Intelligent Green Technologies for Sustainable Smart Cities

Scrivener Publishing
100 Cummings Center, Suite 541J
Beverly, MA 01915-6106

Publishers at Scrivener
Martin Scrivener (martin@scrivenerpublishing.com)
Phillip Carmical (pcarmical@scrivenerpublishing.com)

Intelligent Green Technologies for Sustainable Smart Cities

Edited by
Suman Lata Tripathi
Souvik Ganguli
Tengiz Magradze
and
Abhishek Kumar

Scrivener Publishing

WILEY

This edition first published 2022 by John Wiley & Sons, Inc., 111 River Street, Hoboken, NJ 07030, USA and Scrivener Publishing LLC, 100 Cummings Center, Suite 541J, Beverly, MA 01915, USA

For more information about Scrivener publications please visit www.scrivenerpublishing.com.

Wiley Global Headquarters
111 River Street, Hoboken, NJ 07030, USA

For details of our global editorial offices, customer services, and more information about Wiley products visit us at www.wiley.com.

Library of Congress Cataloging-in-Publication Data

ISBN 978-1-119-81606-5

Cover image: Pixabay.com
Cover design by Russell Richardson

Set in size of 11pt and Minion Pro by Manila Typesetting Company, Makati, Philippines

Printed in the USA

10 9 8 7 6 5 4 3 2 1

Contents

Preface **xv**

List of Contributors **xvii**

1 An Overview of the Intelligent Green Technologies for Sustainable Smart Cities **1**
Tanya Srivastava, Sahil Virk and Souvik Ganguli
- 1.1 Introduction 2
- 1.2 Case Study 1: Oslo—A Smart City 5
- 1.3 Case Study 2: Chandigarh—A Smart City 5
- 1.4 Features of the Smart Cities 6
- 1.5 Well-Planned Public Spaces and Streets 6
  - 1.5.1 Waste Management 6
  - 1.5.2 Energy Management 7
  - 1.5.3 Good Connectivity 7
  - 1.5.4 Urban Residence 8
  - 1.5.5 Smart Grids 8
  - 1.5.6 Smart Governance 8
- 1.6 Intelligent Green Technologies 9
- 1.7 Global and National Acceptance Scenarios 13
- 1.8 Conclusions 15
- References 15

2 Artificial Intelligence for Green Energy Technology **19**
Shanthi Jayaraj and Meena Chinniah
- 2.1 Introduction 19
- 2.2 Solar Energy and AI 20
- 2.3 AI Transforms Renewable Energy 23
- 2.4 IBM Solution Using AI 24
- 2.5 Hydrogen Vehicles 24
- 2.6 Wind Energy and AI 25
- 2.7 Renewable Energy Industry in India 29

2.8 Conclusion 30
References 30
Website Reference 31
Abbreviations 31

3 Effective Waste Management System for Smart Cities 33
G. Boopathi Raja
3.1 Introduction 34
3.2 Literature Survey 36
3.3 Waste Management in India 37
3.4 Existing Methodology 40
3.4.1 IoT-Based Smart Waste Bin Monitoring and Municipal Solid Waste Management System 40
3.4.2 IoT Enabled Solid Waste Management System 41
3.4.3 Smart Garbage Management System 41
3.5 Proposed Framework 42
3.5.1 System Description 42
3.6 Functionality of the Proposed System 44
3.6.1 Sensing Module 44
3.6.2 Storage Module 46
3.6.3 User Module 47
3.7 Workflow of the Proposed Framework 48
3.8 Conclusion and Future Scope 49
References 50

4 Municipal Solid Waste Energy: An Option for Green Technology for Smart Cities 53
Soumitra Mukhopadhyay
4.1 Unavoidable Impacts of Nonrenewable Energy 53
4.2 Municipal Solid Waste Energy as Clean Energy for Smart Cities 55
4.2.1 Renewable Energy Options 55
4.2.2 Municipal Solid Waste as Renewable Energy Option for Smart Cities 56
4.2.3 Why Is MSW Energy Renewable? 58
4.2.4 Various Waste to Energy Technologies 58
4.3 Waste to Energy Technologies (WTE-T) 59
4.3.1 Incineration 59
4.3.2 Pyrolysis 61
4.3.3 Gasification 63
4.3.4 Anaerobic Digestion 65
4.3.5 Landfill with Gas Capture 66

4.3.6 Microbial Fuel Cell (MFC) 68
4.4 Integrated Solid Waste Management Systems (ISWM-S) for Smart Cities 69
4.5 Conclusion 70
References 70

5 E-Waste Management and Recycling Issues: An Overview 73
Simran Srivastava, Sahil Virk, Saumyadip Hazra and Souvik Ganguli
5.1 Introduction 73
5.2 Global Status of E-Waste Management 75
5.3 Industrial Practices in E-Waste Management 77
5.4 Recycling of E-Waste 79
5.5 E-Waste Management Benchmarking 81
5.6 Future of E-Waste Management 82
5.7 Conclusions 83
References 84

6 Energy Audit and Management for Green Energy 89
Arjyadhara Pradhan and Babita Panda
6.1 Introduction 89
6.2 Types of Renewable Energy 91
6.2.1 Solar Energy 91
6.2.2 Wind Energy 91
6.2.3 Biomass 92
6.2.4 Geothermal Energy 92
6.2.5 Ocean Energy 93
6.3 Energy Management 93
6.3.1 Types of Energy Management 94
6.3.1.1 Demand Side Management 94
6.3.1.2 Implementation of DSM 95
6.3.1.3 Supply Side Management 96
6.3.2 Ways to Improve Energy Management 97
6.4 Energy Audit 97
6.4.1 Types of Energy Audit 98
6.4.2 Preliminary Energy Audit 98
6.4.3 Detailed Energy Audit 98
6.4.4 Data Analysis 100
6.4.5 Detailed Steps in Energy Audit 100
6.5 Energy Audit in Solar Plant 101
6.5.1 Technical Inspection Steps of Solar Power Plant 103

6.6 Energy Conservation 104
6.6.1 Energy Conservation Methods 104
6.6.2 Case Study 105
6.7 Conclusion 108
References 108

7 A Smart Energy-Efficient Support System for PV Power Plants 111
Salwa Ammach and Saeed Mian Qaisar
7.1 Introduction 112
7.2 Literature Review 118
7.2.1 Solar Tracking System 119
7.2.2 Solar Cleaning Mechanisms 120
7.2.3 Hotspots Detection 123
7.3 Proposed Solution 131
7.3.1 Solar Tracking 131
7.3.2 Cleaning System 136
7.3.3 Hotspots 136
7.3.4 Modeling and Simulation 136
7.3.5 Limitations 137
7.3.6 Hypothesis 137
7.4 Conclusion 138
References 138

8 A New Hybrid Proposition Based on a Cuckoo Search Algorithm for Parameter Estimation of Solar Cells 143
Souvik Ganguli, Shilpy Goyal and Parag Nijhawan
8.1 Introduction 144
8.2 Modelling of an Amended Double Diode Model (ADDM) and the Objective Function 145
8.3 Proposed Work 149
8.4 Results and Discussions 149
8.5 Conclusions 161
References 162

9 Supervisory Digital Feedback Control System for An Effective PV Management and Battery Integration 165
Amal E. Abdel Gawad, Nehal A. Alyamani and Saeed Mian Qaisar
9.1 Introduction 166
9.2 Literature Review 173
9.2.1 GHI in the Middle East 173
9.2.2 Types of PV Systems 173

9.2.3 Solar Tracking Systems 176
9.2.4 Charger Controller 179
9.2.5 Series Regulator 179
9.2.6 Shunt Regulator 180
9.2.7 Pulse Width Modulation 180
9.2.8 Maximum Power Point Tracker Charger Controller 181
9.2.9 Reducing the Charging Time 182
9.2.10 Dust Remover 183
9.3 Proposed Solution 185
9.3.1 Single Axis Solar Tracking System 186
9.3.2 Supervisory Digital Feedback Solar Tracker Control System 186
9.3.3 Database-Based Digital Solar Tracker Control System 187
9.3.4 Soiling Treatment Module 187
9.3.5 PV-to-Battery Switching Module 187
9.4 Discussion 189
9.5 Conclusion 191
References 191

10 Performance Analysis of Tunnel Field Effect Transistor for Low-Power Applications 195
Deepak Kumar, Shiromani Balmukund Rahi and Neha Paras
10.1 Introduction 196
10.1.1 Limitation of Conventional MOSFET 199
10.1.2 Subthreshold Slope Devices 199
10.2 TFET Structure and Simulation Setup 201
10.3 TFET Working Principle 203
10.3.1 Transport Mechanism in TFET 205
10.3.1.1 Band to Band (BTB) Tunneling Transmission 205
10.3.1.2 Kane's Model 208
10.4 Subthreshold Swing (SS) in Tunnel FETs 209
10.5 Performance of Hetrojunction Tunnel FET 214
10.5.1 Transfer Characteristics Analysis of TFET Devices 214
10.5.2 Frequency Analysis of TFET Devices 219
10.6 Conclusion 221
References 222

11 Low-Power Integrated Circuit Smart Device Design 227
Shasanka Sekhar Rout, Salony Mahapatro, Gaurav Jayaswal and Manish Hooda
11.1 Introduction 228

11.2 Need of Low Power 229
11.3 Design Techniques of Low Power 230
11.3.1 Power Optimization by IC System 230
11.3.2 Power Optimization by Algorithm Section 231
11.3.3 Power Optimization by Architecture Design 231
11.3.4 Power Optimization by Circuit Level 231
11.3.5 Power Optimization by Process Technology 231
11.4 VLSI Circuit Design for Low Power 232
11.4.1 Power Dissipation of CMOS Inverter 232
11.4.1.1 Static Power 232
11.4.1.2 Dynamic Power 233
11.4.1.3 Short Circuit Power Dissipation 233
11.4.1.4 Other Power Issue 233
11.4.2 Capacitance Estimation of CMOS Logic Gate 234
11.5 Circuit Techniques for Low Power 234
11.5.1 Static Power Technique 234
11.5.1.1 Self-Reverse Biasing 234
11.5.1.2 Multithreshold Voltage Technique 235
11.5.2 Dynamic Power Technique 235
11.6 Random Access Memory (RAM) Circuits for Low Power 236
11.6.1 Low-Power Techniques for SRAM 236
11.6.2 Low-Power Techniques for DRAM 237
11.7 VLSI Design Methodologies for Low Power 237
11.7.1 Low-Power Physical Design 237
11.7.2 Low-Power Gate Level Design 237
11.7.2.1 Technology Mapping and Logic Minimization 238
11.7.2.2 Reduction of Spurious Transitions 238
11.7.2.3 Power Reduction by Precomputation 238
11.7.3 Low-Power Architecture Level Design 238
11.8 Power Reduction by Algorithmic Level 239
11.8.1 Lowering in Switched Capacitance 239
11.8.2 Lowering in Switching Activities 239
11.9 Power Estimation Technique 239
11.9.1 Circuit Level Tool 239
11.9.2 Gate Level 240
11.9.3 Architectural Level 240
11.9.4 Behavioral Level 240
11.10 Low-Power Flood Sensor Design 240
11.11 Low-Power VCO Design 241

11.12 Low-Power Gilbert Mixer Design 241
11.13 Conclusion 243
References 243

12 GaN Technology Analysis as a Greater Mobile Semiconductor: An Overview 247
Biyyapu Sai Vamsi, Tarun Chaudhary, Deepti Kakkar, Amit Tiwari and Manish Sharma
12.1 Introduction 248
12.2 Research and Collected Data 250
12.3 Studies Reviewed and Findings 255
12.4 Conclusion 266
References 266

13 Multilevel Distributed Energy Efficient Clustering Protocol for Relay Node Selection in Three-Tiered Architecture 269
Deepti Kakkar, Gurjot Kaur and Aradhana Tirkey
13.1 Introduction 270
13.1.1 Overview 270
13.1.2 Routing Challenges and Design Issues 271
13.1.3 Heterogeneous Wireless Sensor Networks (HWSNs) 272
13.1.3.1 Clustering in WSN 273
13.1.4 Relay Node Selection Scheme 274
13.1.5 Genetic Algorithm 275
13.1.6 Problem Definition and Motivation 275
13.1.7 Proposed Work 276
13.2 Implementation of Proposed Relay Node Selection Based on GA 276
13.2.1 Network Model 276
13.2.2 Heterogenous Network Model 277
13.2.3 Radio Energy Dissipation Model 279
13.2.4 GA-Based Relay Node Selection 279
13.2.5 Steady State Phase or Data Communication Phase 282
13.3 Results of Simulation For Energy Consumption, Lifetime and Throughput of Network 282
13.3.1 Simulation Setup 282
13.3.2 Comparison of Residual Energy Consumption 284
13.3.3 Comparison of Lifetime of Network 284
13.3.4 Comparison of Throughput at BS 286
13.4 Conclusion and Future Scope 287
References 288

14 **Privacy and Security of Smart Systems** 291
K. Suresh Kumar, D. Prabakaran, R. Senthil Kumaran and I. Yamuna
14.1 Smart Systems—An Overview 291
14.2 Security and Privacy Challenges in Smart Systems 292
14.2.1 Botnet Activities in Smart Systems 294
14.2.2 Threats of Nonhuman-Operated Cars 294
14.2.3 Privacy Issues of Virtual Reality 294
14.3 Case Studies—Security Breaches in Smart Systems 294
14.3.1 Breaching Smart Surveillance Cameras 295
14.3.2 Hacking Smart Televisions 295
14.3.3 Hacked Smart Bulbs 295
14.3.4 Vulnerable Smart Homes 296
14.3.5 Identity Stealing using Smart Coffee Machines 296
14.4 Existing Security and Privacy Protection Technologies 296
14.4.1 Cryptography 297
14.4.2 Biometric 299
14.4.3 Block Chain Technology 301
14.5 Machine Learning, Deep Learning, and Artificial Intelligence 301
14.5.1 Machine Learning in Smart Systems 301
14.5.2 Genetic Algorithm 302
14.5.3 Deep Learning in Smart Systems 303
14.5.4 Artificial Intelligence in Smart Systems 303
14.6 Security Requirement for Smart Systems 303
14.6.1 Thwarting of Data Leakage and Falsifications 304
14.6.2 Identification and Prevention of Device Tampering 304
14.6.3 Light Weight Encryption Algorithm for Authentication 304
14.6.4 Access Restrictions to Users 305
14.6.5 Incident Response for Entire Systems 305
14.7 Instruction to Build Strong Privacy Policy 305
14.7.1 Privacy Policy 305
14.7.2 Definition 306
14.7.3 Key Reasons Why There Is a Need for Privacy Policy 306
14.8 Role of Internet in Smart Systems 306
14.8.1 Home Automation 307
14.8.2 Agriculture 307
14.8.3 Industry 308

14.8.4 Health & Lifestyle 309
14.9 Frameworks, Algorithms, and Protocols for Security Enhancements 310
14.9.1 Framework for the Internet of Things by Cryptography 311
14.9.2 Protocols for Security Enhancements 312
14.10 Design Principles of Privacy Enhancing Methodologies 312
14.11 Conclusion 313
References 314

15 Artificial Intelligence and Blockchain Technologies for Smart City 317
Jagendra Singh, Mohammad Sajid, Suneet Kumar Gupta and Raza Abbas Haidri
15.1 Introduction 318
15.2 Standard for Designing Smart City and Society 322
15.2.1 Scalability 322
15.2.2 Intelligent Health Care 322
15.2.3 Flexible and Interoperable 322
15.2.4 Safeguard Infrastructure 322
15.2.5 Robust Environment 323
15.2.6 Distribution and Sources of Energy 323
15.2.7 Intelligent Infrastructure 323
15.2.8 Choice-Based Backing System 323
15.2.9 Monitoring of Behavior 323
15.3 Blockchain and Artificial Intelligence 323
15.4 Contributions and Literature Study 324
15.5 Conclusion 328
References 329

16 Android Application for School Bus Tracking System 331
S. Sriram
16.1 Introduction 331
16.2 Application Methods for Access 332
16.2.1 Driver Portal Screen 333
16.2.2 Parent Portal Screen 334
16.2.3 Teachers Portal Screen 334
16.3 GPS Data Processing Methodology 335
16.4 GPS Working Process 336
16.5 System Implementation 336
16.6 Result and Discussion 336

16.6.1 Reasons to Utilize Android Application for School Bus Tracking System 337
16.6.1.1 Perfect Child Security 337
16.6.1.2 Elaborate Operational Efficiency 337
16.6.1.3 Valid Timely Maintenance 338
16.6.1.4 Automating Attendance Management 338
16.6.1.5 Better Staff Management 338
16.6.1.6 Addressing Environmental Concerns 338
16.7 Conclusion 338
References 339

About the Editors **341**

Index **343**

Preface

Smart cities beautify urban areas. The creation of the smart city relies heavily on the deployment of technology. To further social and environmental improvements, cities have created and advanced their smart policies. A smart city does improve the quality of life. It does, however, also require additional power resources, which highlights green or renewable energy. Green energy is a form of energy generated by natural resources like solar, wind, fossil fuel, etc. Wise utilization of green energy and avoidance of wastage is an integral part of sustainable smart cities. To make energy available round the clock, following several strategies can boost sustainability: smart management of the energy, conversion of waste to efficient energy, and efficient auditing of energy.

Energy conservation or savings is just as important as energy that has been generated. In planning the architecture of smart cities for the upcoming technical era, we need to replace the existing infrastructure in metropolitan areas all around the world with more efficient, smart, and environmentally sustainable ones. Artificial intelligence finds a place in almost every activity of day-to-day life, and the adoption of intelligent technologies helps to make the city smarter. AI and blockchain technologies help to manage the network of smart city architecture towards developing sustainable ecosystems. The integration of electronic equipment in day-to-day life is unavoidable, including maximum information available on the devices. The security of processed and processing information is perhaps the most important aspect of the smart city to avoid unauthorized access to these devices. This book targets future technologies to make the city smarter either by efficient and reliable utilization of energy sources or enhancing the privacy of the information.

List of Contributors

J. Shanthi
Avinashilngam Institute for Home Science and Higher Education for Women, Coimbatore, India

C. Meena
Avinashilngam Institute for Home Science and Higher Education for Women, Coimbatore, India

Tanya Srivastava
Thapar Institute of Engineering and Technology, Patiala, Punjab, India

Soumitra Mukhopadhyay
Doosan Power Systems India Pvt. Ltd, India

G. Boopathi Raja
Velalar College of Engineering and Technology
Erode, Tamil Nadu, India

Arjyadhara Pradhan
Kalinga Institute of Industrial Technology, Bhubaneshwar

Babita Panda
Kalinga Institute of Industrial Technology, Bhubaneshwar

Salwa Ammach
Effat University, Jeddah, Saudi Arabia

Saeed Mian Qaisar
Effat University, Jeddah, Saudi Arabia

Gurjot Kaur
NIT, Jalandhar, India

Aradhana Tirkey
NIT, Jalandhar, India

S. Sriram
Sri Ramakrishna Polytechnic College, Coimbatore, Tamil Nadu, India

Sahil Virk
Thapar Institute of Engineering and Technology, Patiala, Punjab, India

Souvik Ganguli
Thapar Institute of Engineering and Technology, Patiala, Punjab, India

Simran Srivastava
Thapar Institute of Engineering and Technology, Patiala, Punjab, India

Saumyadip Hazra
Thapar Institute of Engineering and Technology, Patiala, Punjab, India

Shilpy Goyal
Thapar Institute of Engineering and Technology, Patiala, Punjab, India

Parag Nijhawan
Thapar Institute of Engineering and Technology, Patiala, Punjab, India

Deepak Kumar
University of Petroleum and Energy Studies, Dehradun, India

Shiromani Balmukund Rahi
Mahamaya College of Agriculture Engineering and Technology
Akabarpur Ambedkar Nagar, Uttar Pradesh, India

Neha Paras
Dr. B R Ambedkar National Institute of Technology, Jalandhar, India

K. Suresh Kumar
IFET College of Engineering, Villupuram, Tamil Nadu, India

D. Prabakaran
IFET College of Engineering, Villupuram, Tamil Nadu, India

R. Senthil Kumaran
IFET College of Engineering, Villupuram, Tamil Nadu, India

I. Yamuna
IFET College of Engineering, Villupuram, Tamil Nadu, India

Jagendra Singh
Bennett University, Greater Noida, India

Mohammad Sajid
Aligarh Muslim University, Aligarh, India

Suneet Kumar Gupta
Bennett University, Greater Noida, India

Raza Abbas Haidri
Khwaja Moinuddin Chishti Language University, Lucknow, India

Shasanka Sekhar Rout
GIET University, Gunupur, India

Salony Mahapatro
Asiczen Technologies, Bhubaneswar, India

Gaurav Jayaswal
Semi-Conductor Laboratory, Department of Space, Punjab, India

Manish Hooda
Semi-Conductor Laboratory, Department of Space, Punjab, India

Biyyapu Sai Vamsi
Dr. B R Ambedkar National Institute of Technology, Jalandhar, India

Tarun Chaudhary
Dr. B R Ambedkar National Institute of Technology, Jalandhar, India

Deepti Kakkar
Dr. B R Ambedkar National Institute of Technology, Jalandhar, India

Amit Tiwari
Dr. B R Ambedkar National Institute of Technology, Jalandhar, India

Manish Sharma
Dr. B R Ambedkar National Institute of Technology, Jalandhar, India

Amal E. Abdel Gawad
Effat University, Jeddah, Saudi Arabia

Nehal A. Alyamani
Effat University, Jeddah, Saudi Arabia

1

An Overview of the Intelligent Green Technologies for Sustainable Smart Cities

Tanya Srivastava[1], Sahil Virk[2] and Souvik Ganguli[2*]

[1]Department of Computer Science and Engineering, Thapar Institute of Engineering and Technology, Patiala, Punjab, India
[2]Department of Electrical and Instrumentation Engineering, Thapar Institute of Engineering and Technology, Patiala, Punjab, India

Abstract
By 2050, 68% of the world population is projected to live in urban areas by the United Nations (UN). With such a rapid surge of people moving into urban areas, the sustainability of many city's social, economic, and environmental practices have become difficult to keep up with. Cities of the future cannot be built in the same way cities of today have been built. Now, urban areas have to be built around incorporating new technologies, which will positively affect the social, environmental and economic terms of the city. The smart city is used as an umbrella term for such cities. In scheming the architecture of smart cities for the upcoming technical era, we need to replace the existing infrastructure in metropolitan areas all around the world with more efficient, smart, and environmentally sustainable ones. In this paper, we explore certain examples of smart cities around the world and try to interpolate the features of one. We also mention smart green technologies and their examples for smart city infrastructure development and try to understand the global acceptance of the concept of smart cities.

Keywords: Intelligent green technologies, sustainability, smart cities, waste management, energy management

**Corresponding author*: souvik.ganguli@thapar.edu

Suman Lata Tripathi, Souvik Ganguli, Tengiz Magradze, and Abhishek Kumar (eds.) Intelligent Green Technologies for Sustainable Smart Cities, (1–18) © 2022 Scrivener Publishing LLC

1.1 Introduction

Energy represents around 70% of the world Gross Domestic Product (GDP) [1]. The rise in the number of the global population has put immense pressure on the available natural resources and city infrastructure. Worldwide, around 56% of the world's population live in urban areas. This percentage is even higher in certain continents as shown in Figure 1.1.

With such high population pressure, energy production and the consequent utilization have immense amounts of greenhouse gas emissions. These greenhouse gases, like carbon dioxide, nitrous oxide, methane, and ozone, have not only resulted in an accelerated rate of climate change but also burgeoned air pollution problems in highly packed cities and megacities. Around 12.6 million people die because of environmental pollution every year. Atmospheric pollution causes severe lung and heart-related diseases [1]. Similarly, with the increase in the chlorofluorocarbons (CFCs) environment, ozone layer depletion has also advanced, which has caused ultraviolet rays to come in contact with people causing skin damage and cancer. In the last century, we have put an enormous amount of carbon dioxide in our environment as seen in Figure 1.2. Urbanization plays a major role in energy consumption and carbon dioxide emissions.

Soil and water pollution are the major reasons behind more than 100 illnesses. Therefore, with the decrease in global natural resources, there is a complete imbalance in the ecological system. Hence, we need to quest ways that will decrease our dependency on conventional resources and invest our finances in planning a more sustainable and safe future for ourselves,

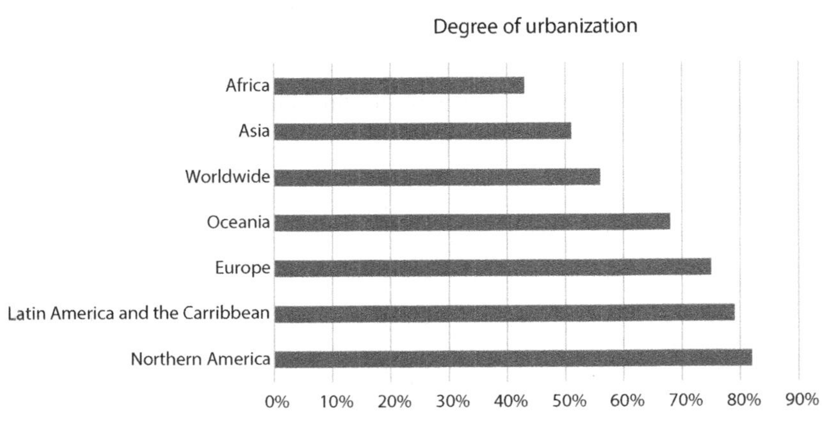

Figure 1.1 The degree of urbanization continent wise based on the year 2020.

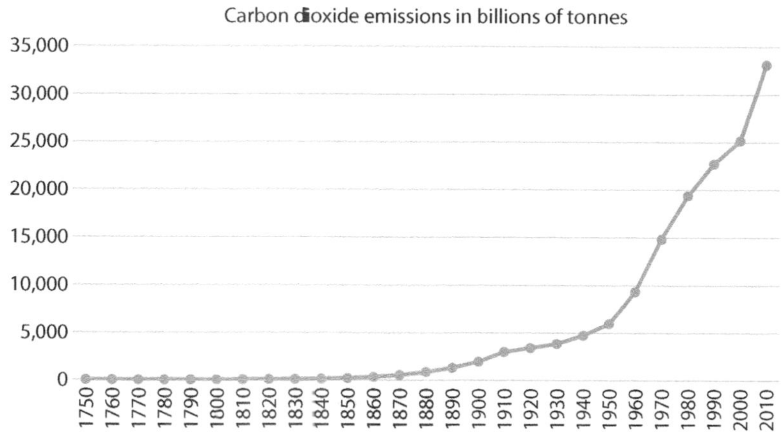

Figure 1.2 The worldwide carbon dioxide emissions from the years 1750–2010.

as well as the upcoming generations. As we have established, urbanization and the projected growth and acceleration of the same would be responsible for negatively impacting our environment. Thus, in the 17 sustainable development goals (SDG) set up by the United Nations General Assembly in 2015, sustainable cities and communities (SDG-11) are listed as one of the SDGs. Others include clean water and sanitation (SDG-6), affordable and clean energy (SDG-7), industry, innovation and infrastructure (SDG-9) and climate action (SDG-13). Smart cities are a way to meet these goals collectively.

According to the United Nations "Smart city planning improves the lifestyle of its people, provides healthy, safe, stimulating and dynamic environment for its citizens to grow." According to Navigant Research's long-term forecasts, the demand for smart cities will expand steadily, reaching a size of 25.3 billion euros in 2023. Smart cities are a modern and growing technology that enhances the lifestyle of the expanding urban population and provides them with better services. Citizen participation also plays a vital role in the planning of a smart city as they pool their ideas and have a lot of experience which could aid in the development of better services and plans. By listening to them, the chances of failure of projects could be reduced and the potential problems could be addressed at an early stage [2].

As mentioned before, the "Smart City" has certain objectives, such as reduction of the negative impact of all the human activities in urban areas and also improving the life quality of its residents. A major problem that experts face while describing the notion of a smart city is a lack of a universally accepted definition for the same; however, the effect of smart city on domains, like transport, buildings, natural resources, and people provides a starting point for exploring the idea. Experts agree that the inclusion of ICT technologies is essential to the concept of a smart city. Smart cities encompass a variety of smart technologies, especially ones applied in homes and public spaces. These places provide a place for commencing smart initiatives which when applied on a larger scale, starts to resemble a smart city. Neirotti *et al.* [3] took this opportunity to highlight the trends in smart city development and discussions as well as identifying hard and soft application domains.

Application domains have been classified into "hard" and "soft" domains. They also further described hard domains like the one where smart technology visions can be physically and quantifiably applied, whereas soft domains are where the ideas of smart cities can be fostered in people and culture. Hard application domain includes energy grids, public lighting, management of water and natural resources, waste management, environmental resources, transport, residential and office spaces, healthcare, and public security. The soft domain includes education, culture, social inclusion, and welfare, public administration, the economy and the government. Subdomains of them are thus listed in Figure 1.3.

Due to concerns about the deteriorating health of our environment, people have become more environmentally sensitive than ever before. Especially in urban cities, residents have become aware that their actions have consequences on the environment. Even after this, residents are still fairly new to the smart technologies that can help reduce this negative impact on the environment. As resources like water and electricity are still available at nominal rates in major metropolitan cities, people thereby rarely take steps to reduce their intake. In the study by Bhati *et al* [4], it was pointed out that there is a gap between today's consumers' understanding

Natural resources and Energy	**Transport and mobility**	**Living**	**Government**	**Buildings**	**Economy and people**
Smart grids	City logistics	Entertainment	E-government	Facility management	Innovation and entrepreneurship
Public lighting	Info-mobility	Hospitality	E-democracy	Housing quality	Cultural heritage management
Renewable energy	People mobility	Pollution control	Procurement	Building services	Digital education
Waste management		Public safety	Transparency		Human capital management
Food and agriculture		Healthcare			
		Social inclusion and culture			
		Public spaces			

Figure 1.3 The domains and subdomains of a smart city.

of smart technology design in smart homes and its integration into people's life. Their paper highlighted the outlook of Singapore's residents on smart technology and how it can be utilized to save resources.

The objective of the paper was to bring light to the famously constructed smart cities at the international and national levels, discussing their designing features and how we can incorporate these learning at local levels and move toward the goal of sustainable development. Besides, the primary and essential features of smart cities have been discussed.

1.2 Case Study 1: Oslo—A Smart City

Oslo, the capital of Norway has targeted to reduce 95% city emissions by 2030. It was also regarded as the eco capital of Europe in 2019. Oslo has only one municipality. The city aims to reduce carbon emissions by promoting electric vehicles and making public transport more attractive. The public electric buses, as well as ferries, operate on renewable hydroelectric power. The government has posed a ban on cars in the cities and encouraged the citizens to use cycles and pave for pedestrian walking. They employ simple waste management strategies, which have given marvellous results in reducing the wasted food as well as in the recycling process. Every household must segregate glass, plastic, metal, and food before dumping them into the garbage bins. The plastic waste is recycled and the food waste goes to the biological treatment plant which produces biogas and biofertilizer and is then delivered to the farms [5]. Datek Light Control system is used in the lighting infrastructure of Oslo city to enhance the quality of light by adding smart capabilities to the lighting infrastructure. It adds sensors for automatic light control and decreases light pollution [6].

1.3 Case Study 2: Chandigarh—A Smart City

After India's independence, Chandigarh was the first city that was smartly planned by French architect Le Corbusier [7]. The city has been divided into various sectors promoting neighborhood planning. Roads in the union territory are all covered with the trees along and are named according to the direction or purpose and not according to the name of the individuals. It has also emerged as the city of elites. The capital of Punjab and Haryana is a green city and is striving to achieve goals of zero emissions in the coming years. Punjab University is one of the good internationally recognized universities in the city [8].

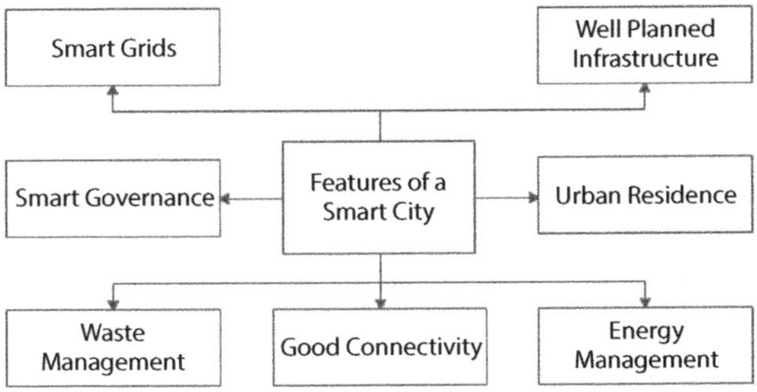

Figure 1.4 The features of smart cities.

1.4 Features of the Smart Cities

The main features of a smart city are shown in Figure 1.4 and explained further in this section.

1.5 Well-Planned Public Spaces and Streets

More social interaction is possible with a good quality environment [9]. This smart public space should be designed for people-to-people interaction, gathering activities, for expressing their views on lifestyles and their cultures. Also, the tradition of asking the people for directions of places should not be lost while designing a plan for the smart city [10]. Planning is not limited to the social impact here. Due to climate change, places are experiencing changing weather patterns and therefore, their infrastructure suffers as they were not built keeping that in mind. As climate change worsens, cities have to think about the long-term effects that their infrastructure may be facing in the next few decades because of climate change.

1.5.1 Waste Management

If we have to reduce and minimize the waste generation of cities, then it would require probably three steps:

(i) Prevention of waste
(ii) Proper waste collection
(iii) Proper value recovery from collected waste

The waste can be utilized by the remanufacturing companies and the waste recovery systems. Urban Mining is a suitable example of such a system. A smart city would use new technologies to better handle the waste. These can be innovated by individuals, enterprises, and public services. Waste collection, recycling, and disposal are done in such a way that minimizes the harmful impact of poor waste management on individuals and the environment as a whole. Waste storage, treatment, recycling, and regeneration are also included.

1.5.2 Energy Management

To design a smart energy management system for a smart city, one has to consider all the five main-energy related areas-generation, storage, infrastructure, facilities, and distribution [11]. Exploiting regenerative or inexhaustible natural resources, such as solar and wind power, are also crucial for smart city energy management. Countries use the smart street lighting system and other IoT (Internet of Things) based applications in the cities for reducing energy consumption.

1.5.3 Good Connectivity

A good transport infrastructure reduces noise pollution and the risk of accidents, as well as traffic congestion. It is an important part of urban planning. The spread of electric vehicles would improve the carbon dioxide emissions in the atmosphere [12]. Here, the technological innovations involving hybrid and electrical cars play a gigantic role in reducing carbon emissions. Taking traffic and energy demand into account when optimizing logistics and transportation in urban environments is pertinent to the existing and future traffic problems. Providing users with real-time, multimodal traffic and transportation statistics and providing environmentally efficient fuels and advanced propulsion technologies to ensure long-term public transit will be the goals of a smart city community.

Promotion of public transports and providing smart parking technology can help in reducing the sheer number of vehicles on the road. Casini [13] pointed out the objectives of achieving good connectivity in a smart city. The author included points like

- Inclusion of early warning signs which include smart parking and conveying traffic conditions in real time to people. This information would go a long way to help people have a more productive life.

- Digitizing the public transport system. This makes public transport attractive to the public while also providing them transparency about the wait timings and distances through a common application.
- Promoting pedestrian spaces and walkways and encouraging people to walk instead of opting for a relatively less environment-friendly way of commuting would have a positive impact on the environment as well as the public's health.

1.5.4 Urban Residence

Urban planning in the smart city supports a sustainable environment [14]. The problem of burgeoning population and urbanization could be effectively dealt with the smart city residence planning [15]. Aspects of a residential building's quality of life, such as convenience, illumination, and heating, ventilation, and air conditioning (HVAC) have to be considered and discussed in great detail to fully optimize smart city residences. It covers anything that has to do with the degree of satisfaction that residents experience while residing in a smart city home.

1.5.5 Smart Grids

It is an interconnected network of microgrids. They are integrated into the current power system to increase the current capacity and efficiency and are a primary source for the charging stations for electric vehicles [16]. They also contribute to exchanging of information with administrations which ultimately help in regulating and optimizing energy flows. Regulating costs in the smart grid systems could include swapping out the inefficient lighting systems with the new generation of LED lights. Also, automating the street lights to adjust their intensity would save a lot of resources. In a smart city, electricity networks would have to account for the behavior of all linked users to provide reliable, cost-effective, and stable electricity. Smart grids should be able to self-heal and withstand device failures. Automated grids use ICT to transport electricity and allow knowledge sharing about usage between utilities and consumers, to lower prices and improve energy supply system efficiency and accountability.

1.5.6 Smart Governance

It increases the citizens' engagement and it is the intelligent and efficient use of information and communication technologies in the decision-making process of the state through collaboration, information sharing, transparency as well as openness [5]. With the advent of machine learning technologies and the numerous applications for which they can be used, it is important now to collect meaningful data. All devices in an urban area network should be linked and used to transmit data, with appropriate permissions, to a collective database for the city i.e., Big Data. This data could help not only in optimizing city resources later but also help in developing applications and academic research for smart cities. Smart cities need to be data-driven. From [13], we highlight the key tools in smart governance implementation.

- Digital democracy means how governmental processes can be impacted by information technologies like social media. This helps in clear communication and transparency between the public and the authorities.
- Open governance is a term used to describe accessibility for the public to look into the government's decisions and processes. The main objective here is to make urban administrative authorities responsible to the public and its needs which, therefore, builds trust.
- Citizen empowerment is to make citizens conscious of their actions and help by participating in major decisions taken up by authorities. Here, the opportunities given to the citizen to participate is crucial. This opens up the responsibility of maintaining and developing the city not only in the hands of the concerned governmental bodies but also the citizens.

To covert cities into smart cities, such measures must be carried out in a systematic and coordinated manner while making use of the infrastructures already and being developed and promoting solution interoperability and scalability.

1.6 Intelligent Green Technologies

Buildings have a huge impact on the environment. Urban areas use up to two-thirds of the world's energy and are responsible for more than 70% of the greenhouse gas emissions emitted by us. With the growing shortage of freshwater, 96% of the urban world has to employ improved methods in other to obtain drinking water. According to the Paris Agreement, the collective goal of all the 197 countries that signed it is to limit global warming to well below 2 degrees Celsius above pre-industrial levels. For us to achieve this target, drastic measures would have to be taken. Our buildings and city infrastructure has to start being aligned with our environmental goals. Seeing how buildings cause extreme levels of environmental strain, they provide the starting point for applying intelligent green technologies

In 2010, Taiwan highlighted the importance of promoting intelligent green buildings. Combined with the Information and Communication Technology (ICT) infrastructures, these intelligent green buildings can help tackle many environmental problems that plague our generation. They also create economic, cultural and social growth in a community. They can reduce carbon emissions, save energy and help in the overall aesthetic appeal of any city. Kuo *et al.* [17] came up with standards and norms for green buildings after probing cases from 1988 to 2014. The Taiwanese government highlighted a hierarchy of promoting different policies concerned with green building materials, green intelligent buildings, an ecological community comprising of the green intelligent buildings linked together and lastly, an intelligent green city. The same is shown in Figure 1.5.

Similarly, in an article published by Everett [18], the author attempted to define the concept of green buildings and examine their performance. They pointed out that green buildings provide excellent learning environments for college students as it sensitizes the students towards sustainability. "Intelligent" buildings, according to them, simply is a building that can reduce capital cost by minimizing wasteful infrastructure duplication while decreasing the carbon footprint. For example, in the earlier days, the wiring in buildings used to be excessive. Every system (CCTV, ventilation,

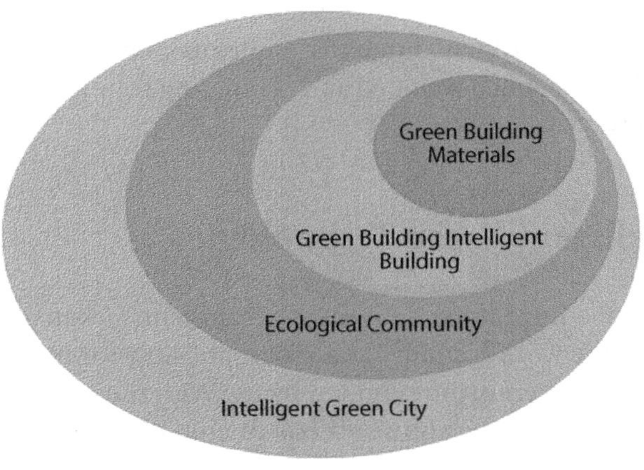

Figure 1.5 The hierarchy of steps for building an intelligent green city.

data, phones) had an isolated set of wires whereas all of that could later be integrated into a single Internet Protocol (IP). This reduced the carbon footprint and also minimized the cost.

There has been a plentitude of definitions but they all encompass a set of common words such as sustainable, responsive, dynamic, healthy, flexible, adaptable and technologically aware. Automation provides the key to implement the expected behaviors of an "intelligent green" building. It automates energy dispersion and optimizes all the resources available to it. It also makes the building flexible to the needs of its occupants.

The main factors that measure the success of a green building technology are cost minimization, optimization of resources, effects on the environment and its acceptance. Intelligent Building Automation Systems (IBASs) have been having a great impact in improving the energy profile of a building. In the recent past, multiuse components, such as cross-functional sensors and control systems, have been integrated with a unified system architecture to further enhance cost savings.

An interesting concept of urban heat island (UHI) comes into play when we talk about southeastern cities like Singapore, Hong Kong and Kuala Lumpur. UHI is a side effect of rapid urban developments in megacities wherein the city is significantly warmer than its surroundings due to human activities. This creates a vicious cycle where inhabitants use more energy to operate air conditioners and refrigerators which contributes to more greenhouse gas emissions, making the city even warmer. This effect would need to be dealt with in a smart city as this exacerbates the problem of global warming. Heat mitigation is thus an important aspect of smart city development in southeastern places. Aflaki *et al.* [19] reviewed the strategies for combating UHI like urban vegetation. They outlined multiple approaches to studying UHI and its impacts as well as the emerging trends in this subject. Singapore has been known to provide cooling to dozens of its buildings through a common underground district cooling system. This system pumps cold water from the main center through all the interconnected buildings in the area (which includes the iconic Marina Bay Sands hotel complex). This has led to a reduction of costs in electricity by up to 40% when compared to utilizing traditional air conditioning systems and emission savings for the same is equivalent to removing 10,000 cars off the road. Some heat mitigation methods include vegetation, urban geometry, water bodies and features and materials and surfaces. Planning transport routes and roads along with lining facades with solar panels also helps in reducing the effects of UHI. Security is also a part of a smart building. Supervised access with biometrics and radio frequency identification (RFID) access cards is essential and should be integrated with all buildings.

Another aspect of utilizing such green technologies is intracity transportation. WHO had pointed out that road traffic crashes killed about 85,000 people in the WHO European Region in 2013. It remains one of the leading causes of death for people aged 5 to 29. Transportation also contributes to enormous amounts of pollution which cause up to 500,000 premature deaths per year in Europe. Noise pollution is also a cause of worry in major cities, where people residing near public transport hubs or construction sites are exposed to noise levels well above 65 dB daily. All these factors add up and demand a change in how we treat transportation in a city.

Smart cities would need to combat such problems efficiently. So, in 2020, Ladha *et al.* [20] proposed an IoT-based framework for an intelligent green public transport system (IIGPTS). They realized that public transports were environmentally draining, dangerous and overall unappealing to the public because of the congestion and pollution. It was clear that the toxic emissions and load management had to be analyzed and balanced. Their system thus collected data through sensors. Passenger density, fuel consumption, emissions and routing paths were measured. Then, simulations were run and it was seen that IIGPTS could handle the massive load with minimal delay by using dynamic routes.

On the same note, traffic handling is also an essential aspect of a smart city. With 1.2 billion vehicles on the road and the number reaching 2 billion within the next 15 years, the cities of tomorrow need to be able to make road travels time efficient. Joseph *et al.* [21] in 2020, outlined the various deep learning methodologies in green and intelligent transport management. They pointed out that with the prodigious amount of data through

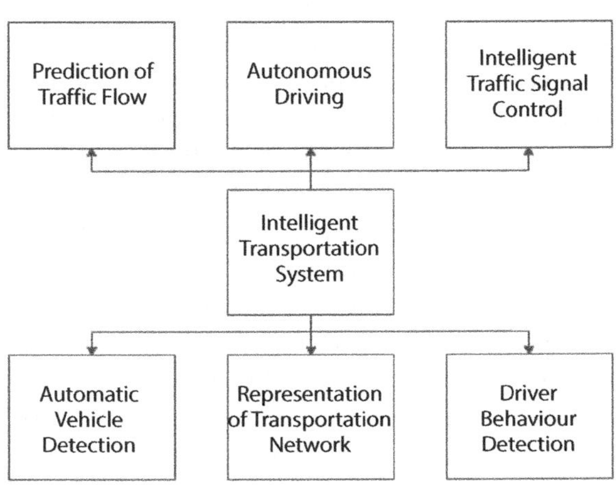

Figure 1.6 Block diagram representation for an intelligent transportation system.

CCTVs, GPS and on-road sensors, machine learning models were failing to build a consistent and robust model for predicting the traffic dynamics. An overall view of the intelligent transportation system covered by [21] is shown in Figure 1.6.

1.7 Global and National Acceptance Scenarios

Smart city development requires planning, clear intention and massive financial investments and support. All of this is only possible if there is acceptance and longing for such infrastructure. For the past several decades, the concept of smart cities has been taken up in a nonspecific and universal manner with little to no regard for social, cultural, and regulative dynamics. As outlined by Sepasgozar *et al.* [22], certain tasks are needed to be done for smart city development. The first task is selecting the appropriate technologies. With the advent of new technologies and innovative data analyzing techniques, there are many to choose from. The second task is adapting the chosen technology and the third relates to managing the acceptance of the technology chosen.

Kshetri *et al.* [23] looked into the role of formal and informal institutions, as both inhibitors and facilitators, in the development of South Korea's new Songdo smart city. Ubiquitous cities are now emerging in Singapore, Japan, and some European countries. These cities use ubiquitous computing to integrate social and information systems to use them for the good of the city. These digitally integrated cities pose as a key foundation toward smart city development. For the analysis of the institutional support for Songdo, Kshetri *et al.* [23] divided them into three categories, namely regulative, normative, and cognitive. Regulative institutions relate to formal constraints, such as laws and rules by the constitution. Normative represents the moral implications and cognitive means recognition or individual nature of reality.

There were many regulative hurdles to jump through for the development of this ubiquitous city, one of them being a regulation that resulted in failing to open an international school in the city. Many families had come in hopes of sending their children to these schools but the Korean government bureaucracy failed to help the smart city attract residents. A similar incident concerning the attraction of domestic companies also hurt the economy of this smart city. The Korean government failed to provide proper incentives for attracting domestic companies whereas the same was provided for foreign companies. The foreign companies also left as there were no accounting, legal, and other such services being provided locally.

The government revised certain bills concerning the protection of personal information which posed a legal obstacle for the development of technological services in New Songdo. Other legal relaxations also helped the smart city development. Furthermore, strict recycling regulations and enforcement were made a law for the citizens.

The normative and cognitive obstacles were relatively few. The cultural environment of South Korea is very conducive to ubiquitous technology implementation. Also, owning or living in an area with high-level integrated technology is a status symbol for South Korean people. As most of the technology used in New Songdo was developed in the US, highly educated and well-travelled South Korean citizens were immediately attracted. Privacy concerns regarding movement tracking of citizens are less of a priority for Asia than the US and Europe. The Asian culture of minimizing their negative impact on the society, especially for the environment, helps in the pitch for New Songdo. Thus, we can say the cultural and social dynamics of South Korea were close to ideal for the adoption and curiosity about an upcoming smart, ubiquitous city.

While institutions have had a major impact on the acceptance of smart technologies, it is ultimately the resident's demand that influences the first steps an administration would take for developing a smart city. To gauge the factors that determine the acceptance of smart city technologies, Habib *et al.* [24] in 2019 attempted to empirically determine them. The questionnaire used for the same focused on seven factors as shown in Figure 1.7. These factors were shown to majorly influence people's opinions but two factors, namely security and privacy came out on top in terms of importance.

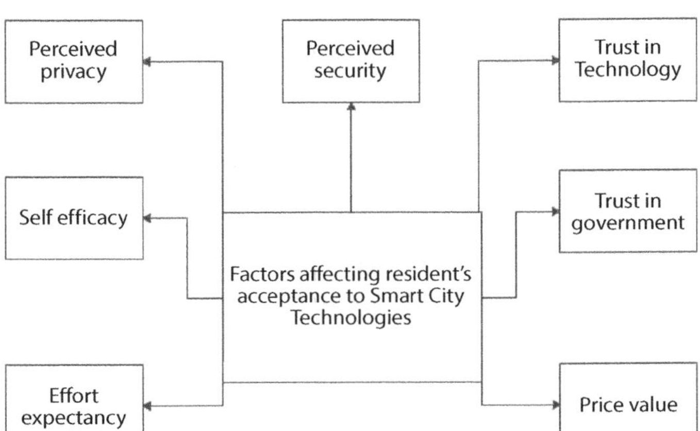

Figure 1.7 The factors affecting resident's acceptance of smart city technologies.

Another point that lowers people's demand for smart cities and smart home technologies is the gap between user expectations and functions. For example, China is a huge market for smart home technologies but yet its involvement in homes is still at a lower level than expected. Ji *et al.* [25] stated that the reason is for the same is that difference in expectations and end product. Similarly, Gimpel *et al.* [26] provided a comprehensive adoption model for smart energy technology for policymakers to implement. This provided policymakers help prepare the people for smart city technology adoption.

1.8 Conclusions

The key to the acceptance of intelligent green technologies is to not think of them as a tradeoff between comfort and sustainability, but rather a new generation of architecture, which only increases the satisfaction of all stakeholders. It also lies in the realization that intelligent green buildings are an investment that may require a high amount of capital for its construction but it balances that out with its cost-cutting and resource optimization throughout its lifetime. In the end, these buildings save resources, even if it does not seem like that initially. This high return of investment is a great incentive for authorities to look into such options.

References

1. Pilipczuk, O., Sustainable smart cities and energy management: The labor market perspective. *Energies*, 13, 22, 6084, 2020.
2. Stupar, A., Jovanović, P., Vojvodić, J.I., Strengthening the social sustainability of super-blocks: Belgrade's emerging urban hubs. *Sustainability*, 12, 3, 903, 2020.
3. Neirotti, P., De Marco, A., Cagliano, A.C., Mangano, G., Scorrano, F., Current trends in Smart City initiatives: Some stylised facts. *Cities*, 38, 25–36, 2014.
4. Bhati, A., Hansen, M., Chan, C.M., Energy conservation through smart homes in a smart city: A lesson for Singapore households. *Energy Policy*, 104, 230–239, 2017.
5. Desouza, K.C., Hunter, M., Jacob, B., Yigitcanlar, T., Pathways to the making of prosperous smart cities: An exploratory study on the best practice. *J. Urban Technol.*, 27, 3, 3–32, 2020.
6. Salha, R.A., El-Hallaq, M.A., Alastal, A.I., Blockchain in smart cities: Exploring possibilities in terms of opportunities and challenges. *J. Data Anal. Inf. Process.*, 7, 3, 118–139, 2019.

7. Singh, K. and Sharma, N., Smart cities in India: Key areas and challenges-case study of Chandigarh City. *Int. J. Manage. Soc. Sci.*, 4, 1, 386–396, 2016.
8. Sharma, A.K. and Gupta, J.K., Strategies for Making Cities Sustainable and Livable in India, *Proceedings of International Convention on Performance of Built Environment*, College of Architecture - Trivandrum, 2007.
9. Heryanto, B., Jamila, R.F., Dharmaraty, A.P., Ruang Kebersamaan: Suatu Kajian Sinergitas Dan Kreativitas Di Ruang Publik Kota. *J. Pembang. Wil. Kota*, 16, 3, 231–244, 2020.
10. Kurniawati, W. and Prihantini, P., Smart City and Shifting Meaning of Public Space, in: *IOP Conference Series: Earth and Environmental Science*, 2019, March, vol. 248, No. 1, IOP Publishing, p. 012012.
11. Singh, R. and Chen, G., *A Community Energy Operations and Planning System: Concept, Use cases, Metrics, and Benefits*, Lawrence Berkeley National Lab.(LBNL), Berkeley, CA (United States), 2020.
12. Carnis, L., Smart cities and transport infrastructures topical collection, Eur. *Transp. Res. Rev.*, 10, 2, 1–3, 2018.
13. Casini, M., Green technology for smart cities, in: *IOP Conference series: earth and environmental science*, 2017, August, vol. 83, No. 1, IOP Publishing, p. 012014.
14. Anthopoulos, L.G. and Vakali, A., Urban planning and smart cities: Interrelations and reciprocities, in: *The Future Internet Assembly*, 2012, May, Springer, Berlin, Heidelberg, pp. 178–189.
15. Bakri, M. and Kasim, A.A., The urban planning concept based on smart city approach. *Int. J. Livable Space*, 3, 2, 63–70, 2018.
16. Nagel, E., Kranz, J., Sandner, P., Hopf, S., How blockchain facilitates smart city applications–development of a multi-layer taxonomy, in *Twenty-Seventh European Conference on Information Systems (ECIS2019)*, Stockholm-Uppsala, Sweden, pp. 1–17, 2019.
17. Kuo, C.F.J., Lin, C.H., Hsu, M.W., Analysis of intelligent green building policy and developing status in Taiwan. *Energy Policy*, 95, 291–303, 2016.
18. Everett, R., The 'building colleges for the future'program. Delivering a green and intelligent building agenda. *New Rev. Inf. Netw.*, 14, 1, 3–20, 2008.
19. Aflaki, A., Mirnezhad, M., Ghaffarianhoseini, A., Ghaffarianhoseini, A., Omrany, H., Wang, Z.H., Akbari, H., Urban heat island mitigation strategies: A state-of-the-art review on Kuala Lumpur, Singapore and Hong Kong. *Cities*, 62, 131–145, 2017.
20. Ladha, A., Bhattacharya, P., Chaubey, N., Bodkhe, U., Iigpts: IoT-based framework for intelligent green public transportation system, in: *Proceedings of First International Conference on Computing, Communications, and Cyber-Security (IC4S 2019)*, Springer, Singapore, pp. 183–195, 2020.
21. Joseph, S.I.T., Velliangiri, S., Chandra, C.S., Investigation of deep learning methodologies in intelligent green transportation system. *J. Green Eng.*, 10, 931–950, 2020.

22. Sepasgozar, S.M., Hawken, S., Sargolzaei, S., Foroozanfa, M., Implementing citizen centric technology in developing smart cities: A model for predicting the acceptance of urban technologies. *Technol. Forecasting Soc. Change*, 142, 105–116, 2019.
23. Kshetri, N., Alcantara, L.L., Park, Y., Development of a smart city and its adoption and acceptance: The case of new songdo. *Commun. Strateg.*, 96, 113, 2014.
24. Habib, A., Alsmadi, D., Prybutok, V.R., Factors that determine residents' acceptance of smart city technologies. *Behav. Inf. Technol.*, 39, 6, 610–623, 2020.
25. Ji, W. and Chan, E.H., Between users, functions, and evaluations: Exploring the social acceptance of smart energy homes in China. *Energy Res. Soc. Sci.*, 69, 101637, 2020.
26. Gimpel, H., Graf, V., Graf-Drasch, V., A comprehensive model for individuals' acceptance of smart energy technology—A meta-analysis. *Energy Policy*, 138, 111196, 2020.

2

Artificial Intelligence for Green Energy Technology

Shanthi Jayaraj* **and Meena Chinniah**

Department of Physics, Avinashilingam Institute for Home Science and Higher Education for Women, Coimbatore

Abstract
Artificial intelligence can mimic the output patterns of all human experiment with solar and wind energy the need of the hour. It can be used to grapple with the power challenges of the current global demand. Unpredictable situations with regard to energy crisis can be solved by ensuring complete safety measures. As people become more inventive and depend on green energy, artificial intelligence can support people with what best can be done. To improve efficiency and reliability of all equipments, artificial intelligence helps in the maintenance in splendid ways. Though artificial intelligence never talks back, is always attentive, and like Alexa or Siri, is the best perfect companion for human beings. Not everyone is Edison, who held 2,332 patents worldwide (1093 in the US alone) and hence the authors aim to address the ways in which artificial intelligence can help in green energy.

Keywords: artificial intelligence, solar energy, wind energy generation, fuel cell hydrogen vehicle, photovoltaic cell

2.1 Introduction

The artificial intelligence (AI) technology is essential for utilizing renewable energy to manage power demand. The electricity supply and demand in real-time, customer's energy consumption are balanced by this AI. Solar and wind generation aids in compensating loss of base load generation

**Corresponding author*: shanthi_phy@avinuty.ac.in

Suman Lata Tripathi, Souvik Ganguli, Tengiz Magradze, and Abhishek Kumar (eds.) Intelligent Green Technologies for Sustainable Smart Cities, (19–32)

from coal and gas turbines. Weather data from local weather locations, cloud records, satellite image, and sensor networks are all used to produce renewable energy power. Solar horizontal irradiance, wind speed is essential for maintaining grid stability. Sometimes additional power is needed during festival holiday, big events, in such case predictions with large data sets becomes remarkable. Energy traders need more accurate forecasting to maintain the output and rebalance their position. AI helps in dealing with diverse data more efficiency.

In a perfect world, electricity produced from a local grid on solar PV systems could be used to increase reliability by incorporating AI. Hence, energy traders can make better financial return. The algorithm of AI can absorb the data from smart meters and predict the consumption accurately. Hence, consumers can take caution in minimizing energy consumption, so that electricity bills can get lowered [1].

AI can be used to optimize solar home and battery UPS to meet energy demand for timely heating and cooling production. It also helps customers control energy costs without interrupting heating ventilating and air conditioning (HVAC) systems operations. It can also forecast energy use in buildings 24 hours in advance. AI can play an important role with storage batteries, which can be quickly activated during peak power demand. Accurate demand forecasting is also important, and AI will help with this because of its ability to boost economic load transfer and demand-side control. Smart meter implementation has increased, allowing request data to be delivered to service providers. AI algorithms will internalize data that is transmitted hourly and forecast network load and usage patterns correctly [2, 3].

2.2 Solar Energy and AI

During the month of August (2020), the Fraunhofer ISE achieved 25.9% of energy efficiency for III–V group solar cell deposited on silicon substrate. The modified version of this two terminal architecture device was built using III–IV group semiconductor layers connected to the silicon substrate. Around 35.9% power efficiency rate with the use of semiconductor material like gallium-indium-arsenide-phosphide (GaInAsP) for middle cell was observed. The measured power conversion value under the terrestrial conditions has set new world record and demonstrates the potential of silicon based photovoltaics. The successful improvements in Silicon triple junction solar cells enhances the lifetime of the charge carriers and thus achieve a higher cell voltage. The applications of such PV cells are drones and satellites, where low weight and high efficiency are more essential [4]. There are many benefits when it comes to PV solar energy systems because they reduce pollution, the required infrastructure expenditures are a one-time investment and require regular maintenance costs, but the fuel cost is nothing. Demand for renewable energy will increase in the near future; hence, it is crucial to apply emerging technologies, such as AI and ML, to increase productivity and overcome deficiencies. The constructed solar PV system in our university is shown in Figure 2.1.

300 kwp solar PV system has been installed in our institution. 60 kwp installed in library building rooftop is shown in Figure 2.1. The yield is good and reduces the power bill for our institute from 2015, the year of inception. Solar yield (kwh) for 2018–2019 and 2020–2021 is also shown in the bar graph in Figure 2.2. We also have a single windmill installed. The output being tied to the electricity grid by our institute aids in green energy measures. This measure being followed by many educational institutions, industries will definitely reduce carbon emission.

The team at IBM Research Australia is familiar with the challenges of unique home PV profile building in last five years. AI programs developed by IBM have resulted in 30% improvement in accuracy of weather forecasting. In a partnership financed by the State of Victoria, a project with two business partners, installer and electrify (provider of energy storage), and cloud-based applications will better be orchestrated to get the most

Figure 2.1 Constructed solar panels in our institute.

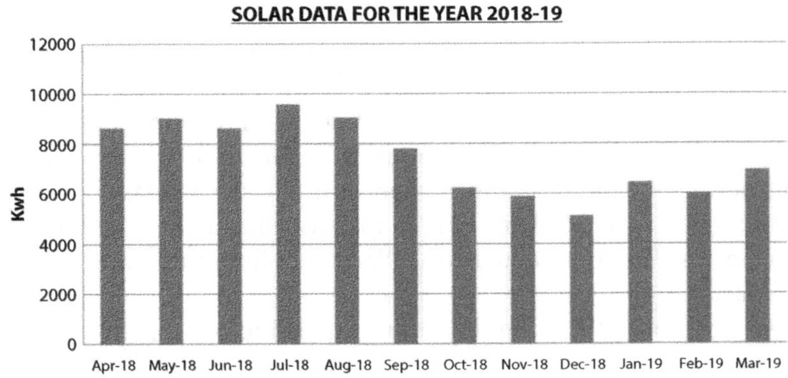

	Apr-18	May-18	Jun-18	Jul-18	Aug-18	Sep-18	Oct-18	Nov-18	Dec-18	Jan-19	Feb-19	Mar-19
Kwh	8623.95	9021.576	8616.485	9570.166	9035.585	7797.417	6227.425	5887.056	5104.255	6441.778	6004.392	6934.3

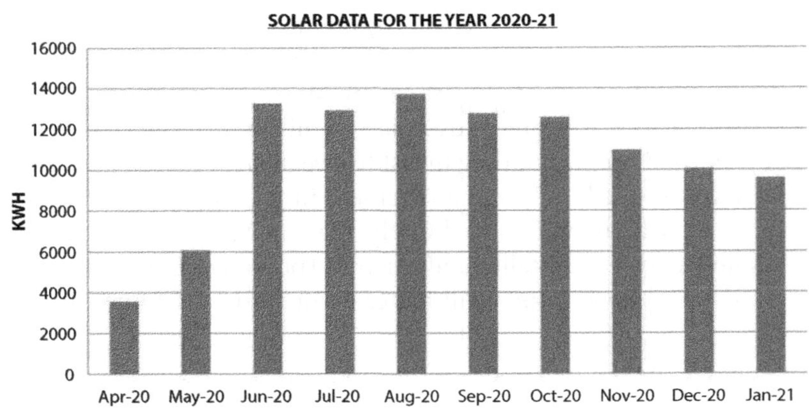

	Apr-20	May-20	Jun-20	Jul-20	Aug-20	Sep-20	Oct-20	Nov-20	Dec-20	Jan-21
Kwh	3561.467	6071.191	13272.412	12909.71	13715.22	12780.93	12577.64	10977.43	10058.63	9597.906

Figure 2.2 Solar yield (Kwh) for the year 2018-2019 and 2020–2021.

production from solar panels. This is where forecasting becomes essential as RE system owners aim for good returns taking into account anticipated demand and an understanding of the electricity tariff structure. Some retailers are even offering discharge to the grid during peak hour's demand as well.

- *Using AI systems, the Microsoft, Google, and Tesla* organizations have explored some efforts for *earth-friendly environment*. The AI system helps to improves energy efficiency and supply energy for remote communities.

- The smart grid extensively employs artificial intelligence techniques to coordinate and monitor portions of the adjacent power grid in order to generate the precise amount of energy needed. Standard power grids that have a certain amount of electricity can be avoided [5].

2.3 AI Transforms Renewable Energy

AI weather data algorithms ensure that power grids are used optimally by adjusting activities to weather conditions at any moment. When spikes in demand occur, this could supply electricity below a fixed point as consumers monitor energy prices and without describing procedures, according to Josh Lehman, senior director of product marketing at US Stem Energy Storage.

The clients' ability to consider their habits and achieve greater stability in smart grids is attributed to prior experience creating energy in AI algorithms, according to Panorama of Critical Resilience. AI will also help energy suppliers with their processes and maintenance. The real time of mechanical failure can be automatically detected by AI algorithms, which helps to improve the reliability and efficiency in power system.

Scientist Tesla invented hundreds of technologies that play a big part in our daily life like; wireless transmission, Neon & Fluorescent lights, robotics and so on. Tesla recently created advanced environment framework to help its energy hardware devices. It is used in over 40 countries. Tesla created tools for energy goods and increased efficiency at gigawatt-scale sites. Tesla's well-known scheme for the largest energy storage product has supported solar power plants, truck charging stations, and utility scale power stations. Autonomous is Tesla's suite of optimization software tools that includes deep learning, forecasting, optimization, and real-time management algorithms that are used to minimize energy bills, participate in demand response, micro-grid control, and wholesale energy market offerings. Tesla's self-driving algorithms simplify the sending of energy reserves in order to optimize economic profit. With features including Virtual Machine Mode, Tesla has also built applications to allow more renewable energy production on the power system.

With artificial intelligence, image recognition can play a significant role in justifying open source data. ENIAN, a tech corporation based in the United Kingdom, owns one of the world's biggest clean energy initiative directories, as well as publicly accessible reports on power plants and grid grids. The organization runs a similar code through the sets of data, and

AI is used to determine how the wind farm can work. The AI examines millions of collections of wind farm data to determine which photos and coordinates align.

2.4 IBM Solution Using AI

IBM is developing the next generation of energy-efficient supercomputers. In reality, IBM systems ranked six of the top ten in this year's Green 500, which ranks the energy performance of the world's best supercomputers. The first three systems mentioned below will minimize energy consumption in a novel and groundbreaking way. Rensselaer Polytechnic's AiMOS machine, ranked third on the list and the most efficient modern supercomputer in the semi-annual TOP500 ranking, has achieved a balance between two separate computational architectures - CPUs and GPUs - to optimize energy and energy usage. This machine was created in collaboration with IBM, Rensselaer, and the New York State Partnership in December 2019. The software is as follows: The Massachusetts Institute of Technology developed the carbon calculator (Satori, a modern IBM POWER9 supercomputer), which measures the carbon footprint of specific tasks. It provides researchers with new perspectives into specific programs running on a mainframe. The most efficient supercomputer in the world was created by Oak Ridge National Laboratory. The automated self-regulated water cooling system, which can recognize temperature variations and configure the settings, was uniquely attached to this.

IBM and Raise Green revealed collaboration with every developer to launch their own neighborhood solar projects for renewable energy sources in order to create a Raise Green program solution built on the IBM Cloud using Red Hat Open Shift. Raise Green approached IBM Global Business Services (GBS) with the need to develop parties to the Originator Engine - a groundbreaking digital platform that clients, developers, and ecosystem collaborators would use to build, fund, and operate solar photovoltaic firms.

2.5 Hydrogen Vehicles

Hydrogen is the lightest and most abundant element. It is non-toxic and can be easily produced from either Natural gas or from Water. Hydrogen fuel cells can last up to 50,000 hours. Japanese car manufacturer TOYOTA is developing a hydrogen engine for motorsports. Mirai was one of the world's first hydrogen fuel vehicle developed by Toyota in 2014. It has no harmful emissions, and the only byproduct is water vapor. Alstom, the French train manufacturer has developed world's first hydrogen train. Alstom CoradiaiLint fuel cell hydrogen train is tried and has shown sustainability and economic viability.

US-based oil and gas companies Chevron and Toyota Motor signed a memorandum of understanding to collaborate on commercially viable, large-scale businesses in hydrogen. The focus of the strategic collaboration will be on fuel cell electric vehicles and joint research and development activities in hydrogen-powered transportation and storage. Duke Company is planning to reduce carbon dioxide emissions by 50% before 2030. It has also partnered with Siemens energy and Clemson University to study the use of hydrogen for energy storage in its power plant in South Carolina. To produce hydrogen from water on large scale, scientists need to find more efficient ways as well as cheaper methods

2.6 Wind Energy and AI

Different AI approaches, such as organized data processing, information mining capabilities, and deep learning methods, are supporting the AI-powered smart energy grid in the energy market. For the development of AI, it needs to add all levels of energy grids. The AI tools help in improving the energy production and design of infrastructure improvement. In European Union countries, there is a growing interest in the renewable energy sector. An analysis of economic facts can lead to important conclusions regarding the RE sector efficiency improvement, carbon emission reduction and green city. Also, the fields of AI, with advanced machine learning act as a solution provider for power grids. The use of AI tools to enhance the sector's RE efficiency towards economic efficiency and sustainability is currently an urgent need. AI services directed at RE us for societal need, and a multi-faceted research initiative is the central theme of future research work. The analysis of the implications for the economy from adopting RE measures poses capacities to reframe the major cities is a challenge for the researchers. This will contribute for studying the contribution of AI to establish green measures [6–8].

Variable wind or solar radiation patterns are causing problems [8] for energy providers as RE accounts for a larger share of electrical systems. As a factor, they need weather conditions (renewables) to know how much electricity they can pump through into network in the coming era. Capacity planning has become critical for power grid and activity planning, as it is needed for the proper operation of the power grid and the optimum use of RE systems. Load and green energy forecasts are critical for effective energy management. The real power grid is not equipped to deal with variable power sources or fluctuating load capability, as in RE. Currently, the situation in the clean energy market is evolving, and new smart algorithms for development and transportation are being investigated. New algorithms necessitate less processing time and greater precision. It also makes use of the hybridization of two or more algorithms to solve the shortcomings of a single algorithm [9]. Networks are becoming increasingly complex, which necessitates the exploration of AI data and the use of advanced algorithms in conjunction with weather forecasting in order to realize the maximum investment value of RE and forecast capacity provided by all network output components. Since the initial investments in renewable energy are high, mass output of renewable energy has a lower return. Since AI allows systems to capture and interpret vast data base obtained from power grid, timely decision-making algorithms to achieve the right resource distribution are critical in the current energy shortage situation. Forecasting, modeling, and inception of RE systems based on the data set helps to recognize. To avoid power outages, demand and supply must be managed efficiently within smart grids. It can also help in making decisions based on previous practice in order to use available evidence in similar ways [10]. Benefits are risks and related costs that suggest the system's integrity. Use of such AI would assist in going beyond conventional factors of production when taking into account the effects on the climate, human possibly the best, and economic growth [11]. This technique is particularly evident when AI is used in the real estate market. AI would reshape the manufacturing and labor markets by transforming the use of technology, expanding job opportunities, and promoting lifelong learning. Given this viewpoint, the labor market poses a new threat to the RE industry in terms of expertise and labor shortages. Established economies are advantageous in terms of investment and labor market because they have a highly educated workforce .Regardless of improvements in workplace composition, RE offers tremendous potential for creating new employment prospects, which is a current political issue in many nations. So AI means improved decision complexity and tempo that exceeds the human aspect, remains found in the operations center, capacity to modify, schedule, superintend, execute laws, safety process guidelines, and so on. Created in the field of renewable energy creation (production), service, oversight, and delivery that will provide expanded potential for human capital use and produce higher income for the society [8].

Wind energy forecasting assists wind farms in organizing their schedule plans by predicting energy generation. Wind speed variations make it risky and difficult for wind turbines to use massive amounts of wind power in their power plants. Wind energy forecasting is a powerful tool for maintaining the security and safety of energy networks. The productivity of a wind farm is determined by wind velocity, temperature, and location. The equation for wind energy, $P = 0.5\ Av^3$, where A is area and v is wind velocity. Any variation in wind speed will cause a cubic variation in wind power. Furthermore, in a real wind farm, this relationship is more complicated since different wind speeds and paths are used to maximize wind power. Wind speed forecasting cannot ensure the dependability and efficiency of wind energy. As a result, upgrading wind energy analysis is critical. Using manufacturers' power curves to add wind velocity to power systems is a popular technique, but the best results are difficult to achieve. The best solution is to produce power curves based on real wind speeds at various axle heights. Wind energy is difficult to integrate into power systems because of its randomness and interference, which limits grid efficiency and power quality. Wind and energy speed forecasting will effectively reduce energy system risks from unreliable wind causes. Wind energy probabilistic forecasting is critical in wind farm use and system activity. It may reflect the unpredictability of predictive results and help in risk evaluation. Wind energy and its volatility can be used to construct models that forecast wind speed and electricity production in an energy system. Accurate wind speed and power estimation will minimize voltage and frequency fluctuations induced by shifts in wind power, as well as unreasonable motion due to a sudden drop in wind power due to extreme wind velocity in traditional power systems. Similarly, choices such as load distribution, blade shape variance, and other control behaviors are made in a wind farm control scheme enrich wind promptness estimate. Furthermore, on big modern wind turbines, the power generating firm must derive the best possible power from the wind speed. Different control mechanisms are often applicable to various operating procedures. The predictive control model is an efficient way to manage all aspects of a wind farm. Predicts future outputs based on past data and inputs. As predictive models, traditional models such as state equations, transition functions, linear stable models, and nonlinear models can be used. When predictive control is used in the model, it is predicted that wind and power velocity prediction can be built on a wide scale.

In terms of wind speed and power prediction, AI algorithms have demonstrated their dominance, but many issues remain, such as uncertain models, slow convergence speed, local preference, number of samples needed, and so on. To solve it, more efficient AI algorithms must be proposed. Support vector machine (SVM) is a machine learning algorithm based on strict mathematical learning theory. It improves generalizability by eliminating structural risks and solves many problems such as thin, linear, and high-dimensional samples. Many of artificial neural network (ANN)'s drawbacks, such as uncertain network mechanisms, interpretability, and curved surface, can be resolved. As a result, SVM is a useful instrument for predicting wind and energy velocity. When opposed to conventional ANNs, they use more opaque layers to solve difficult challenges and achieve good performance. Because of their distinctive deep frameworks, their teaching and predictions of power consumption is preferred. Deep learning models are expected to outperform traditional models in terms of wind velocity and power prediction capacity and efficiency. Other approaches ELM, and data processing are also likely to produce a variety of predictive problems. The hybrid models on which they are based willfully use the benefits of each variable to solve a wide range of issues and improve overall predictive efficiency. It should be remembered that regardless of the algorithm used, predictive precision is the most critical factor. There are two methods that combine AI algorithms to improve predictive accuracy. And that is to use other algorithms to better boost the main algorithm's predictive ability. The other option is to employ other algorithms to reduce the descriptive challenge of the initial wind speed or power results, allowing the main algorithm to have a better predictive accuracy. With these approaches, we think AI algorithms may be more beneficial in the future for wind direction and energy forecast.

The configuration of ANN is neuron-dependent and complex. Via a training phase, it will understand and arrange the special framework adaptively. ANN's fault tolerance means that it can handle missing or spammy results. Parallel computation takes advantage of its simplification and high performance. Thus, ANN will show strong generalization potential and efficiency in wind farms for complex dynamic problems. There is, however, no theoretical basis for strengthening its composition. Manual tuning is typically used to evaluate network configurations and parameters, such as the number of training tests, neurons, and hidden nodes. Although the deep learning network has excellent learning capacity and predictability, a huge number of hyper parameters must be changed, and the training process is largely slow due to its complicated nature and large number of samples. Another issue is ANN over fitting as a result of the training process. A high training accuracy or a long training time often improves the fit of training samples while decreasing predictive efficiency. As a result, the required preparation process must be investigated.

SVM is a machine learning algorithm that uses nonlinear mapping to transform a nonlinear input region into a region with high dimensional characteristics in order to achieve a superlative level. It is a mathematical model and mechanism is entirely different from those of ANN. Maps native feature space to high dimensional space using kernel functions. As a result, SVM can overcome dimensional problems while still ensuring consistency and reliability. Furthermore, because of its compact structure and statistical learning theory, it can model complex nonlinear boundaries and come to terms with the adequacy query. Other characteristics such as high performance, ease of use, and low computing cost have also led to its adoption by many researchers for a variety of problems in wind farms. It does not, however, provides theoretical criteria for choosing kernel functions. SVM output is highly influenced by various kernel functions and parameters. Furthermore, SVM is better suited for analyzing small sample sizes because sample size significantly increase computing time.

2.7 Renewable Energy Industry in India

Due to the growing economy, electricity consumption by consumers is expected to reach 15,280 TWh in 2040. Most of the demand will come from the real estate and transportation sectors. This sector is more attractive to local and foreign investors due to the government's participation in the ambitious green energy goals. The policy in support of the Indian renewable energy industry is (1) The Indian government's target is to achieve 227 GW of electric power generation by 2022; (2) the government's aim is to achieve 523 GW of installed by 2030 and (3) High efficiency solar modules PV and industrialize over 5 years by the PLI scheme.

The largest project to generate TATA renewable energy (solar and wind energy) near international borders. In 2020, a 41,500-capacity park MW obtained a permit from the Ministry of Defense and the Government Revenue Administration devoted to the Khavda grounds in Kutch. The joint infrastructure for the proposed project is developed by Gujarat Power Corporation (GPCL). As per a top member in the state energy industry, the plant is also one of the projects of national significance, and as a result, the state legislature has been proactive in completing the project before the 2022 deadline. The Gujarat government anticipates a huge investment of about Rs 1.35 crore here, according to the official, who added that major companies such as Solar Energy of India (SECI), NTPC, GSEC, Suzlon, and Adani Group have already been allocated land. He mentioned that Adani is projected to receive the lion's share of the overall property, amounting to approximately 20 thousand hectares. Adani Group has announced a Rs 30,000 crore investment in the integrated field by the end of 2024. Meanwhile, Tata Power has reported possession of a grant letter (LOA) from Gujarat Urja Vikas Nigam Limited (GUVNL) to build 100 MW solar projects at Gujarat's Dholera Solar Park. GUVNL will be powered by Tata Group under a Power Purchase Agreement (PPA) that will be in effect for 25 years from the date of commercial operation.

2.8 Conclusion

Renewable energy surely will reduce the carbon emission as all the governments in the entire globe are concentrating on installation of solar PV systems, wind farms, and electric vehicles use every year. Consumption of electricity from the grid increase can be reduced with green energy measures taken care of by allocating more funds. AI plays a pivotal role to balance the electricity supply and the demand needs of the consumer. It can also help in optimal energy usage and pollution reduction in the digital age. AI tools have to be explored in a systematic way by major research institutes, and awareness should also be instilled.

References

1. Jha, S.K., Bilalovic, J., Jha, A., Patel, N., Zhang, H., Renewable energy: Present research and future scope of artificial intelligence. *Renewable Sustainable Energy Rev.*, 77, May 2016, 297–317, 2017, doi: 10.1016/j.rser.2017.04.018.
2. Mocanu, E., Nguyen, P.H., Gibescu, M., Kling, W.L., Deep learning for estimating building energy consumption. *Sustain. Energy Grids Netw.*, 6, 91–99, 2016, doi: 10.1016/j.segan.2016.02.005.
3. Yang, X. *et al.*, Development status and prospects of artificial intelligence in the field of energy conversion materials. *Front. Energy Res.*, 8, July, 1–12, 2020, doi: 10.3389/fenrg.2020.00167.
4. Das, U.K. *et al.*, Forecasting of photovoltaic power generation and model optimization: A review. *Renewable Sustainable Energy Rev.*, 81, August 2017, 912–928, 2018, doi: 10.1016/j.rser.2017.08.017.
5. Kuo, P.H. and Huang, C.J., A green energy application in energy management systems by an artificial intelligence-based solar radiation forecasting model. *Energies*, 11, 4, 1–11, 2018, doi: 10.3390/en11040819.
6. Hong, T. and Fan, S., Probabilistic electric load forecasting: A tutorial review. *Int. J. Forecast.*, 32, 3, 914–938, 2016, doi: 10.1016/j.ijforecast.2015.11.011.
7. Voyant, C. *et al.*, Machine learning methods for solar radiation forecasting: A review. *Renewable Energy*, 105, 569–582, 2017, doi: 10.1016/j.renene.2016.12.095.

8. Diamantoulakis, P.D., Kapinas, V.M., Karagiannidis, G.K., Big data analytics for dynamic energy management in smart grids. *Big Data Res.*, 2, 3, 94–101, 2015, doi: 10.1016/j.bdr.2015.03.003.
9. Sinha, S. and Chandel, S.S., Review of recent trends in optimization techniques for solar photovoltaic-wind based hybrid energy systems. *Renewable Sustainable Energy Rev.*, 50, 755–769, 2015, doi: 10.1016/j.rser.2015.05.040.
10. Xiao, L., Wang, J., Dong, Y., Wu, J., Combined forecasting models for wind energy forecasting: A case study in China. *Renewable Sustainable Energy Rev.*, 44, 271–288, 2015, doi: 10.1016/j.rser.2014.12.012.
11. Mosavi, A., Salimi, M., Ardabili, S.F., Rabczuk, T., Shamshirband, S., Varkonyi-Koczy, A.R., State of the art of machine learning models in energy systems, a systematic review. *Energies*, 12, 7, 1–32, 2019, doi: 10.3390/en12071301.

Website Reference

1. Clean Power Professionals Group, *The Rise of Artificial Intelligence in Renewable Energy* [online], 2020, https://energycentral.com/c/cp/rise-artificial-intelligence-renewable-energy-.
2. Kumar, V., *It's Time to Prioritize Energy Efficient Green Artificial Intelligence* [online], 2020, https://Www.Analyticsinsight.Net/Time-Prioritize-Energy-Efficient-Green-Ai/.
3. Murleedharan, S., *Role of Artificial Intelligence in Environmental Sustainability* [online], 2021, https://www.ecomena.org/artificial-intelligence-environmental-sustainability/.
4. *Hydrogen train coradiailint completes successful tests* [online], 2020, http://www.alstom.com/press-release-news/2020/3/alstoms-hydrogen-train-coradia-ilint-completes-successful-tests.
5. Business world, *Artificial Intelligence and the Power Sector Promising Future* [online], 2021, http://www.businessworld.in/article/Artificial-Intelligence-And-The-Power-Sector- Promising-Future/24-01-2021-369309/.
6. Chen, C., *Google, Microsoft, IBM Are Using AI to Predict Climate Outcomes* [online], 2020, https://www.greenqueen.com.hk/google-microsoft-ibm-are-using-ai-to-predict-climate-outcomes-ccnow/.

Abbreviations

AI- Artificial Intelligence
HVAC- Heating Ventilating and Air conditioning
ANN- Artificial Neural Network
RE- Renewable Energy
SVM- Support Vector Machine

3

Effective Waste Management System for Smart Cities

G. Boopathi Raja

Department of ECE, Velalar College of Engineering and Technology, Erode, Tamil Nadu, India

Abstract

Increasing the volume of wastages has become an unavoidable risk in developing nations due to an unprecedented rise in the rate of population. Nowadays, handling the wastages of any metropolitan cities has become a challenging task. Proper handling and disposal of these wastages are essential for better livelihood. The present scenario in the handling of waste management includes the usage of numerous sensors that indicate the level of the trash bin. From the existing system, several problems have been identified that lead to raise the accumulation of wastages and slow down the disposal mechanism. This is due to the lack of intelligent mechanisms used in municipal waste management. Furthermore, several other issues arise due to a lack of proper knowledge in the waste management system, such as pollution, harmful diseases, etc. In the proposed system, it is capable of collecting wastages effectively based on the detection of moisture level, hazardous gas, and fire in the waste material. This setup was handled and monitored by a control unit, which is located in the municipal offices of each smart city. The important feature of this framework is to collect the wastages on time to reduce the accumulation of waste materials.

***Keywords*:** Disposal mechanism, hazardous materials, solid waste, waste management

Email: g.boopathiraja14@gmail.com

Suman Lata Tripathi, Souvik Ganguli, Tengiz Magradze, and Abhishek Kumar (eds.) Intelligent Green Technologies for Sustainable Smart Cities, (33–52) © 2022 Scrivener Publishing LLC

3.1 Introduction

In the last two decades, waste production in major cities has increased dramatically. According to research undertaken around the world, annual solid waste production is projected to hit around 3.40 billion tonnes by 2050, resulting in an estimated cost of $635.5 billion in urban waste management [1]. Many aspects contribute to making waste disposal more difficult in smart cities. For example, rapid demographic growth, population shifts from small to large cities, economic progress, rising rates of goods use, regional area, system management, collection and disposal of wastages, inadequate knowledge on forecasting the collection of waste materials, and a lack of utilization of smart technology to help the waste management in municipalities and corporation [2]. Inadequate waste management in communities causes significant economic and human life losses. For example, unexpected fires have been recorded during the wastage disposals in open yards located in congested parts of cities, resulting in significant losses in both human life, as well as economy. It is also noted that any form of waste that can easily catch fire, such as solid, liquid, or containerized gas, can result in harm, illness, economic ruin, and environmental damage [2, 3].

Also, hazardous waste may cause a major problem in collecting the wastages from the bin. Today and future challenges are to find a suitable way to handle the hazardous wastes and it is difficult to introduce an efficient methodology to decompose those waste materials. These materials are harmful to both humans and the environment. There are several solutions available to assist communities in properly disposing of household hazardous wastes, protecting the environment, and keeping homes clean. For certain hazardous wastes, recycling services are eligible.

It is necessary to find a way to handle their waste has changed dramatically over the past two decades. The recycling procedure must be effective and adapt to the changes according to the amount of waste materials generated, it may be increased dramatically along with the hazardous waste. Many toxic chemicals or goods containing dangerous elements may be found in ordinary homes. Hazardous wastes can be unsafe to dispose of by daily garbage pickups. Several factors influence household radioactive wastages. Some of them are listed as follows: paints with a solvent foundation, lithium-ion batteries (for example car, mobile phone, or regular household batteries), chemicals for cleaning and polishing, compact fluorescent globes (CFLs), chemicals in a swimming tub or a spa bath, neon lamps, barometers, thermostats, and thermometers, pharmaceuticals (all medicines), computer hardware that is no longer in use, oils for automobiles (for example from cars or mowers), petrol and kerosene, pesticides and other contaminants used in the greenhouse.

Two main reasons play an important role in the contribution of large-scale production of wastages. They are urban population growth and average mean living standards [4]. Hence, these two aspects must be regulated, it is also critical to develop an effective infrastructure, procedures, and approaches to help solid waste management schemes in municipalities, as well as corporation peoples. It must also support the other incidents in a smart city, such as early fire detection. This can be achieved and implemented only with the help of advancements in technologies, innovations in waste management. Internet of Things (IoT) is considered an unavoidable technology and became trending in several fields that can help with these problems. IoT is not a separate framework or device; however, it is an integrated term that encompasses a wide range of mechanical devices, electrical appliances, instruments, modern cars, structures, and all other devices that people have handled use regularly [5, 6]. Many of these tools and equipment are wired to the internet so that they can exchange data with one another, and they are classified as IoT [7, 8]. IoT allows people to access the operation of these machines directly using a previously installed network infrastructure, allowing for a more straightforward introduction of wirelessly operated networks with real-world life experience.

This chapter explains the procedure of building an interconnected physical infrastructure with the utilization of knowledge about engineering backgrounds and utilizing the features of wireless sensor networks. These networks will help smart cities handle themselves better. The proposed scheme is built based on waste bin location recognition using a global positioning system (GPS). The concept used in the proposed framework consists of a small prototype interfaced by Wi-Fi or IoT module. This prototype also includes several sensors that measure the volume of waste materials filled in containers, humidity, temperature, and detection of fire. It also can replay captured data to a control center through a wireless medium. The data collected from each bin is then used to forecast the amount of bin filled and the overall quantity of waste disposal on a scheduled basis.

The projected flow of this chapter includes an automated infrastructure framework for the efficient recycling of waste materials using IoT-based bins and a reduction in the current municipal system's maintenance costs. Flame sensors are built into the device to detect fires that occur in the bins. The detection of fire is being implemented to prevent major economic and human casualties. Furthermore, with the implementation of a smart city, potential waste disposal is expected to benefit the waste management industry. By making good use of waste bins, the scheme would minimize the amount of manpower used in municipal administration. Individuals will only go and gather waste as bins are sent out.

The major contributions of this chapter are discussed as follows: The proposed model would provide urban solid waste management with cost-effective, reliable, and IoT-based garbage collection solutions. By sending out routine evacuation warnings, it would increase air conditions and thereby mitigate health concerns. It will save money and thereby increased the income. It would aid in the reduction of road congestion and noise caused by public transportation. It would minimize city management's manpower commitment by ensuring the trash cans are used effectively.

The outline of the chapter is described as follows. Section 3.2 covers a wide range of similar work. Section 3.3 discusses the waste management schemes followed in India. Section 3.4 analyzes the existing methodology followed in smart cities. Section 3.5 explores the engineered framework and the specifics of its operation. The results and interpretation of the method and bins distribution model are discussed in Section 3.6. Section 3.7 contains the forecast for potential waste. Finally, as a result, Section 3.8 summarizes this research.

3.2 Literature Survey

Several IoT-based waste management systems were successfully implemented and followed in practice in many smart cities to address several issues. Each of these systems has its own merits and demerits. According to the literature, the most important problem for the smart city is the implementation of efficient solid waste management. To address these issues, researchers have used a variety of approaches and strategies, including solid waste management [9]. For solid waste monitoring and processing, weight, humidity, temperature, and toxic sensors are used in the studies of Tao and Xiang [10], Anghinolfi *et al.* [11], and Hannan *et al.* [12].

With the support of modern IT technology, Hannan *et al.* proposed an effective management framework for recycling collection content and the disposal of solid waste in municipalities [12]. A model for waste management, transportation, recycling, and disposal was built in this report. The findings of this study show that the built framework aids municipal authorities in making use of the data produced at each level of waste monitoring and processing. Finally, the scheme accomplished the goal of intelligent waste recycling of collection materials.

Mamun *et al.* introduced an effective monitoring system for measuring the waste material loading amount of bins [13]. This device collects waste information from bins on time using a controller unit, transmitter and receiver unit, and communication module such as RF (radio frequency) unit, and/or IR (infrared) sensor. Sensors in this device detect the amount of waste in the bins and warn the controller. These alerts are encoded by the microcontroller and sent to the main control unit.

Khedikar *et al.* presented an alternate novel approach for collecting data about waste materials from bins [14]. The system implemented by using this technique consists of a GSM module and ultrasonic sensors. The purpose of an ultrasonic sensor is to sense the amount of wastages accumulated in the bin. Then the processor compares it to a predetermined threshold standard. If the present volume of waste material accumulated in the bin exceeds the predefined threshold value, the sensor continuously updates the status to the Arduino Uno. Then, the Arduino processor forwards this information to the GSM module. Finally, the information can be shared with the concerned authorities that are wirelessly connected to the smartphones.

Anjomshoaa *et al.* introduced a new methodology in that the Raspberry Pi is used by an ultrasonic sensor to determine the exact ratio of waste content [15]. This machine allows sensors and humans to communicate over the internet to collect information about how much garbage is filling up the bins. Besides, Kasliwal Manasi *et al.* propose a system for collecting knowledge about solid and liquid waste materials from containers [16]. The WSN and IoT are two major components used in this scheme. Different sensors are connected to smart bins, and WSN was used to communicate with the bins and the server. The waste collector agent (WCA) is crucial in this situation because it receives all of the warnings provided by the sensors. A geographical information system (GIS) is proposed in Asansol (an Indian city) that particularizes the plans for smart waste generation, storage, and transportation [17]. Hong *et al.* propose a real-time smart dumpster built on a GSM module that effectively gathers waste from the city's targeted areas [18].

3.3 Waste Management in India

The Union Ministry of Environment, Forests, and Climate Change is in charge of waste management in India. The Solid Waste Management (SWM) Rules, which replaced the Municipal Solid Waste (Management and Handling) Rules, which had been in effect for 16 years, were published in 2016.

Each year, 62 million tonnes of municipal solid waste (MSW) are produced in urban India (which has a population of 377 million people).

A total of 43 million tonnes (70%) are processed, 11.9 million tonnes (20%) are treated, and 31 million tonnes (50%) are disposed of in landfills.

Urban metropolitan solid waste generation is expected to rise to 165 million tonnes in 2030 as a result of changing usage habits and fast economic growth. Table 3.1 and Figure 3.1 describe the composition of waste materials in South Asia in terms of percentage [19]. From Table 3.1, it is clear that the major components of wastages include food.

Table 3.2 and Figure 3.2 describe the world-leading top 5 municipal solid waste generators annually in terms of percentage [19]. In the survey, India was placed in the top position in the generation of municipal waste and also it may suffer from an exponential rise of solid waste every year. Hence, it is necessary to find a suitable and efficient way to handle the huge generation of solid waste.

Table 3.3 and Figure 3.3 describe the top 10 Indian cities in the generation of waste materials annually in terms of lakh tones [19]. According to the report of Time of India, Indian capital Delhi bagged the first position with approximately 30 lakh tonnes of waste generation annually. From the survey, it is clear that the city with a dense population is the major contributor to waste generation. Hence, it is important to implement an effective waste management system in those cities.

Therefore, an efficient waste management framework has been implemented with an integration of IoT, artificial intelligence, and blockchain technology. The utilization of these technologies is essential for handling the huge amount of waste generation in future [20, 21].

Table 3.1 Composition of waste in South Asia [19].

Composition of waste in South Asia (in %)	
Food and green	57
Paper and cardboard	10
Plastic	8
Glass	4
Metal	3
Rubber and leather	2
Wood	1
Other	15

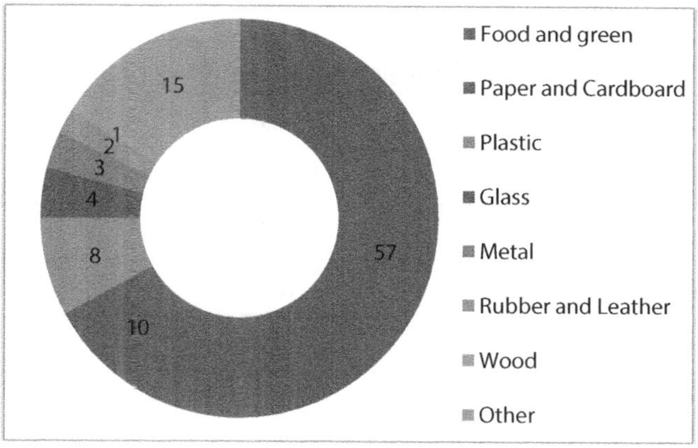

Figure 3.1 Different composition of waste materials in South Asia (in %) [19].

Table 3.2 Top 5 Municipal solid waste generators [19].

Top 5 municipal solid waste generators annually (in million tonnes)			
Country	**2016 estimate**	**2030 projected**	**2050 projected**
India	277.1	387.8	543.3
US	263.7	311	359.9
China	220.4	295	335.8
Brazil	79.1	96.7	114.3
Indonesia	65.2	87.9	118.6

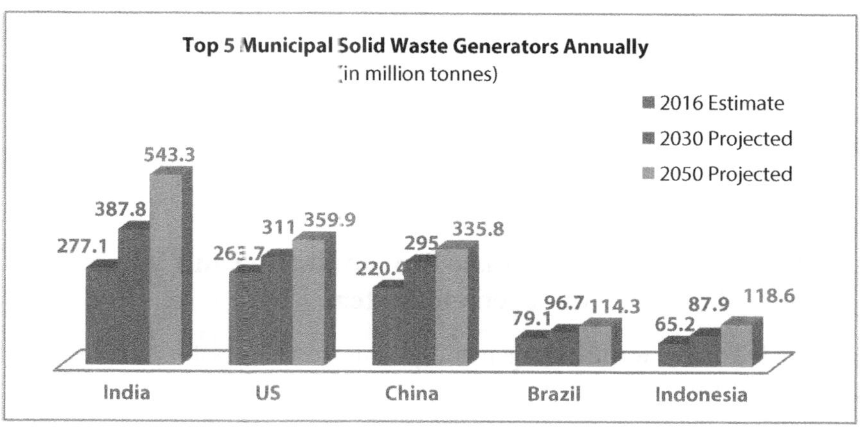

Figure 3.2 Top 5 annual municipal solid waste generators [19].

Table 3.3 Annual waste generation in few important Indian cities [19].

Top 10 Indian Cities for waste generation annually (in lakh tonnes)	
Delhi	30.6
Mumbai	24.9
Chennai	18.3
Hyderabad	16.4
Bangalore	12.8
Ahmedabad	12.1
Pune	6.2
Surat	6.2
Kanpur	5.5
Jaipur	5.4

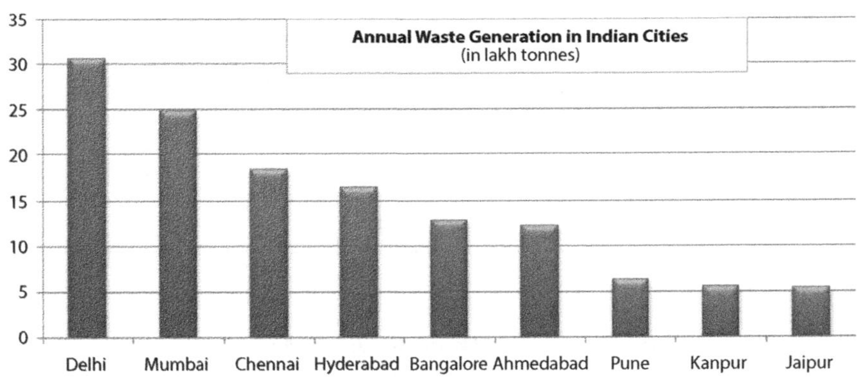

Figure 3.3 Annual waste generation in few important Indian Cities [19].

3.4 Existing Methodology

3.4.1 IoT-Based Smart Waste Bin Monitoring and Municipal Solid Waste Management System

Several IoT-based waste management systems were successfully implemented in many of the cities. One such system was described below. The waste bin monitoring and municipal solid waste management system were successfully implemented in the municipality of Najran city of Saudi Arabia [22]. The purpose of this system is to collect a huge amount of waste materials collected at an exponential rate. Also, this framework was extended to provide the additional service to detect the flaming by providing safety to human lives.

3.4.2 IoT Enabled Solid Waste Management System

It uses a hybrid network architecture to properly manage garbage bins in public spaces and residential regions. End sensor nodes, such as the public bin level monitoring unit (PBLMU) and the home bin level monitoring unit (HBLMU), are used to monitor garbage bins in public areas and home areas, respectively [23].

PBLMUs are deployed using a LoRaWAN networking architecture, whereas HBLMUs are deployed using Wi-Fi-based communication since garbage bins are linked to residences. The PBLMUs transmit unfilled level and geo-location data from garbage bins to the LoRaWAN gateway at a frequency of 915 MHz. The LoRaWAN gateway collects data from the PBLMUs and uploads it to a server for storage and display, whilst the HBLMUs have a Wi-Fi module to connect to home networks and send data to the central monitoring station. The data from the PBLMUs and HBLMUs is published into the server using the MQTT broker, which uses a publish-subscribe communication paradigm.

MQTT protocol is particularly appealing for IoT-based remote monitoring systems because of its low power consumption, quick data transfer, lightweight nature, and ease of installation. The authorized individuals may monitor and evaluate the unfilled levels and related geo-locations of all the trash bins for effective garbage collection using the sophisticated GUI.

3.4.3 Smart Garbage Management System

The identification system, the automatic lid system, the display system, and the communication system are all part of this system. The Arduino Uno microcontroller is used to synchronize all four systems, which are explained in below.

The identification system, automatic lid system, microcontroller, display system, and communication system are all synchronized using an Arduino application. The front of the waste bin is equipped with an ultrasonic sensor. The ultrasonic sensor's transmitter generates an ultrasonic sound that is audible to the human ear, while the receiver collects the reflected sound waves from solid objects. As a result, if any people are spotted within a specified range, the lid will automatically open, thanks to a servo motor installed in the waste bin's upper half. After the lid has been opened, waste may be placed into the bin, and the lid will remain open until the staff are within detecting range. Finally, the lid will automatically close if the workers depart the detecting range.

External causes, such as animals and unexpected weather conditions, might cause littering of garbage, thus a closed container is given here to avoid meddling. An ultrasonic sensor installed inside the waste bin monitors the garbage level continually. The ultrasonic sensor sends out ultrasonic sound, which is reflected by the rubbish in the bin. Between sending ultrasonic sound and receiving reflected sound waves, there is a temporal delay. The percentage of the waste bin that has been full is computed using this time gap. With the aid of the LCD in front of the waste bin, the computed numeric data is presented.

When the waste bin is completely full, the display will show 100% full. The automatic lid mechanism will be disabled as a result of this. The lid will not open if employees are identified within the detecting range, and a message will be sent to the appropriate authorities through GSM to collect the rubbish.

3.5 Proposed Framework

The idea behind the proposed framework on effective waste management is the utilization of cloud computing and IoT concepts. Smart bin is placed in different places of a city based on the daily accumulation of waste materials and population. Sensors are fixed with each bin, and these sensors are responsible for updating the status periodically about the collection of waste materials in the cloud. Based on the status updated, the decision will be taken by the authorities of municipalities. Figure 3.4 shows the overview of the proposed methodology.

3.5.1 System Description

With the support of appropriate hardware components, such as a controller, various sensors, IoT modules, and software/programming tools, an IoT-based intelligent waste collection and monitoring system was developed. This system is designed to assist the municipality with solid waste management for the waste collection produced at a faster rate. Furthermore, this device is capable of detecting fire (flaming) within the bins, which may save lives and reduce financial losses.

The proposed framework is focused on the data about the collection of waste materials from smart bins in an area. The level of waste materials filled in the bins was continuously monitored and reported based on the three different specified fill-up thresholds. The below are the three stages that have been identified:

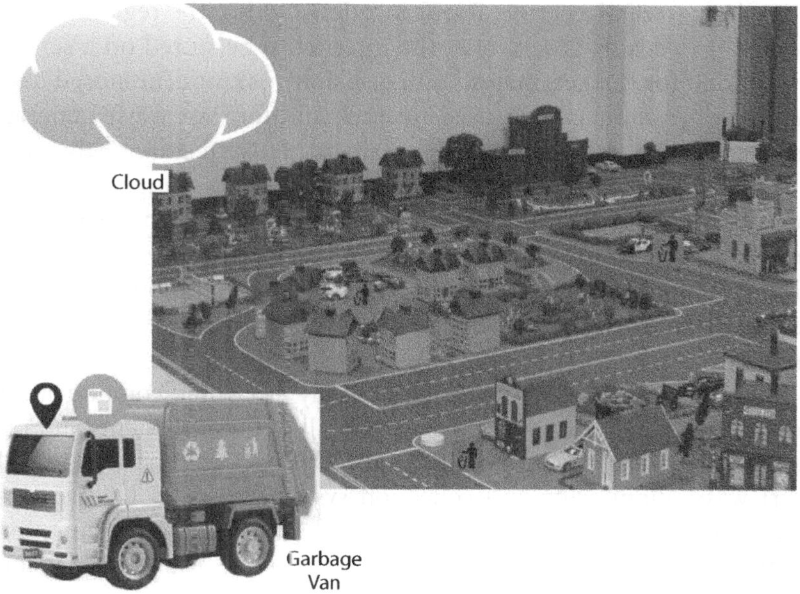

Figure 3.4 Overview of proposed methodology.

a. Full Level

The system sends a warning notification to the authorities of municipalities and truck drivers once it receives a "full alert" message from the sensor attached to the smart bin.

b. Half Level

This is another loading level required for the system to predict the time at which the bin fills up completely. It provides the exact status of the bin and based on this municipality authorities and truck drivers going to take the decision. Suppose if the truck driver collects the waste materials from the other bin, he may also collect the waste materials from this bin only if he knows the exact status. This position aids in the reduction of waste material collection truck transportation costs in terms of both time and diesel. It also reduces the need for manpower.

c. Empty Level

Whenever the waste bin is empty or being empty, the sensors placed in the bin sends the information to the truck driver about the empty level of bins. It avoids the unwanted transportation of the truck driver.

The bins were placed in various locations within the region. The data gathered by sensors is sent over the internet to be stored on a server or in the cloud for further analysis and decision-making. The stored data in the cloud or server is then used to track and forecast waste generation forms and areas in the future. The main attractive feature of the system is that it will benefit from past experiences and exchange information not only for the regular updates of the volume of bins but also from the forecast information of waste generation in the future. The suggested solution also aids in the management of road congestion and provides a cost-effective strategy, as well as other aspects that are impossible to foresee by humans.

The proposed device is made up of many electronic components and sensors that are all linked together. When all of these elements and modules are combined, the proposed design achieves the desired results. The smart waste monitoring and collection system based on IoT were built completely by the integration of several additional components. These additional requirements are described below. The framework includes three categories of modules:

1. Hardware-based module,
2. Software-based module, and
3. External components for communication between hardware and software, such as link wires and a breadboard.

3.6 Functionality of the Proposed System

Figure 3.5 illustrates the structure and outline of the proposed scheme, which are formed by an integration of three modules, which are described in the following:

3.6.1 Sensing Module

The sensing module used in this framework includes four major components that integrate to develop a system. These four elements are listed and described as follows:

- Microcontroller
- Sensors
- Network Interface
- Battery Power

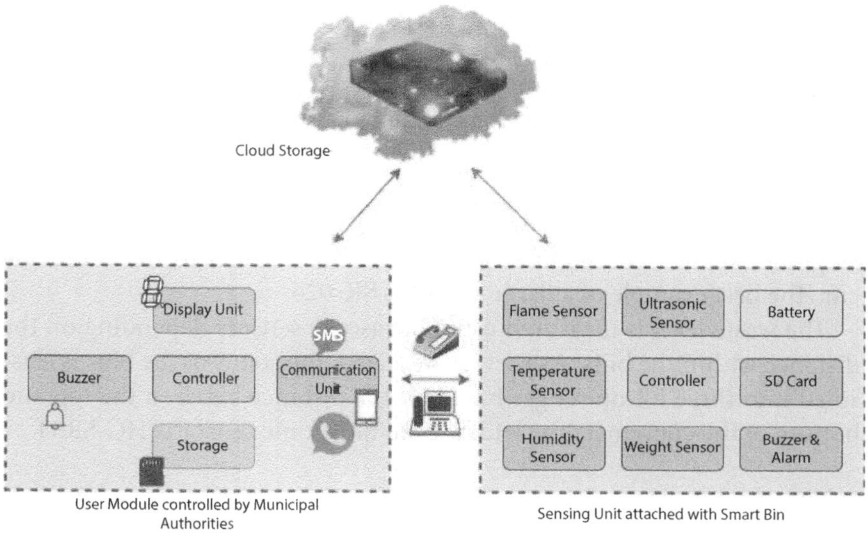

Figure 3.5 Block diagram of the proposed framework.

a. Microcontroller

In this project, all other ATmega328p-equipped module is controlled and monitored in the form of the network by an Arduino Uno microcontroller. The clock speed is about 16 MHz and can operate up to 1.5 Mips per MHz. It consists of 2 kilobytes of RAM and 32 kilobytes of flash memory. This device may operate at a voltage range of 2.5 to 5.5 V and consumes relatively little electricity (40–50 mA). The memory, synthesis, low-cost requirements, and low power consumption may influence the selection of microcontroller. The Arduino Uno is capable of receiving data from sensors and transmitting it to a cloud or master node via a network interface card.

b. Sensors

In the proposed prototype, four different types of sensors are used for their respective purposes. They are listed as below:

i. Weight sensor
ii. Humidity sensor
iii. Temperature sensor
iv. Flame sensor
v. Ultrasonic sensor

Weight sensors are preferred to determine the weight of actual waste collected from every bin to forecast potential waste production in the region. The examples of weight sensors are SEN-10245 and SEN-10245.

The temperature and humidity of the smart bins can be continuously tracked and recorded with the help of Temperature and Humidity sensors. The purpose of preferring this sensor is for the early detection of a fire or other unfavorable events. DHT-11, DHT-21, and HW-505 belong to this category.

The fire inside the bins is detected using flame sensors. We set the wavelength to 760 nm to 1100 nm and the detection distance to 100 cm to 200 cm. The best example of a flame sensor is SK-026.

The waste level is measured using ultrasonic sensors depending on the distance within the bins. As previously stated, the status of the bins is classified into three fill-up ranges based on the distance between the top and bottom of the container. The widely used ultrasonic sensor is HC-SR04.

c. Network Interface

The data is collected by the sensing module via various sensors, and it is then transmitted to the server via a wireless connection. The designed prototype uses used Wi-Fi as a network interface. Wireless Extension Module GSM and GPRS are included in the SIM900A V4.0 Kit. This configuration provides the highest level of ATmega328p microcontroller compatibility while being cost-effective. The planned prototype was completed for less than 50 dollars.

d. Solar Battery

The battery is an important component for any system since it extends the system's life cycle. In this proposed prototype, two 12-V batteries were used and solar-based recharging was allowed. Energy use has a significant impact on wireless infrastructure, data sensing, and forwarding speeds.

3.6.2 Storage Module

In the new methodology, the storage module is made up of three submodules that function together as a single unit. The prediction algorithm, distance measurement model, and database are the three major components required to construct a storage module. These sub-modules are discussed as follows:

a. Prediction Algorithm

   The future waste generation on a timely basis, that is, month or year was predicted by a model based on the proposed algorithm to take the right decisions and plans for the smart city's future waste management. The predictive model is used to estimate waste generation in the future based on data obtained from several nearby bins over a specified time frame. The predictive model is based on empirical evidence and forecasts future events.

b. Distance Calculation Model
In this study, the gap among bins can be determined and positioned at different locations in appropriate areas using a distance measurement model. This model aids in determining the best location for bin installation in the lane, taking into account the population density and waste generation ratios in the city.

c. Database
All of the data obtained from different sensors must be stored in the storage module, such as a centralized server or cloud. This database organizes all of the information so that it can be used in several different ways. The database includes the stored data such as information received from sensors, a field where the waste bin is placed, etc. is preferred to forecast overall waste collection monthly and annually, as well as estimate bin levels.

3.6.3 User Module

The user module designed in this system is made up of two submodules that work together to form a single integrated unit. It consists of the display unit and waste data collection unit. Both submodules are described as follows:

a. Display
The current state of the bins is shown on an LCD and broadcast across a variety of platforms, including cell phones, tablets, navigation systems with data connections, and other wired display units. The truck driver of the waste materials collecting vehicle can observe the current state of the bins and take appropriate action.

b. Waste Data Collection
The waste data is processed on a predetermined date by the waste truck. When the waste truck received a warning from the bins or according to the schedule, it went to certain locations. GPS locations of bins and the record about the visited routes are additional details that waste trucks must hold in their records.

3.7 Workflow of the Proposed Framework

The overall flow of the proposed system for detecting the amount of waste materials accumulation and fire (flaming) within the bins is seen in Figure 3.6.

The power supply initializes the connected elements. Both related sensors check the status of bins regularly. The coordinates of the exact position

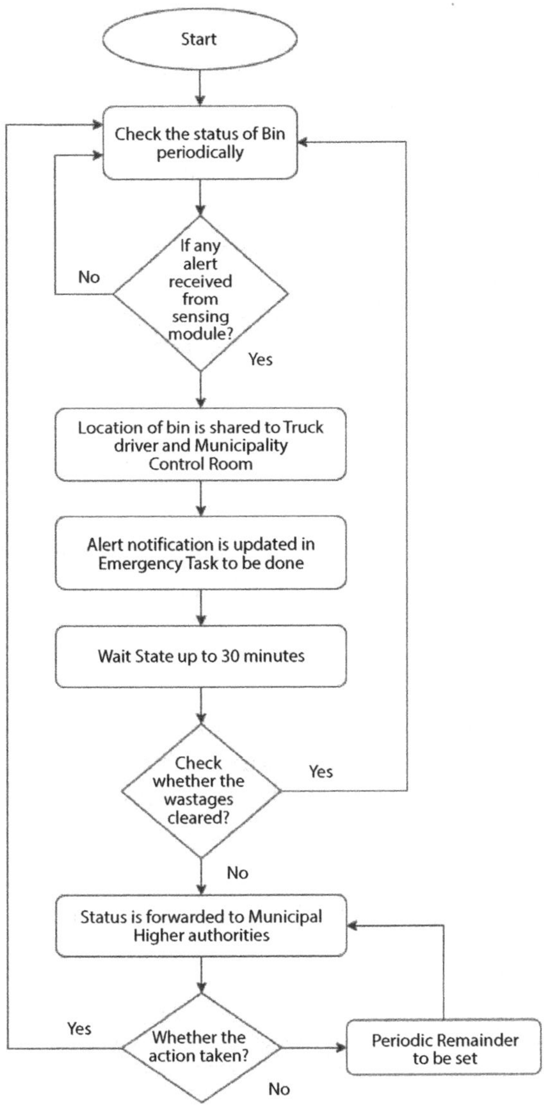

Figure 3.6 Flow diagram of the proposed methodology.

of waste bins were obtained with the help of GPRS based on the feedback or any other notification received from the sensors attached with the bin. Also, location maps are submitted to the corresponding truck drivers, the control room associated with the solid waste management board at the municipalities, and other related waste trucks so that appropriate and timely steps can be taken.

After waiting the prescribed amount of time, the proposed scheme would re-check the waste-filled state of bins to see whether they are empty/clear of waste, and then report the current situation to municipal administration, together with any remedies given. If the state of the bins is not scheduled to resolve within the agreed time, the system will send signals to a higher authority, who will take action.

3.8 Conclusion and Future Scope

This framework describes the innovative procedure that has been followed in the monitoring and collection of waste materials in a smart city. An IoT sensing prototype is preferred to monitor the bin levels and transmit the data to a cloud or server through internet services. For more retrieval and analysis, the data stored in the database and/or server, as well as in the cloud is used. The entire allocated region for collecting the waste materials was split into smaller areas based on the obtained information. The optimized path for each area was found and waste collecting truck/vehicle is allocated to an individual zone that comprises few nearer areas. The chapter is mostly concerned with the system's reliability and maintenance costs. The findings show that waste obtained using an IoT-based approach is more efficient than conventional approaches. The real-time filling status tracking increases average waste collection quality by collecting bins on the same day they become complete and lowering emissions ratios in the environment. In terms of cost-effectiveness, IoT-based solutions are marginally more expensive at first due to rollout and servicing, but the operating costs are much smaller than non-IoT approaches. The findings of this analysis suggest that the device is capable of accurately and efficiently forecasting future waste generation. The potential directions of work include analyzing the outcomes based on finding the optimized path of waste materials collecting truck and reduction in the cost of system implementation, including construction and maintenance costs, which are two possible paths for the work. Furthermore, potential measures may include using publicly accessible GIS data to aid in the improvement of the IoT base system's performance without relying on assumptions.

The proposed scheme is limited to the collection of waste materials and also there is a restriction in continuous monitoring of waste bins. In the future, along with the collection of waste materials with continuous monitoring, the services can be extended to the removal of blocks in sewages due to the accumulation of excess waste materials. For the implementation of future work, it is necessary to integrate gas sensors along with actual setup to detect hazardous gases such as methane ammonia, carbon dioxide, sulfur dioxide, hydrogen sulfide, and nitrous oxides.

References

1. Kaza, S., Yao, L.C., Bhada-Tata, P., Van Woerden, F., *What a Waste 2.0: A Global Snapshot of Solid Waste Management to 2050*, Urban Development, Washington, DC: World Bank, 2020. https://openknowledge.world bank.org/handl e/10986 /30317. Accessed 24 Apr 2020.
2. Kansara, K., Zaveri, V., Shah, S., Delwadkar, S., Jani, K., Sensor based automated irrigation system with IOT: a technical review. *Int. J. Comput. Sci. Inf. Technol.*, 6, 6, 5331–5333, 2015.
3. Tapashetti, A., Vegiraju, D., Ogunfunmi, T., IoT-enabled air quality monitoring device: a low cost smart health solution, in: *2016 IEEE Global Humanitarian Technology Conference (GHTC)*, IEEE, pp. 682–685, 2016.
4. Kelly, S., Suryadevara, N., Mukhopadhyay, S., Towards the implementation of IoT for environmental condition monitoring in homes. *IEEE Sens. J.*, 13, 10, 3846–3853, 2013, https://doi.org/10.1109/jsen.2013.2263379.
5. Kim, K., Interacting socially with the Internet of Things (IoT): Effects of source attribution and specialization in human-IoT interaction. *J. Comput.-Mediat. Commun.*, 21, 6, 420–435, 2016, https://doi.org/10.1111/jcc4.12177.
6. Schaffers, H., Komninos, N., Pallot, M., Trousse, B., Nilsson, M., Oliveira, A., Smart cities and the future internet towards cooperation frameworks for open innovation, in: *The Future Internet FIA 2011 Lecture Notes in Computer Science*, vol. 6656, J. Domingue, *et al.*, (Eds.), Springer, Berlin, 2011.
7. Pavithra, D. and Balakrishnan, R., IoT based monitoring and control system for home automation, in: *2015 Global Conference on Communication Technologies (GCCT)*, IEEE, pp. 169–173.
8. Mitton, N., Papavassiliou, S., Puliafito, A., Trivedi, K., Combining cloud and sensors in a smart city environment. *EURASIP J. Wirel. Commun. Netw.*, 2012, 1, 1–10, 2012, https://doi.org/10.1186/1687-1499-2012-247.
9. Hu, L. and Ni, Q., IoT-driven automated object detection algorithm for urban surveillance systems in smart cities. *IEEE Internet Things J.*, 5, 2, 747–754, 2018, https://doi.org/10.1109/jiot.2017.2705560.
10. Tao, C. and Xiang, L., Municipal solid waste recycle management information platform based on internet of things technology, in: *Proceedings of*

IEEE International Conference on Multimedia Information Network Security, pp. 729–732, 2010.

11. Anghinolfi, D., Paolucci, M., Robba, M., Taramasso, A., A dynamic optimization model for solid waste recycling. *Waste Manage.*, 33, 2, 287–296, 2013, https://doi.org/10.1016/j.wasman.2012.10.006.
12. Hannan, M., Arebey, M., Begum, R., Basri, H., Radio Frequency Identification (RFID) and communication technologies for solid waste bin and truck monitoring system. *Waste Manage.*, 31, 12, 2406–2413, 2011, https://doi.org/10.1016/j.wasman.2011.07.022.
13. Mamun, M.A., Hannan, M.A., Hussain, A., Basri, H., Wireless sensor network prototype for solid waste bin monitoring with energy efficient sensing algorithm, in: *Proceedings of IEEE International Conference Comput. Science and Engineering*, pp. 382–387, 2013.
14. Khedikar, A., Khobragade, M., Sawarkar, N., Waste management of smart city using IOT. *Int. J. Res. Sci. Eng.*, 3, 2, 6–10, 2017.
15. Anjomshoaa, A., Duarte, F., Rennings, D., Matarazzo, T., deSouza, P., Ratti, C., City scanner: Building and scheduling a mobile sensing platform for smart city services. *IEEE Internet Things J.*, 5, 6, 4567–4579, 2018, https://doi.org/10.1109/jiot.2018.2839058.
16. Kasliwal Manasi, H. and Suryawanshi Smitkumar, B., A novel approach to garbage management using Internet of Things for smart cities. *Int. J. Curr. Trends Eng. Res. (IJCTER)*, 2, 5, 348–353, 2016.
17. Yadav, V., Karmakar, S., Dikshit, A., Vanjari, S., A feasibility study for the locations of waste transfer stations in urban centers: a case study on the city of Nashik, India. *J. Cleaner Prod.*, 126, 191–205, 2016, https://doi.org/10.1016/j.jclepro.2016.03.017.
18. Hong, I., Park, S., Lee, B., Lee, J., Jeong, D., Park, S., IoT-based smart garbage system for efficient food waste management. *Sci. World J.*, 2014, 1–13, 2014, https://doi.org/10.1155/2014/646953.
19. https://timesofindia.indiatimes.com/india/in-30-years-india-tipped-to-double-the-amount-of-waste-it-generates/articleshow/74454382.cms.
20. Raja, G.B., Impact of Internet of Things, artificial intelligence, and blockchain technology in Industry 4.0. In: Kumar, R., Wang, Y., Poongodi, T., and Imoize, A.L. (eds), *Internet of Things, Artificial Intelligence and Blockchain Technology*. Springer, Cham, 2021, https://doi.org/10.1007/978-3-030-74150-1_8.
21. Raja, G.B., Impact of IIOT in future industries: Opportunities and challenges. In: A.K. Rana, A.O. Salau, S. Sharma, S. Tayal, and S. Gupta (Eds.), *Internet of Things: Energy, Industry, and Healthcare*, 1st Edition, CRC Press, Taylor & Francis, USA, 2021.
22. Ali, T., Irfan, M., Alwadie, A., Glowacz, A., IoT-based smart waste bin monitoring and municipal solid waste management system for smart cities. *Arab. J. Sci. Eng.*, 45, 12, 10185–10198, 2020, https://doi.org/10.1007/s13369-020-04637-w.

23. Vishnu, S., Ramson, S.R.J., Senith, S., Anagnostopoulos, T., Abu-Mahfouz, A.M., Fan, Z., Srinivasan, S., Kirubaraj, A.A., IoT-enabled solid waste management in smart cities. *Smart Cities*, 4, 1004–1017, 2021, https://doi.org/10.3390/smartcities4030053.
24. Sohag, M.U. and Podder, A.K., Smart garbage management system for a sustainable urban life: An IoT based application. *Internet Things*, 11, 2020, 100255, 2020, https://doi.org/10.1016/j.iot.2020.100255.

4

Municipal Solid Waste Energy: An Option for Green Technology for Smart Cities

Soumitra Mukhopadhyay

Doosan Power Systems India Pvt. Ltd., India

Abstract

Due to environmental issue and depletion of fossil fuel resources, every country is trying to shift to renewable energy sources to obtain considerable amount of power. Out of various renewable energy options available, solar source of energy is a promising one. But unavailability of this energy round the clock, power obtained from municipal solid waste (MSW) is a promising option for smart cities, which many countries have already started to use, and energy from MSW is renewable in nature. Since MSWs are inevitable to generate, these can be utilized through a systematic procedure to generate power; otherwise it would have been dumped as landfill, producing harmful greenhouse gas (GHG) methane (CH_4), which is much more dangerous than carbon dioxide. There are various waste to energy technologies (WTE), out of which incineration, pyrolysis, gasification, anaerobic digestion, landfill with gas capture and microbial fuel cell are established methods. Smart cities should explore one or multiple of these options to generate renewable energy and implement it through integrated solid waste management system (ISWM-S), which involves waste segregation, collection, transportation, and processing of biodegradable waste to energy.

Keywords: Municipal solid waste, smart city, renewable energy, waste to energy technology, integrated solid waste management system

4.1 Unavoidable Impacts of Nonrenewable Energy

In the present times, a much talked subject is renewable energy options (or renewables) from which power can be generated. These resources are

Email: soumitra1995@gmail.com

Suman Lata Tripathi, Souvik Ganguli, Tengiz Magradze, and Abhishek Kumar (eds.) Intelligent Green Technologies for Sustainable Smart Cities, (53–72) © 2022 Scrivener Publishing LLC

such that either they are available in an abandoned quantity or unlimited in nature or can be replenished very easily.

Renewable energies are taken as opposite to fossil fuel energies or non-renewable sources though fossil fuels take a major share in the power generation in the world till now. The most commonly used nonrenewable resources for power generation are coal, oil, and natural gas, which are having several disadvantages:

A) The stocks of above fossil fuels are limited in the human timescale. As per BP Statistical Review of World Energy 2016, the reserves-to-product (R/P) ratio of coal, natural gas and oil are 114, 52.8, and 50.7 years, respectively. The reserves-to-product ratio means the number of years of production left for the known reserves and annual production levels. These values may change if new reserves are discovered, or there is a change in annual production.

B) As per Statistical Review of World Energy 2020, the fossil fuels still supply 84% of the total energy consumption of the world, and it has also been established from several studies that primary energy demand shows a rising trend, both for developed as well as emerging economies. In such a situation, if fossil fuels are used in the same pace, following major issues will be faced by the mankind:
(i) Global Warming and Air Pollution: Burning of fossil fuels contribute to the emission of greenhouse gases, which increases the temperature of mother earth and air pollution. Even the fossil fuels emit harmful air before they are burned. From active oil and natural gas wells and from transport and processing facilities, the carcinogenic benzene and formaldehyde are emitted.
(ii) Acid rain: Acid rain is a chemical phenomenon while the pollutants sulphur dioxide and nitrogen oxides chemically react with water, oxygen and other chemicals high in the atmosphere. These pollutants are produced by burning of fossil fuels in thermal power plants. Acid rain is a serious issue, which damages trees and crops, architecture, statues, rivers, and wildlife.
(iii) Mercury emissions: Due to combustion of coal, mercury is generated, which is highly toxic and damages the environment. It has been observed that the largest emitters of Hg from coal combustion are China, India, and America.

Asia contributes about 54% of the global Hg emission, with China being the largest Hg emitting country [1].

(iv) Land use and the impact on wildlife: It has been observed that wherever fossil fuel reserves have been found, primary focus is the extraction of fossil fuels and due to extraction, production and transportation of fossil fuels, local wildlife is disturbed. Again due to extraction process, where surface mines are used, scars are left upon the landscape.

4.2 Municipal Solid Waste Energy as Clean Energy for Smart Cities

4.2.1 Renewable Energy Options

So to produce sustainable energy with lower emissions, achieve sustainable growth, improve energy security, improve access to energy, and above all, to avoid catastrophic climate change, every country is focussing to a rapid and global transition to renewable energy technologies, which are solar, wind, biomass, waste, hydropower, etc. They are also called green energies or clean energies. Among various renewable energy options available, solar energy has become very much popular due to several reasons:

(i) Abundance in nature and clean source: Solar energy is easily available, accessible, unlimited in human life time and abundant in nature. It does not release harmful greenhouse gas and pollutants — like carbon dioxide, sulphur dioxide, etc. into the air.

(ii) Reduces carbon footprint: Since solar power is a very clean source of energy, it helps in carbon reduction goals for every country, which is a vital step towards combating climate change. Cutting down the carbon footprint will also help to protect wildlife habitats, communities and economies that are affected by warming climate.

(iii) Reduction in cost of solar system installation: The cost of installation of solar systems has dropped drastically in recent years. Now, rooftop solar panels have become very much popular and can easily be installed on roof. It usually takes only a day or two for a home to be set-up with rooftop solar and running on solar energy. In addition, net-metering system has been adopted by many governments to attract people towards installation of this system.

(iv) Reduction in electricity cost: By generating own power through solar panels, electricity cost/bill is also reduced.

But the major disadvantages of solar energy are:

(i) Weather dependency: Solar power is fully dependent on weather; maximum efficiency and power are obtained from solar system when it is sunny.
(ii) High storage cost: Though battery storage cost is reducing, but this is one of the biggest challenges the world is facing with solar power storage. The batteries used for the storage purpose are still quite costly, and they are also large. The more the power required, the larger the battery required.

4.2.2 Municipal Solid Waste as Renewable Energy Option for Smart Cities

Municipal solid waste (MSW), commonly known as trash or garbage, consists of everyday items discarded by people they use. These come from our homes, hospitals, schools, offices, etc. Though the composition of municipal solid waste varies greatly from municipality to municipality, wastes from different municipalities consist of following compositions [2].

(i) Biodegradable waste: green waste, paper, food and kitchen waste, etc.
(ii) Recyclable materials: paper, cardboard, glass, bottles, jars, tin cans, aluminium cans, tires, batteries, etc.
(iii) Inert waste: dirt, rocks, debris from construction.
(iv) Electrical and electronic waste: electrical appliances, bulbs, washing machines, TVs, computers, mobile phones, watches, etc.
(v) Composite wastes: tetra pack food and drink cartons, waste plastics, such as toys.
(vi) Hazardous wastes: paints, chemicals, tires, batteries, fertilizers, and pesticides.
(vii) Biomedical wastes: expired pharmaceutical drugs, etc.

Now, every government is trying to convert urban localities into smart cities without harming the environment to make these urban areas more liveable and truly sustainable and municipalities across the globe are joining the smart city movement. However, municipal solid waste (MSW) has become a very crucial issue nowadays for the smart cities. The World Bank reported that while 1.3 billion tonnes of waste is generated per year in 2012 all over the world, this amount will increase to 2.2 billion tonnes per year by 2025 and by 2050, annual waste generation will increase by approximately 70% to 3.4 billion metric tons [3]. The damps of solid wastes will seriously damage the environmental conditions in smart cities if these are not properly handled and properly taken care of. City managers, who are responsible for protecting public health and overall liveability, are very much worried about the effective waste collection and treatment. It comprises of generation of waste, its storage, collection, transfer and transportation, processing and finally its disposal. Many IoT-based smart systems have been proposed and developed for storage, collection, transfer, etc. of solid wastes [4] but it has been found that MSW is mostly dumped on land or landfilled on riverbanks, or into the sea. For technoeconomic reasons, landfilling is the most suitable option. In developing countries, including India, open dumping is practiced. The solid wastes handled inefficiently may create serious negative environmental impacts for the smart cities like land and water pollution, infectious diseases, obstruction of drains and loss of biodiversity. In addition, Methane generated from the dumps is a non-CO_2 gas that contribute to global climate change and global warming with a potential much higher than CO_2.

So to minimize environmental impacts, Solid Waste Management is defined as the discipline associated with the control of generation, collection, storage, transfer and transport, processing and disposal of solid wastes in a manner so that public health, engineering, economics, conservation and aesthetics are maintained in the best possible way, which is the primary requirement of smart cities. The 7R principle of solid waste management is

(i) Refuse
(ii) Reduce
(iii) Repurpose
(iv) Reuse
(v) Recycle
(vi) Rot and
(vii) Rethink

4.2.3 Why Is MSW Energy Renewable?

Though recycling and reuse are in practice, municipal solid waste (MSW) is a locally available one of the largest renewable energy sources, which produce low carbon dioxide (CO_2) emissions. It has been observed that about 80% of the carbon in MSW is biomass derived and thus can be considered as a renewable source [5]. While MSW is managed in a better way in developed countries, it can be used in a sustainable way in smart cities to gain advantages from following issues:

(i) Greenhouse gas emission
(ii) Acid rain
(iii) To reduce depletion of fossil fuels, natural gas etc.
(iv) Water and land contamination

4.2.4 Various Waste to Energy Technologies

Most of the MSW can be converted to heat and power through different waste to energy technologies (WTE-T) thereby minimizing environmental effects associated with the landfilling or open dumping. For example, major cities in Malaysia generate MSW of 1.7 kg/person/day, and it has energy potential of 640 kW/day through the process of incineration [6]. As volume and weight of waste are reduced considerably while power is generated through WTE-T, requirement of land for dumping residual waste is also minimized. Various methods are practiced for Waste to Energy technologies, as shown in Figure 4.1.

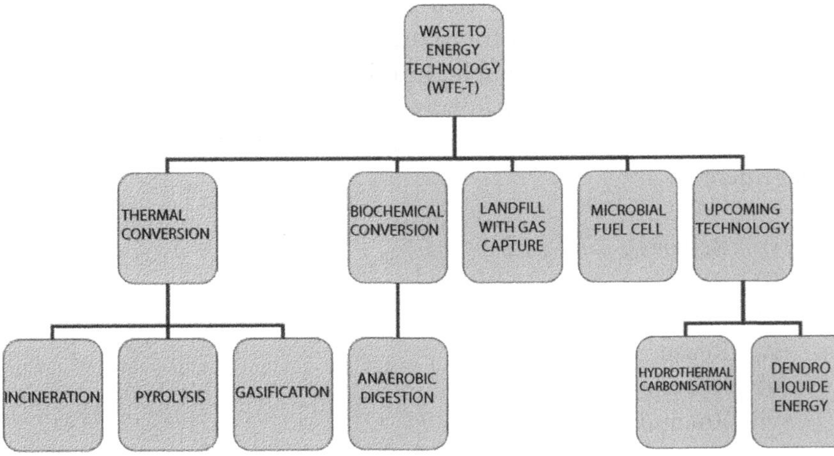

Figure 4.1 Various methods for waste to energy technologies (WTE-T).

Composition of wastes is influenced by various factors like culture, climate, economic development, etc. While it has been observed that low-income countries have the highest proportion of organic wastes, high-income countries dispose of paper, plastics, and other inorganic materials as solid wastes [3]. Since MSW composition varies upon various factors, a specific classification is required for the research of thermal conversion of MSW components. Zhou *et al.* [7] had taken 26 kinds of MSW component samples, and MSW components were classified according to the proximate and ultimate analyses, heating value results, as well as thermogravimetric (TG) characteristics. Generally, combustible MSW were classified into six physical groups: wood waste, food residue, textiles, paper, plastics, and rubber. From these groups, typical components can be selected for thermal conversion in waste-to-energy research. As already explained, the composition of MSW is highly variable, particularly between developed and developing countries, mainly dependent on which income-group the country is falling. The residue has a significant calorific or heat value making it suitable for energy recovery options. Typically, MSW has 8 to 12 MJ/kg of energy, while received energy from coal is 25 to 30 MJ/kg, i.e., a tonne of MSW has about one-third of the calorific value of coal and can produce about 600 kWh of electricity [5].

4.3 Waste to Energy Technologies (WTE-T)

In the smart cities, huge wastes generated can be converted to energy/power in multiple ways. The commonly used established methods have been described below in brief.

4.3.1 Incineration

Incineration is simply the burning of wastes. It is a thermal decomposition of organics in the presence of oxygen at higher temperatures. The process normally requires temperatures between 850°C and 1100°C. The total process has been shown in Figure 4.2. Garbage bunkers are used to store the wastes required to run the incineration plant. First of all, garbage is pretreated and then the waste is fed into the combustion chamber, i.e., boiler. High-pressure steam is generated in the boiler. This steam is then passed through turbine(s) (HP/IP/LP as the case may be) which is coupled with the generator to produce electricity. The waste is converted into ash, which is about 30% to 40% of the waste of the original waste. The process

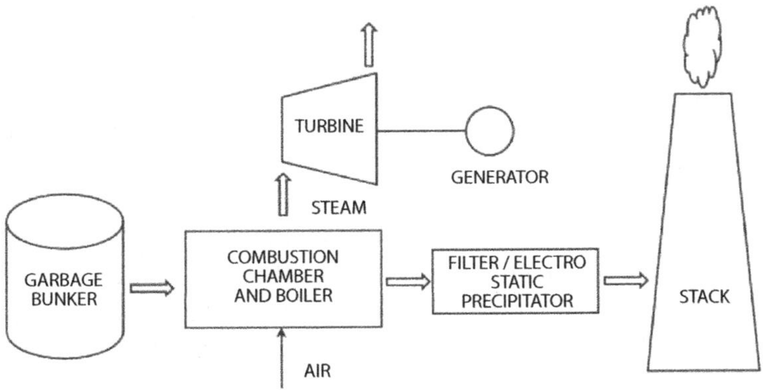

Figure 4.2 Block diagram of incineration process.

of burning produces toxic smoke and flue gas, which are passed through various equipments like electrostatic precipitator (ESP), scrubbers to eliminate hazardous elements and is released into the atmosphere via the stack/ chimney. In addition to the generation of electricity through the incineration process, MSW mass and volume are reduced and the net electrical efficiency is about 15% [8].

Different types of incinerators producing WTE are:

(i) Rotary Kiln Incinerators
(ii) Fluidized Bed Incinerators
(iii) Liquid Injection Incinerators
(iv) Multiple Hearth Incinerators
(v) Catalytic Combustion Incinerators
(vi) Waste Gas Flare Incinerators etc.

Major advantages of incineration:

(i) Heat and Power Availability: From the process of incineration, both steam and power are obtained. This steam can be used as process heat and power is renewable one.
(ii) Reduced Landfill Issue: This process reduces considerable weight and volume of MSW, by 70% and 90% respectively [8], thereby easing out landfilling problem.
(iii) Reduces GHG Emissions: From the open dump site of MSW, methane (CH_4) — a greenhouse gas (GHG) — is emitted. Incineration eliminates this problem. It stabilizes waste by oxidizing organic fraction of the waste stream.

(iv) Reuse of By-product of Incineration: By-product of incineration process can be used as secondary aggregates for parking lots and paved roads.
(v) Toxicity of Hospital Waste Eliminated: High-temperature incineration destroys toxicity of hazardous hospital wastes.

Major dis-advantages of incineration:

(i) Expensive Pollution Control: Incinerators emit environmental pollutants like nitrous oxides, acid gas and particulate matters. These should be avoided to escape through stack during the incineration process and so, pollution control devices are installed, which add cost to the plant. A pollution control device could be an ESP, scrubber, dilution equipment, etc.
(ii) Landfill Pollution: By products after the process of incineration is the pollutants like fly ash, bottom ash, boiler ash, slag, etc., which are about 30% of the MSW, and these are deposited in landfills. These residual ashes contain a wide variety of toxic compounds and elements, which can pollute the groundwater, rivers, and soil.

4.3.2 Pyrolysis

Pyrolysis of MSW is rapidly developing thermal conversion technology, and it is gaining attention due to its high efficiency and good ecofriendly performance. Pyrolysis technology provides an opportunity for the conversion of municipal solid wastes into clean energy by converting urban wastes of smart cities into products, which can be effectively used for the production of electricity, heat and chemicals.

This process consists of both simultaneous and successive reactions when carbon-rich organic material is heated in a nonreactive atmosphere. Simply speaking, it is a thermochemical decomposition of organic product in the absence of oxygen or inert gas environment leading to formation of new molecules. Thermal decomposition is carried between 300°C and 900°C and beyond 700°C. syngas is vital product [9] in the absence of oxygen.

Pyrolysis of municipal wastes begins with mechanical separation of glass, metals, paper, organic-waste, and inert materials prior to processing the remaining waste in a pyrolysis reactor, which acts as an effective waste-to-energy convertor, as shown in Figure 4.3. External heat source is

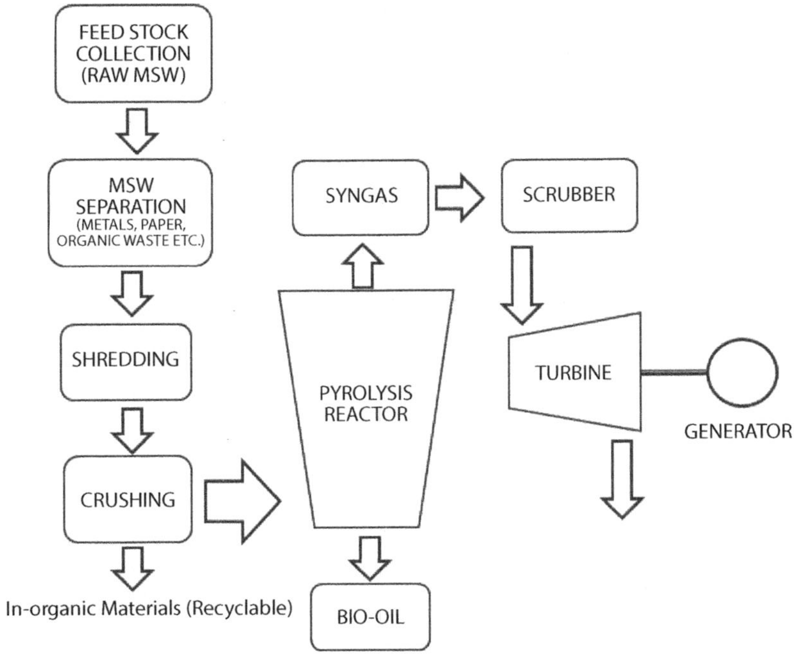

Figure 4.3 Block diagram of pyrolysis process.

required to maintain the high temperature in the process. Produced syngas is passed through scrubber for the purpose of filtering. Then it is passed through gas turbine, which is mechanically coupled with generator to produce electricity.

The main products obtained from pyrolysis of municipal wastes are synthesis gas or syngas with high calorific value, a biofuel or pyrolysis oil, and solid residue char. The type of product generation mainly depends on temperature and heating rate and biomass composition.

Depending on operation conditions, materials input and products, different types of pyrolysis reactors are:

(i) Fixed bed reactors and batch reactor
(ii) Fluidized bed reactors
(iii) Spouted bed reactors
(iv) Rotary kiln reactors
(v) Microwave assisted reactors
(vi) Plasma reactors
(vii) Solar reactors

Major advantages of pyrolysis:

(i) Limited oxygen is used in this process producing few air emissions.
(ii) Syngas contamination is easy to control since it is cleaned before releasing it to the atmosphere
(iii) Pyrolysis produces various useful products
(iv) It is more efficient process than incineration

Major disadvantages of pyrolysis:

(i) It generates toxic residues such as mineral ash, inorganic compound etc. which are harmful to environment
(ii) It generates toxic air emissions too such as acid gases, nitrogen oxides, sulphur dioxide, particulates, etc.

4.3.3 Gasification

Harnessing energy from MSW gasification is gaining the interest of the energy researchers due to its wide availability and environment friendly applications. Regarding the use of gasification, it can be converted into three products: electrical/heat energy, transport fuel, and chemical feedstock. Gasification is the partial combustion of MSW, i.e., carbonaceous (carbon-based) raw material to generate gas after shredding, which is used to reduce the size of feedstock [10].

Gasification is set of heating process of MSW at about 700°C to 1100°C under substoichiometric condition of air supply to produce a low or medium calorific gas. Depending upon the carbon, hydrogen, and oxygen content of feedstock and also the gasifier properties, the product gas composition and calorific value may change. These gases are often referred as syngas. Gasification occurs in a gasifier, which is generally a high-temperature/pressure vessel where oxygen (or air) and steam are directly contacted with the feed material. It causes a series of chemical reactions, which convert the feed to syngas and ash/slag. Composed primarily of the colourless, odourless, highly flammable gases carbon monoxide (CO) and hydrogen (H_2), syngas has a variety of uses. The syngas can be further converted to hydrogen and carbon dioxide (CO_2) by adding steam and reacting over a catalyst in a water-gas-shift reactor. Gasification reduces volume by up to 90% with virtually zero hazardous emissions [11]. The composition of syngas depends on the feedstock characteristics, type of catalyst, type of gasifier, and operating conditions, such as temperature, pressure, residence time, etc. The process flow is shown in Figure 4.4.

A gasification system is generally composed of three stages:

(i) A gasifier producing syngas
(ii) A cleaning system for removal of pollutants and harmful compounds
(iii) An energy recovery system such as a gas engine with sub-system preventing environmental impacts.

Syngas thus produced can be used in a number of ways some of which are noted below:

(i) Syngas can be burned in a boiler to generate steam for power generation or process heating.
(ii) Syngas can be used as a fuel in a gas engine.
(iii) Syngas, after cleaning, can be used in a gas turbine

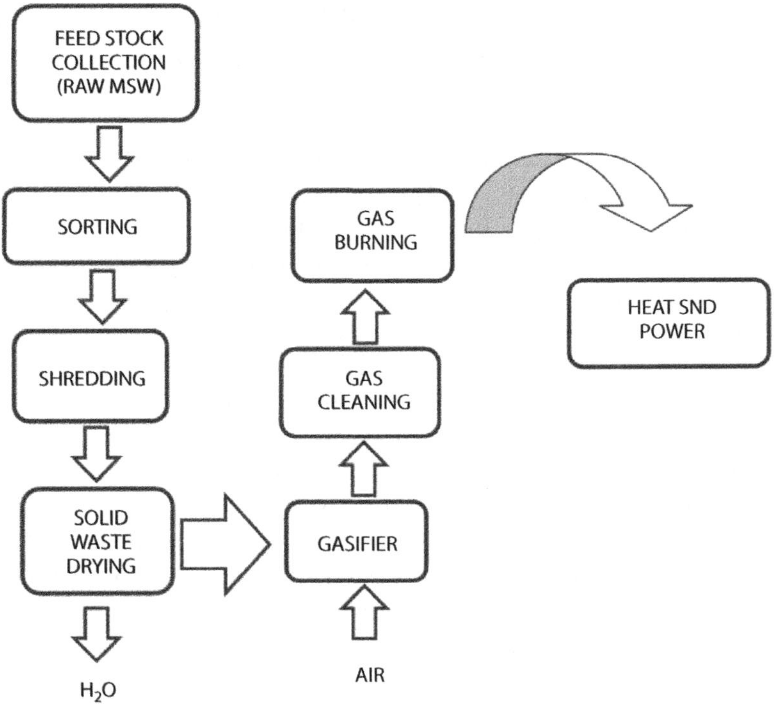

Figure 4.4 Block diagram of gasification process.

Different types of gasification reactors:

- (i) Bubbling fluidized bed gasifiers
- (ii) Circulating fluidized bed gasifiers
- (iii) Dual fluidized bed gasifiers
- (iv) Plasma gasifiers
- (v) Entrained flow gasifiers

Major advantages of gasification:

- (i) Gasification occurs in a low oxygen environment, which limits the formation of large quantities of SOx and NOx.
- (ii) Since stoichiometric/substoichiometric amount of oxygen is required in this process, volume of syngas is low, which helps in using less expensive gas cleaning equipment.
- (iii) The syngas produced can be used with reciprocating engines, combined cycle turbines, fuel cells, etc.

Major disadvantages of gasification:

- (i) MSW has a relatively high ash content, which results in the formation of tars, particulates, halogens, heavy metals and alkaline compounds. These are accumulated in the fluidised bed of gasifier.

4.3.4 Anaerobic Digestion

Anaerobic digestion (AD) can broadly be called a natural bacterial decomposition process of organic matters [12] — such as animal manure, food wastes, etc., in the absence of oxygen. The process reduces the amount of wastes and produces biogas, which is used as an energy source. There are four steps in AD process, such as:

- (i) Hydrolysis: Organic compounds are converted to structural molecules, such as fatty acids, amino acids, and related compounds.
- (ii) Acidogenesis: From the products of hydrolysis, gases such as CO_2, CH_4, and NH_3 are produced with the help of acidogenic bacteria.
- (iii) Acetogenesis: In the acidogenesis phase, simple molecules created, which are further digested by acetogens. This

acetogenesis process produces largely acetic acid, as well as carbon dioxide and hydrogen.

(iv) Methanogenesis: At this stage CO_2, CH_4, and water are produced with the help of methanogens bacteria.

The process flow diagram of the Anaerobic Digestion is shown in Figure 4.5. This process generates three by-products including biogas, bioliquid, and fiber digestate. Biogas, which is a gaseous mixture composed of 60% of methane (CH_4) and about 40% of carbon dioxide (CO_2), with some amounts of hydrogen sulphide and ammonia, can be used in combustion engines, microturbines or in combined heat and power (CHP) plants to generate heat and electricity.

Major advantages of AD:

(i) AD produces by-products, which can easily be captured and used for soil fertilization and the generation of heat and/or electricity
(ii) In comparison to incineration, pyrolysis, and gasification, anaerobic digestion produces the less amount of pollutants
(iii) AD plants are small, which makes it suitable for instalment in any location

Major disadvantages of AD:

(i) Sometimes, AD does not completely digest all the wastes, which may not meet government standards; sometimes, it can emit dangerous element like mercury.
(ii) AD produces methane and carbon dioxide with high concentration and these are toxic greenhouse gases, which are associated with global warming and climate change.

4.3.5 Landfill with Gas Capture

From the municipal waste disposal sites, landfill gas (LFG) is emitted, which primarily consists of methane and carbon dioxide, which are around 98% of total gas emitted and LFG is generated through anaerobic decomposition. Emitted amount of methane from landfill can be reduced to carbon dioxide through the installation of landfill gas capture where combustion of methane takes place to convert it to carbon dioxide. This landfill gas has sufficient heat content of about 18 MJ/m^3 [8], which can be used for energy recovery through heat and electricity.

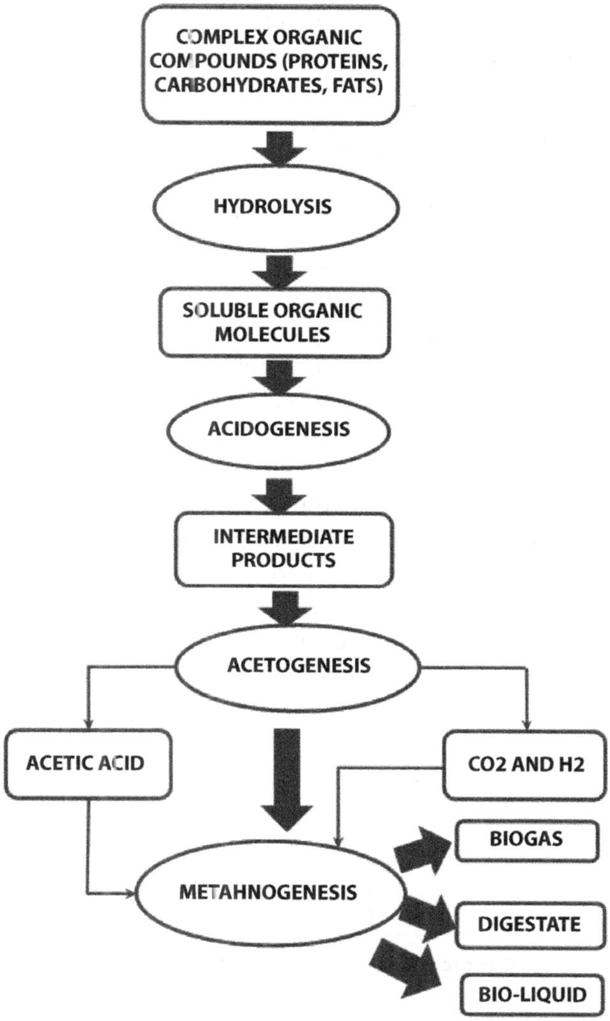

Figure 4.5 Process flow diagram of anaerobic digestion.

A series of wells are made in the landfills, and the gas is extracted through these wells. To make it usable, it is treated first to remove offensive odours, water, particulates, etc. It is then passed to a central point where it can be processed. From this point, the gas can be used in an LFG energy project, thermal uses in kilns, etc.

Major advantages of landfill gas are:

(i) It is an excellent energy source.
(ii) Modern landfills are ecofriendly.

(iii) The system keeps cities, towns, and districts clean.
(iv) Landfills are cheap.

4.3.6 Microbial Fuel Cell (MFC)

The traditional version of fuel cell (FC) which generates energy uses metal catalysts but Microbial Fuel Cell (MFC) is a device that converts chemical energy into electrical energy through the catalytic activities of microorganisms.

In its construction of MFCs, they consist of anode and cathode chambers, separated by a proton exchange membrane (PEM) [13]. Organic substrates are oxidized by the active biocatalyst in the anode and produces electrons and protons [14]. Through the PEM, the protons are transmitted to the cathode chamber, and the electrons are conveyed through the external circuit. In the cathode chamber, protons and electrons are reacted along with reduction of oxygen to water as shown in Figure 4.6.

The main limitations of the MFCs are:

(i) Enough power is not generated by this system.
(ii) Since microbial reactions are slow at low temperatures, MFCs cannot operate at low temperatures

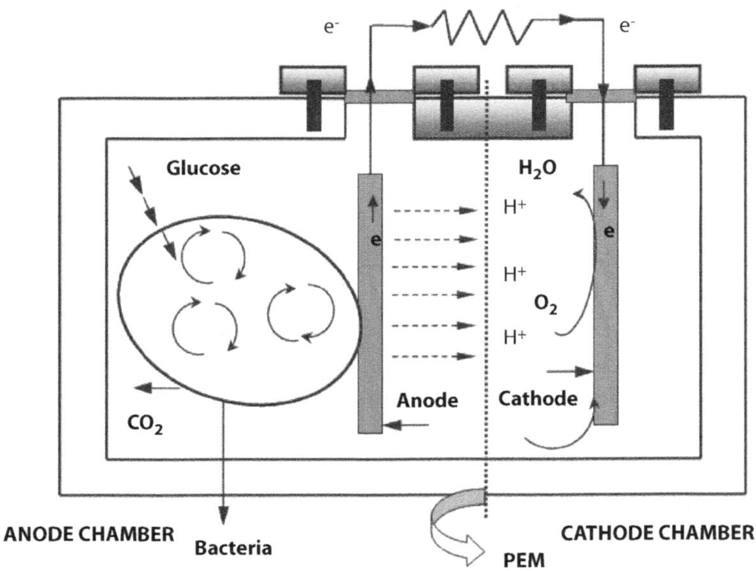

Figure 4.6 Microbial fuel cell.

4.4 Integrated Solid Waste Management Systems (ISWM-S) for Smart Cities

Definitely, there are certain pain points in waste collection in cities and urban localities:

(i) Many people still see the waste management as an expenditure
(ii) Frequency of trash collection is very poor
(iii) Budget allocation by municipalities in waste collection is not sufficient
(iv) Sometimes trash bins are turned upside down or kicked off

Above issues can be found more in developing countries and some developed nations as well but solid waste should be taken as a resource from which energy can be recovered. Smart cities should be encouraged for Integrated Solid Waste Management Systems (ISWM-S) which encourages policy makers to develop and implement policies, strategies and practices for cleaner, safer and efficient use of wastes. In ISWM-S, authority should emphasis on following steps:

(i) Waste segregation at source
(ii) Door to door collection of segregated waste
(iii) Transportation of waste to dumping ground
(iv) Segregation of waste at dumping ground if required and
(v) Processing of biodegradable waste either to compost or energy

Benefits of ISWM-S are as follows

(i) Cleaner and safe neighbourhoods
(ii) Efficient use of resources
(iii) Reduced final weight of waste reducing the cost in waste management
(iv) Better business opportunities and economic growth Local

Many countries/localities have incorporated ISWM-S into their policy. For example, the Lee County has set up WTE plant, which generates 53MW of electric energy through incineration process. In Sheffield, UK, an energy recovery plant is located, which generates electricity for the National Grid and heat for the District Network. In Porto, Portugal, total solid waste production is utilized: gasification plant for organic wastes, an incineration plant for solid wastes, and a composting plant for organic fraction and the remaining pretreated MSW is disposed of into the landfill system [15].

4.5 Conclusion

Solid waste management is burning challenges for the cities. Due to rapid urbanization, improved living standard and changing lifestyle, solid waste generation is increasing day by day. Smart cities should think of it for efficient collection, transportation and utilization for energy generation in one or multiple methods as described earlier. Smart cities should also implement Integrated Solid waste management system to use the solid wastes, which are inevitable to generate, in a better way.

References

1. *Reducing Mercury Emission from Coal Combustion in the Energy Sector in Thailand*, A UN Environment Report, Faculty of Public Health, Thammasat University, Pathumthani, 12120, Thailand, 2018.
2. Municipal solid waste, 2021, https://en.wikipedia.org/wiki/Municipal_solid_waste.
3. Hoornweg, D. and Bhada-Tata, P., What a Waste: A Global Review of Solid Waste Management. Urban Development & Local Government Unit, World Bank, Washington, DC, USA, 2012.
4. Ali, T., Irfan, M., Alwadie, A.S., Glowacz, A., IoT-based smart waste bin monitoring and municipal solid waste management system for smart cities. *Arab. J. Sci. Eng.*, 45, 10185–10198, 2020.
5. *Municipal solid waste and its role in sustainability*, A position paper prepared by IEA Bioenergy, IEA Bionergy, New Zealand, 2003.
6. Kathirvale, S., Yunus, M.N.M., Sopian, K., Samsuddin, A.H., Energy potential from municipal solid waste in Malaysia. *Renewable Energy*, 29, 559–567, 2003.
7. Zhou, H., Long, Y.Q., Meng, A.H., Li, Q.H., Zhang, Y.G., Classification of municipal solid waste components for thermal conversion in waste-to-energy research. *Fuel*, 145, 151–157, 2015.
8. Nixon, J.D., Dey, P.K., Ghosh, S.K., Davies, P.A., Evaluation of options for energy recovery from municipal solid waste in India using the hierarchical analytical network process. *Energy*, 59, 215–223, 2013.

9. Chen, D., Yin, L., Wang, H., He, P., Pyrolysis technologies for municipal solid waste: A review. *Waste Manage.*, 34, 2466–2486, 2014.
10. Belgiorno, V., De, F.G., Della, R.C., Napoli, R.M.A., Energy from gasification of solid wastes. *Waste Manage.*, 23, 1, 1–15, 2003.
11. Kumar, S., Technology options for municipal solid waste-to-energy project, in: *TERI Information Monitor on Environmental Science*, vol. 5(1), pp. 1–11, 2000.
12. Maunder, D.H., Brown, K.A., Richards, K.M., Generating electricity from biomass and waste. *Power Eng. J.*, 9, 4, 188–196, 1995.
13. Ghasemi, M., Daud, W.R.W., Ismail, M., Rahimnejad, M., Ismail, A.F., Leong, J.X., Miskan, M., Liew, K.B., Effect of pre-treatment and biofouling of proton exchange membrane on microbial fuel cell performance. *Int. J. Hydrogen Energy*, 38, 5480–5484, 2012.
14. Antonopoulou, G., Stamatelatou, K., Bebelis, S., Lyberatos, G., Electricity generation from synthetic substrates and cheese whey using a two chamber microbial fuel cell. *Biochem. Eng. J.*, 50, 10–15, 2010.
15. Moya, D., Aldas, C., Lopez, G., Kaparaju, P., Municipal solid waste as a valuable renewable energy resource: a worldwide opportunity of energy recovery by using waste-to energy technologies. *Energy Proc.*, 134, 286–295, 2017.

5

E-Waste Management and Recycling Issues: An Overview

Simran Srivastava, Sahil Virk, Saumyadip Hazra and Souvik Ganguli*

Department of Electrical and Instrumentation Engineering, Thapar Institute of Engineering and Technology, Patiala, Punjab, India

Abstract

E-waste is a rapidly growing subject that has gained the attention in past few years. It is not only harmful in terms of environmental aspects but also in terms of health as well. Along with these negative impacts, a positive impact is the remarkable amount of valuable metals that can be recovered from it. International organisations, which are the concerned authorities of e-waste, have made many rules and regulations for the collection, recycle, disposal, etc., but these efforts are still not implemented to their full potential. Although several types of research have been carried out on different aspects of e-waste, still execution is not up to the mark and the outcome is not satisfactory. This chapter deliberates on the overview of the several issues of e-waste management. The global status of e-waste, industrial practices of e-waste, recycling of e-waste is also reviewed in this chapter. It thereafter examines the future and benchmarking of e-waste management. Another striking issue is that the steps taken toward the control and regulation of management are not sufficient and is not receiving many contributions.

Keywords: e-waste, global status of e-waste management, industrial practices of e-waste management, benchmarking of e-waste management, recycling issues

5.1 Introduction

The use of electronic devices has increased with the growing technology over time and became a requirement for mankind. In addition to the electronic

**Corresponding author*: souvik.ganguli@thapar.edu

Suman Lata Tripathi, Souvik Ganguli, Tengiz Magradze, and Abhishek Kumar (eds.) Intelligent Green Technologies for Sustainable Smart Cities, (73–88) © 2022 Scrivener Publishing LLC

facilities, science is also involved in some problems and the problem here is that these commodities are short in existence and are rapidly diminishing due to rapid change [1]. When these commodities reach near the end of usefulness or lifetime then they are considered as e-waste [2]. Electronic waste is popularly referred to as e-waste in abbreviated form and is the name of both electrical and electronic products toward the termination of their lifetime [3]. It is assumed to be one of the most rapid-growing municipal solid waste [4]. There are concerns about the continuously increasing amounts of e-waste, along with the complex nature of these products and the problems related to their proper processing [5]. Even though e-waste has been producing a large number of problems, it's indeed a large source of useful metals like copper, gold and silver which can be recovered from production [6]. The recovery of these metals may to some degree reduce the overall global demand for new metals supply. The recycling of e-waste also reduces amounts of waste disposal in landfills [7]. It is a well-known fact incineration and landfilling of e-waste are very dangerous for the environment [8]. A linear progression between manufacturing, personalized usage, stored and waste disposal is a typical e-waste product's life cycle [9].

For e-waste, it is not feasible to implement the predictable waste management policies that are well suited to deal with traditional forms of waste because of its characteristics, which contain highly hazardous material and other essential metals that can be reclaimed [10]. The count of environmental regulations and policies that centre the cycle of product production has enhanced in the current times to reduce the impact due to the use of these products on the environment [11]. An appropriate system for the control of e-waste was introduced which included all amendments. Legislation is not beneficial until it is completely applied. The government or authorities cannot do this on their own unless the customers are ready to contribute to the initiative [12]. There are no e-waste collecting centres in India, and no specific details are available regarding the volume of e-waste discarded per year and the environmental risks caused due to them [13]. Using the option of free trade, e-waste produced from manufacturers who are generally developed countries enter easily in developing countries and the situation becomes worse [14]. Nevertheless, this is not the situation in countries such as the USA which is a developed country; still, there is no law for appropriate disposal, treatment and other requirements such as recycling of e-waste. Just 15% to 20% of the total waste is being reprocessed and tracked, while 80% is still going to sites in the US, according to a survey [15].

Managing e-waste effectively is fundamentally centralized and is seen as a critical challenge for the world today. The key problems are either because e-waste management activities are not being undertaken or not enough regulations are formulated. While one of the fastest-developing economic states and one of the major generators of e-waste places viz. India and China may have formulated their law recently, most developing countries still are searching for clear guidance on e-waste. Compliance outcomes are not yet achieved [16]. Programs and legislations have been introduced to increase the possible level of recycling of e-waste, including Expanded Producer Accountability (EPR). Different researches on effective treatment and storage of e-waste were performed. Although management of e-waste has been thoroughly examined from manifold perspectives, however, an overall approach is still required to demonstrate the cost and environmental effects of e-waste management, including waste generation, variations in electronic-waste sources and different types of its processing [17].

In this analysis; e-waste management activities that have been subsequently implemented in different areas responsible for the problem management has been reviewed. Section 5.2 consists of global standards for the management of electronic waste, while section 5.3 discusses the recycling aspects of e-waste. In section 5.4 is offered a forward-looking view of e-waste whereas section 5.5 sets the guidelines for electronic waste management.

5.2 Global Status of E-Waste Management

The use of electronic and electrical equipment is rising globally. E-waste is thus a problem growing globally posing an immense menace to the world's environment and human health [18]. E-waste produced worldwide is approximately 50 million tons a year. This weight is greater than the weight of all the commercially available airlines, and only 20% of it is being correctly reclaimed. Global e-waste production, if current trends continue, is on the way to 120 million tons a year by 2050. Nearly fifty four million tons of e-waste originated worldwide in the year 2019 [19]. Compared to the last five years, there has been an increase of almost 21%. From this total e-waste, the main part was created in Asia, followed by the United States and Europe [20]. The worldwide production of e-waste along with the population increase has been depicted in Figure 5.1.

In the world, the highest quantity of e-waste is being produced by China but the waste generated per person is highest for Germany [21]. Figure 5.2 indicates the top ten e-waste generating nations along with their total share in global e-waste.

For any individual commodity, the contribution made by it in the generation of e-waste annually (W kg/year), depends on its mass M (kg), the

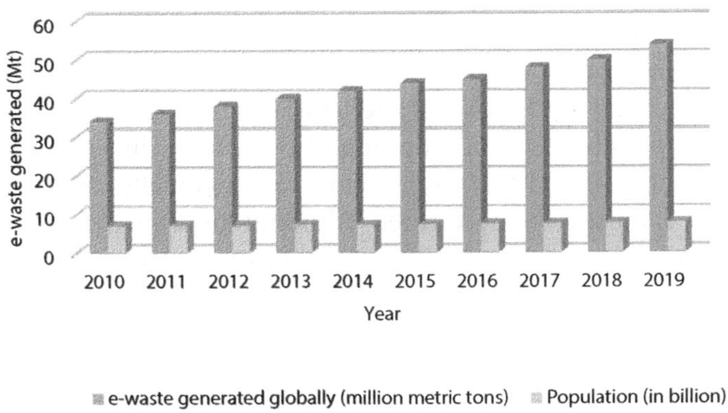

Figure 5.1 The production of e-waste (in million metric tons) in comparison to the increasing world population (in billion).

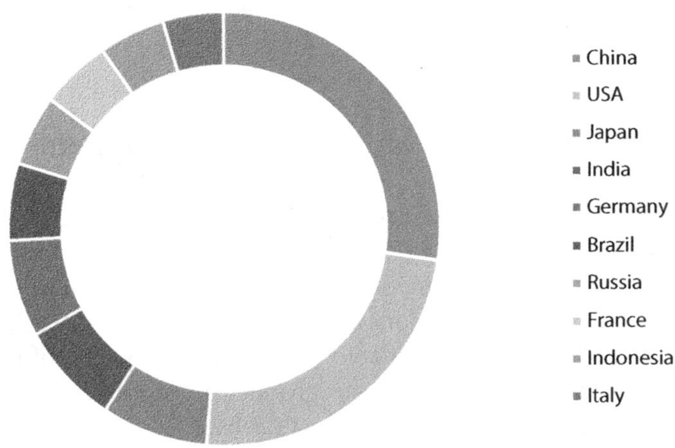

Figure 5.2 Contribution of top ten e-waste generating countries.

total number of that commodity which is in service N and the average life of that commodity. Finally, it can be written with the help of equation (5.1) as given below [22]:

$$W = \frac{MN}{L} \tag{5.1}$$

Classification of e-waste is done in six divisions on the basis of their life span, temperature change devices, monitors, lights, large machinery, small devices, and small stuff. This implies that every group has distinct amounts of waste content, commericial standards and consequences that have a lasting influence on environment as well as health [23].

5.3 Industrial Practices in E-Waste Management

Thanks to the excellent resistance toward corrosion, good electric conductance and high catalytic efficiency, precious metals are commonly found in applications for many fields. But precious metal reserves are getting low [24]. The recycling of e-waste through the provision of appropriate technology is desired because of its inclusion in PCBs and is regarded as a lucrative undertaking in developing countries [25]. E-waste can cause a lot of problems that may degrade the environment during the extraction of useful metals and waste management, if not properly handled, because of hazardous material content. Within the midst of these changes, several countries have enacted legislation aimed at enhancing the necessary recovery, recycle and reuse to limit disposal [26]. Figure 5.3 summarizes step by step the retrieval process.

Electronic and electrical equipment includes a wide variety of useful materials and plastics [27]. In complex electronics, there are up to 60 elements from the periodical table, many theoretically recoverable, although the market established limits them economically [28]. In terms of their merits and demerits, the recovery of metal requires different thermal and chemical treatments [29]. E-waste contains valuable metals, including silver, platinum, palladium, gold and copper, but also hazardous reactive metals, like aluminium and iron, along with recyclable plastics [30].

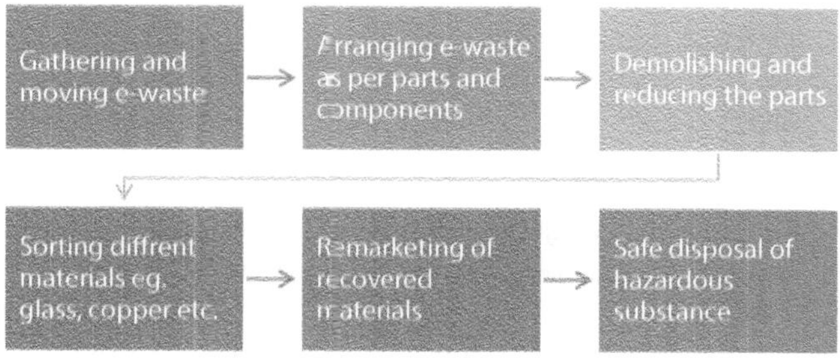

Figure 5.3 Process of recovery of material from e-waste.

E-waste is projected to be EUR 55 million in raw materials [31]. Figure 5.4 represents the metals that are recovered with their income. A lot of e-waste is required to produce 280 grams of gold [32]. The future value of the raw material recovered is calculated in Figure 5.5. The analysis shows that the highest amount of iron is recovered and that the least amount of recovered element is iridium.

The structure for maximizing the approach to counter the e-waste problems would include the longevity of goods and components through the repair and reuse and enhancing the recovery of valuable resources through the simple recycling of products [33].

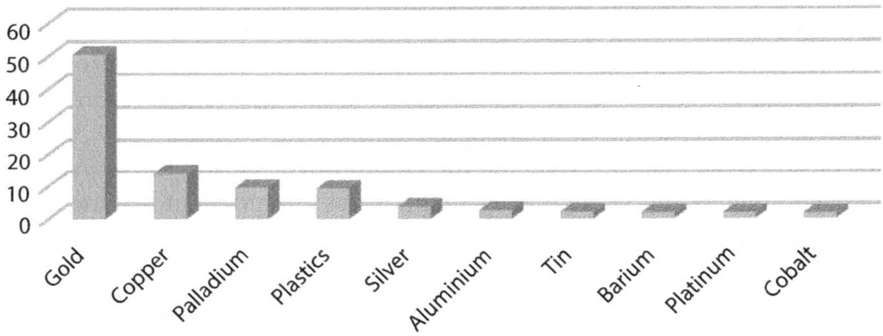

Figure 5.4 Amount of revenue collected from e-waste in percentage.

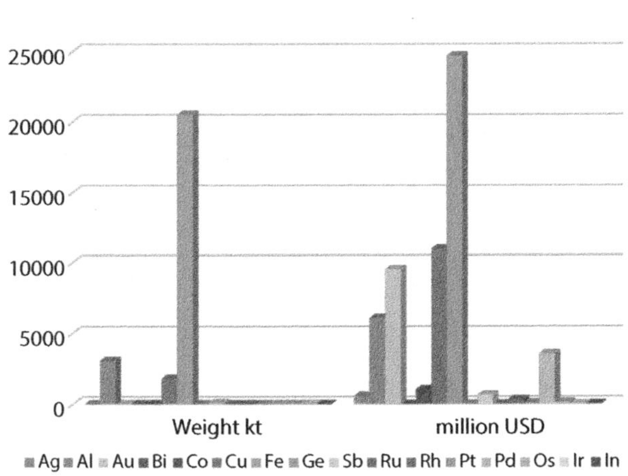

Figure 5.5 Amount of material recovered (in kt) and their value (in million USD).

5.4 Recycling of E-Waste

Reprocessing and reusing of electronic and electrical equipment of any kind which has been discarded or deemed obsolete can be called e-waste recycling. This method is important to address four challenges: loss of rich raw materials, management of solid waste, toxic substances, hazardous residues. For both developed and developing countries, therefore, recycling is required [34]. Globally, however, recycling is small. According to the World Economic Forum, which wrote in one of its reports that the production rate of e-waste is greater than the rate of generation of any other solid waste in the world [45].

The definition of Extended Producers Accountability (EPR) was introduced in the Basel Convention to control e-waste management [36]. The responsibility for the e-waste channelization for the biologically sound treatment of such waste is the responsibility of the manufacturers and importers of electrical or electronic products [35]. Following its guidance and main points, the unique Directive steps for storage, management, recovery and recycling of all electrical and electronic items are based on EPR [36]. The primary objectives of each EPR process are reduction and prevention of waste; the recycled material use in the production; decreased natural resource consumption; environment internalization [37]. In various laws and policies in different countries, the EPR definition is used [31]. In 2019 the national e-waste policies, legislation and regulations protected 71% of the global population, i.e. 78 nations. Since 2014, major improvements have been recorded and before that only 44% of the population was covered [20]. Nevertheless, many aspects of implementing the EPR are difficult to tackle for developing countries, such as fundraising by producers or importers, complex policy opportunities and regional disparities [18]. The life cycle of electrical or electronic devices is illustrated in Figure 5.6 as per EPR guidelines.

Design for environment (DfE) is an additional strategy in the world and is considered as a modern form of emission control and has been receiving considerable interest. In product design, the DfE is the mechanism by which the environmental effect of goods on the market can be greatly reduced. Despite major efforts made in 2018, the year in which 49.8 MMT of e-waste was produced merely 20% of it was recycled [40]. The following Figure 5.7 provides details of the worldwide share of e-waste by the disposal process. Most (80%) of the e-waste is not registered and waste not collected (76%) can be seen in statistics. Europe is the biggest contributor to this amount out of the 20% waste recycled in

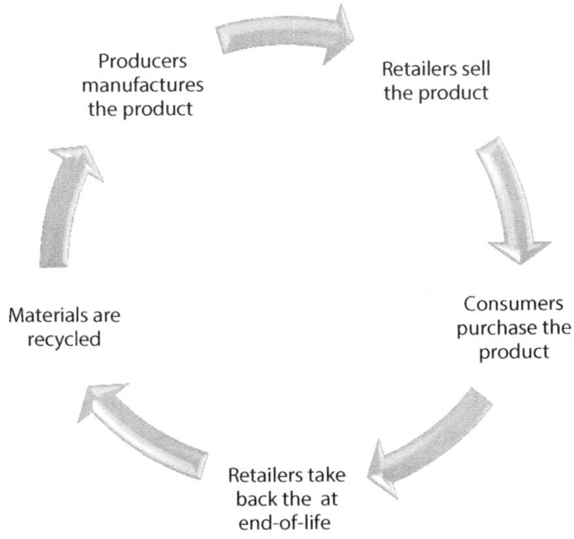

Figure 5.6 The life cycle of electrical or electronic equipment according to EPR.

Figure 5.7 The share of e-waste as per disposal method.

the world. Recycling in Europe accounts for 42.5% and Asia, 11.7% of recorded waste, but only these two continents have an immense recycling gap [20]. Figure 5.8 demonstrates the role of multiple continents in the recycling process.

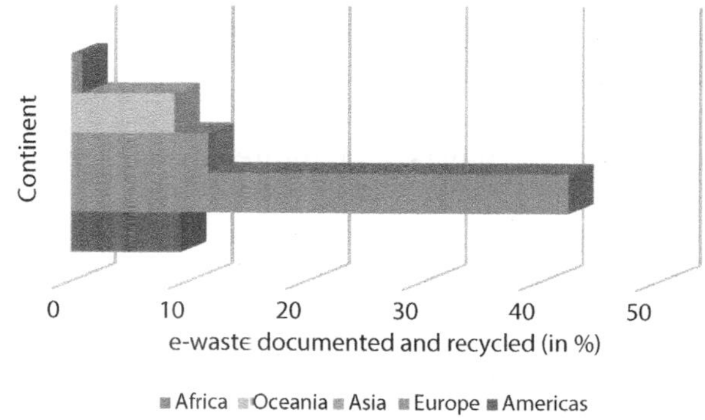

Figure 5.8 The amount of e-waste recycled in continents.

5.5 E-Waste Management Benchmarking

As the fastest developing industry worldwide, the electronic and electrical equipment industry generates a huge amount of e-waste [41]. For all countries, the future trend mentioned in Figure 5.9 is an issue of concern. Therefore, e-waste benchmarking is an important tool for monitoring the quality of management and achieving the performance objective through recognizing past best practices and errors [42]. The process of identification, understanding and performance development can be continuously improved [43]. The e-waste management system is monitored by many national as well as international organizations. Examples of such organisations include the ISWA, the Basel Action Network (BAN), the UN University, and the Ministry of the Environment, Forestry and Climate Change.

The benchmarking process will overcome a range of challenges [44]:

1. The policy of legislation needs to be better and stricter
2. Lack of awareness around e-waste management
3. Installations of recycling are not to the norm
4. Social responsibility of companies is very weak

The main aim of benchmarking solid waste is to compare common elements in solid waste systems and track waste management from production to disposal. The benchmarking process is carried out following the steps as described below:

1. Review of the situation today
2. Description of weaknesses and strengths
3. Role recognition relative to nobles
4. Mutual best practices
5. Round tables, seminars and future knowledge exchange platform
6. The possibility to examine in depth the vulnerabilities found [45].

In developing this framework, the most central, but the challenging aspect is to define a few, yet efficient criteria. A comparison and discussion of best practices between municipalities should be done to encourage others. Indicators on waste management measurement and reporting of progress toward sustainable development are used at international and national levels. Quality is a key variable to be calculated by choosing the correct indicator of the comparison in the public sector benchmarking phase [46].

5.6 Future of E-Waste Management

Figure 5.1 poses a troubling scenario, which is why future generation provision and preparation to reduce risks of e-waste are essential requirements. In planning future effective management strategies, data on the production, disposal and recycling of e-waste is of importance [47]. Therefore, to reduce the risk for the future it is crucial to examine the e-waste management system including the unknown and known faults of waste products and related material, the economical evaluation of the recovery material and the responsibility of authorities [48]. The largely untapped potential for waste to become resources and the environmentally friendly recycling and reuse of some sections of waste streams is one of the biggest unaddressed challenges [49]. The future trend is calculated by analysing the previous patterns in e-waste generation to better assess the situation [50]. Figure 5.9 demonstrates that channelization and management of generated e-waste is a grave issue that must be addressed at the moment since potential data is very frightening. All these developments are opening the way to amazing e-waste volumes, which treatment plants that recyclers ought to handle in the future [51].

Nonetheless, merely research is not sufficient to solve the problem. There needs to be a new system of rules, standards, incentives and investment flows [52]. Hazardous materials such as lead, halogenated flame

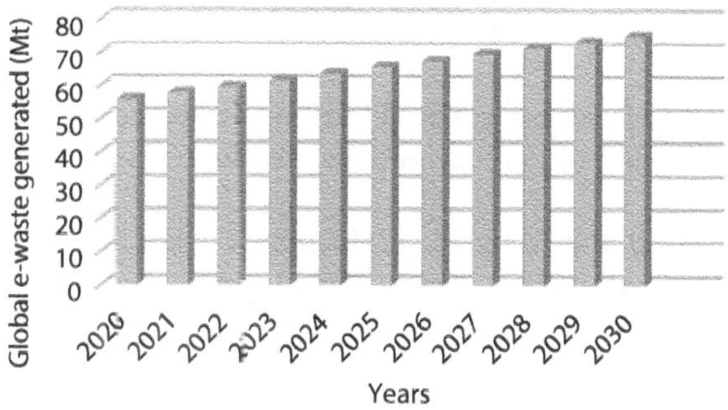

Figure 5.9 The prediction of global e-waste generation in next 10 years.

retardants should be substituted [53]. In addition, it could prove fruitful to invest greater focus for designing future electronic and electrical products to minimize the usage of any harmful metals and compounds and choose plastics that have minimal burning releases of toxic by-products [54].

Eventually, the production and consumer models will determine the size and nature of the future e-waste problem [55]. While certainty surrounds the progress of future technology, the use of electronic and electrical products and thus e-waste generation will certainly grow in the next few decades at least [56]. In addition to all the risk points for e-waste, it is also important to explore how and when a program tailored for current e-waste would be economical and beneficial to the environment [57].

5.7 Conclusions

E-waste seems to be an ever-growing problem if the production rate will increase that the same pace. It is an issue not only for health but also for the environment. It is not that only a few countries are contributing to this trend but all the countries are somehow responsible for the situation. The rules and regulations need to be amended in a way that not only the government and concerned authorities but consumers also participate in the initiative of management of e-waste. All the trends need to be studied carefully for a better interpretation of the future and present situation. The recycling process should be given a boost so that the economic benefits can be received from it. Above all the collection process should be increased so that recycling can be done as much as possible. Future studies should focus on the methods of making the situation better than the predicated one.

References

1. Kang, H.Y. and Schoenung, J.M., Electronic waste recycling: A review of US infrastructure and technology options. *Resour. Conserv. Recycl.*, 45, 4, 368–400, 2005.
2. Sthiannopkao, S. and Wong, M.H., Handling e-waste in developed and developing countries: Initiatives, practices, and consequences. *Sci. Total Environ.*, 463, 1147–1153, 2013.
3. Garlapati, V.K., E-waste in India and developed countries: Management, recycling, business and biotechnological initiatives. *Renewable Sustainable Energy Rev.*, 54, 874–881, 2016.
4. Chen, A., Dietrich, K.N., Huo, X., Ho, S.M., Developmental neurotoxicants in e-waste: an emerging health concern. *Environ. Health Perspect.*, 119, 4, 431–438, 2011.
5. Tsydenova, O. and Bengtsson, M., Chemical hazards associated with treatment of waste electrical and electronic equipment. *Waste Manage.*, 31, 1, 45–58, 2011.
6. Sinha-Khetriwal, D., Kraeuchi, P., Schwaninger, M., A comparison of electronic waste recycling in Switzerland and in India. *Environ. Impact Assess. Rev.*, 25, 5, 492–504, 2005.
7. Kumar, A., Holuszko, M., Espinosa, D.C.R., E-waste: an overview on generation, collection, legislation and recycling practices. *Resour. Conserv. Recycl.*, 122, 32–42, 2017.
8. Man, M., Naidu, R., Wong, M.H., Persistent toxic substances released from uncontrolled e-waste recycling and actions for the future. *Sci. Total Environ.*, 463, 1133–1137, 2013.
9. Kahhat, R., Kim, J., Xu, M., Allenby, B., Williams, E., Zhang, P., Exploring e-waste management systems in the United States. *Resour. Conserv. Recycl.*, 52, 7, 955–964, 2008.
10. Khetriwal, D.S., Kraeuchi, P., Widmer, R., Producer responsibility for e-waste management: key issues for consideration–learning from the Swiss experience. *J. Environ. Manage.*, 90, 1, 153–165, 2009.
11. Nnorom, I.C. and Osibanjo, O., Overview of electronic waste (e-waste) management practices and legislations, and their poor applications in the developing countries. *Resour. Conserv. Recycl.*, 52, 6, 843–858, 2008.
12. Borthakur, A. and Govind, M., Emerging trends in consumers' e-waste disposal behaviour and awareness: A worldwide overview with special focus on India. *Resour. Conserv. Recycl.*, 117, 102–113, 2017.
13. Garlapati, V.K., E-waste in India and developed countries: Management, recycling, business and biotechnological initiatives. *Renewable Sustainable Energy Rev.*, 54, 874–881, 2016.
14. Joseph, K., Electronic waste management in India–issues and strategies, in: *Eleventh international waste management and landfill symposium, Sardinia*, 2007, October.

15. Gollakota, A.R., Gautam, S., Shu, C.M., Inconsistencies of e-waste management in developing nations–facts and plausible solutions. *J. Environ. Manage.*, 261, 110234, 2020.
16. Srivastava, R.R. and Pathak, P., Policy issues for efficient management of e-waste in developing countries, in: *Handbook of Electronic Waste Management*, pp. 81–99, Butterworth-Heinemann, United Kingdom, 2020.
17. Jaunich, M.K., DeCarolis, J., Handfield, R., Kemahlioglu-Ziya, E., Ranjithan, S.R., Moheb-Alizadeh, H., Life-cycle modeling framework for electronic waste recovery and recycling processes. *Resour. Conserv. Recycl.*, 161, 104841, 2020.
18. Chi, X., Streicher-Porte, M., Wang, M.Y., Reuter, M.A., Informal electronic waste recycling: a sector review with special focus on China. *Waste Manage.*, 31, 4, 731–742, 2011.
19. UN report: time to seize opportunity, tackle challenge of E-waste, UN, Switzerland, 2019. Available: https://www.unenvironment.org/news-and-stories/press-release/un-report-time-seize-opportunity-tackle-challenge-e-waste.
20. Forti, V., Balde, C.P., Kuehr, R., Bel, G., The Global E-waste Monitor 2020: Quantities, flows and the circular economy potential. United Nations University/United Nations Institute for Training and Research, International Telecommunication Union, and International Solid Waste Association, Bonn, Geneva, and Rotterdam, 2020.
21. Forti V., Baldé C.P., Kuehr R., Bel G. The Global E-waste Monitor 2020: Quantities, flows and the circular economy potential. United Nations University (UNU)/United Nations Institute for Training and Research (UNITAR) – co-hosted SCYCLE Programme, International Telecommunication Union (ITU) & International Solid Waste Association (ISWA), Bonn/Geneva/Rotterdam.
22. Robinson, B.H., E-waste: an assessment of global production and environmental impacts. *Sci. Total Environ.*, 408, 2, 183–191, 2009.
23. Baldé, C.P., Forti, V., Gray, V., Kuehr, R., Stegmann, P., The global E-waste monitor 2017: quantities, in: *Flows and Resources*, vol. 33(6), 2017.
24. K. Yamada, K. Tayama, N. Ezawa, H. Inoue, M. Ogino, K. Takadate, U.S. Patent Application No. 10/548,440, 2006.
25. Chatterjee, S. and Kumar, K., Effective electronic waste management and recycling process involving formal and non-formal sectors. *Int. J. Phys. Sci.*, 4, 13, 893–905, 2009.
26. Nnorom, I.C. and Osibanjo, O., Electronic waste (e-waste): Material flows and management practices in Nigeria. *Waste Manage.*, 28, 8, 1472–1479, 2008.
27. Pant, D., Joshi, D., Upreti, M.K., Kotnala, R.K., Chemical and biological extraction of metals present in E waste: a hybrid technology. *Waste Manage.*, 32, 5, 979–990, 2012.

28. Debnath, B., Chowdhury, R., Ghosh, S.K., Sustainability of metal recovery from e-waste. *Front. Environ. Sci. Eng.*, 12, 6, 2, 2018.
29. Ding, Y., Zhang, S., Liu, B., Zheng, H., Chang, C.C., Ekberg, C., Recovery of precious metals from electronic waste and spent catalysts: A review. *Resour. Conserv. Recycl.*, 141, 284–298, 2019.
30. Norton, A., de Klerk Batty, J., Dew, D.W., Basson, P., U.S. Patent and Trademark Office, Washington, DC, 2005.
31. Baldé, C.P., Forti, V., Gray, V., Kuehr, R., Stegmann, P., *The Global E-waste Monitor*–2017, pp. 978–92, United Nations University (UNU), International Telecommunication Union (ITU) & International Solid Waste Association (ISWA), Bonn/Geneva/Vienna. ISBN Electronic Version, 2017.
32. Golev, A., Corder, G.D., Rhamdhani, M.A., Estimating flows and metal recovery values of waste printed circuit boards in Australian e-waste. *Miner. Eng.*, 137, 171–176, 2019.
33. Cayumil, R., Khanna, R., Rajarao, R., Mukherjee, P.S., Sahajwalla, V., Concentration of precious metals during their recovery from electronic waste. *Waste Manage.*, 57, 121–130, 2016.
34. Kaya, M., Recovery of metals and nonmetals from electronic waste by physical and chemical recycling processes. *Waste Manage.*, 57, 64–90, 2016.
35. Unep, P., Itu, I., Unido, U., A New Circular Vision for Electronics Time for a Global Reboot, PACE, Switzerland, 2019.
36. Wath, S.B., Vaidya, A.N., Dutt, P.S., Chakrabarti, T., A roadmap for development of sustainable E-waste management system in India. *Sci. Total Environ.*, 409, 1, 19–32, 2010.
37. Osibanjo, O. and Nnorom, I.C., The challenge of electronic waste (e-waste) management in developing countries. *Waste Manage. Res.*, 25, 6, 489–501, 2007.
38. Gaidajis, G., Angelakoglou, K., Aktsoglou, D., E-waste: environmental problems and current management. *J. Eng. Sci. Technol. Rev.*, 3, 1, 193–199, 2010.
39. Dwivedy, M. and Mittal, R.K., Willingness of residents to participate in e-waste recycling in India. *Environ. Dev.*, 6, 48–68, 2013.
40. Tiseo, Waste Management Inc.'s recycled materials by type 2018, 2020, Energy and Environment (Waste Management), USA. Retrieved from: https://www.statista.com/statistics/653044/waste-management-inc-recycling-by-material/.
41. Heeks, R., Subramanian, L., Jones, C., Understanding e-waste management in developing countries: Strategies, determinants, and policy implications in the Indian ICT sector. *Inform. Technol. Dev.*, 21, 4, 653–667, 2015.
42. Anand, G. and Kodali, R., Benchmarking the benchmarking models. *BIJ*, 15, 3, 257–291, 2008.
43. Auluck, R., Benchmarking: a tool for facilitating organizational learning? *Public Adm. Dev.: The International Journal of Management Research and Practice*, 22, 2, 109–122, 2002.
44. Adediran, Y.A. and Abdulkarim, A., Challenges of electronic waste management in Nigeria. *Int. J. Adv. Eng. Technol.*, 4, 1, 640, 2012.

45. Ilić, M. and Nikolić, M., Waste management benchmarking: A case study of Serbia. *Habitat Int.*, 53, 453–460, 2016.
46. Dwivedy, M. and Mittal, R.K., An investigation into e-waste flows in India. *J. Cleaner Prod.*, 37, 229–242, 2012.
47. Kumar, A. and Dixit, G., An analysis of barriers affecting the implementation of e-waste management practices in India: A novel ISM-DEMATEL approach. *Sustainable Prod. Consumption*, 14, 36–52, 2018.
48. Dasgupta, D., Debsarkar, A., Hazra, T., Bala, B.K., Gangopadhyay, A., Chatterjee, D., Scenario of future e-waste generation and recycle-reuse-landfill-based disposal pattern in India: a system dynamics approach. *Environ. Dev. Sustain.*, 19, 4, 1473–1487, 2017.
49. Islam, M.T. and Huda, N., Material flow analysis (MFA) as a strategic tool in E-waste management: Applications, trends and future directions. *J. Environ. Manage.*, 244, 344–361, 2019.
50. Gao, Y., Ge, L., Shi, S., Sun, Y., Liu, M., Wang, B., Tian, J., Global trends and future prospects of e-waste research: a bibliometric analysis. *Environ. Sci. Pollut. Res.*, 26, 17, 17809–17820, 2019.
51. Kumar, A., Holuszko, M., Espinosa, D.C.R., E-waste: an overview on generation, collection, legislation and recycling practices. *Resour. Conserv. Recycl.*, 122, 32–42, 2017.
52. Cucchiella, F., D'Adamo, I., Koh, S.L., Rosa, P., Recycling of WEEEs: An economic assessment of present and future e-waste streams. *Renewable Sustainable Energy Rev.*, 51, 263–272, 2015.
53. Williams, E., International activities on E-waste and guidelines for future work, in: *Proceedings of the Third Workshop on Material Cycles and Waste Management in Asia, National Institute of Environmental Sciences: Tsukuba, Japan*, 2005.
54. Nowakowski, P., A novel, cost efficient identification method for disassembly planning of waste electrical and electronic equipment. *J. Cleaner Prod.*, 172, 2695–2707, 2018.
55. Man, M., Naidu, R., Wong, M.H., Persistent toxic substances released from uncontrolled e-waste recycling and actions for the future. *Sci. Total Environ.*, 463, 1133–1137, 2013.
56. Parajuly, K., Kuehr, R., Awasthi, A.K., Fitzpatrick, C., Lepawsky, J., Smith, E., Zeng, X., Future e-waste scenarios, StEP (Bonn), UNU ViE-SCYCLE (Bonn) & UNEP IETC (Osaka), 2019.
57. Heacock, M., Trottier, B., Adhikary, S., Asante, K.A., Basu, N., Brune, M.N., Chen, A., Prevention-intervention strategies to reduce exposure to e-waste. *Rev. Environ. Health*, 33, 2, 219–228, 2018.

6

Energy Audit and Management for Green Energy

Arjyadhara Pradhan* **and Babita Panda**

School of Electrical Engineering, Kalinga Institute of Industrial Technology, DU, Bhubaneshwar, Odisha, India

Abstract

Modernization is a boon to society and the world has started trending toward smart concepts where technology involves existing systems along with an intermix of information technology. With the rapid industrialization process, the use of conventional sources has increased a lot. This will lead to a shortage of conventional sources like coal and petroleum. Moreover, energy extraction from these sources and their uses in daily activities creates environmental pollution. Thus, the need of the time is to switch to green technology, which is clean and pollution-free. There are different types of renewable sources like solar, wind, biomass, geothermal, and ocean energy. Every industry whether renewable or nonrenewable requires effective energy audit and management. Energy management is the process of planning, organizing, and managing the usage of energy consumption in an organization considering several aspects like time management, system input, cost-benefit analysis, output to bring overall improvement to the company. Similarly, an energy audit is also carried in every organization to have a regular check of energy inflow and outflow and record surplus energy balance with a scope of any improvement in energy usage.

***Keywords*:** Energy, audit, management, nonconventional, conservation, photovoltaic, environment, solar

6.1 Introduction

With the rise of energy consumption and growing demand for electric power every day there is a continuous need for adequate generation and

**Corresponding author*: arjyadhara.pradhanfel@kiit.ac.in

Suman Lata Tripathi, Souvik Ganguli, Tengiz Magradze, and Abhishek Kumar (eds.) Intelligent Green Technologies for Sustainable Smart Cities, (89–110) © 2022 Scrivener Publishing LLC

transmission facilities. Environmental concern about fossil fuels and their constraints have spurred great interest in generating electricity from the nonconventional source of energy [1]. Considering energy efficiency and power quality issues the generation of electrical energy from renewable sources is an important factor for meeting world energy requirements. Take like solar, wind, geothermal, biomass, ocean energy. Solar energy is predominant due to its free availability worldwide. Solar energy can be utilized in two ways i.e. one is concentrated solar power and the other is solar photovoltaic. The solar PV system has various advantages in comparison to solar thermal.

The PV plants are mostly installed in the United States of America and Spain. Even India is also dominant among various renewable energy sources [2]. It has been estimated that the electricity from the PV generation will be contributing to 7% of the world electricity needs by 2030 which will also rise to 25% by the year 2050 [3]. Photovoltaic being one of the fastest-growing technology has an annual growth rate of 35-40%. The present installed power generation capacity of India is 334.40 GW as of 31.01.2018. Of the total installed capacity, 67.8% comprises the conventional power plants and 32.2% comprise nonconventional power plants. Of the total installed power capacity, renewable energy accounted for 18.37% excluding large hydropower. The electricity sector in India is growing at a rapid pace. The present peak demand is about 1,15,000 MW and the installed capacity is 1,52,380 MW using generation from thermal (63%), hydro (25%), nuclear (9%), and renewables (9%) sector. India is one of the countries with a large production of energy from renewable sources. As of March 31, 2020, 35.86% of India's installed electricity generation capacity is from renewable sources, generating 21.22% of total

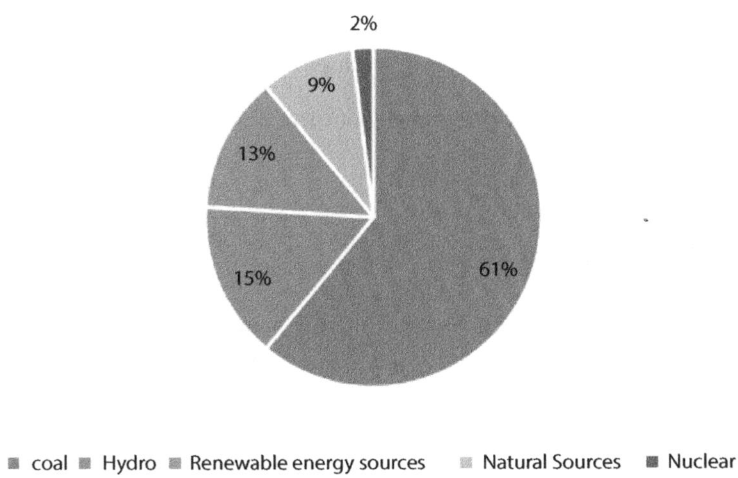

Figure 6.1 Sources of electricity in India by installed capacity.

utility electricity in the country. Figure 6.1 shows the perecentage of installed capacity from various sources in India. From this figure it is clear that the contribution from nuclear is very less in comparison to other sectors.

There are different types of energy problems in India like production of coal base of fossil fuels, fourth largest consumer of oil and petroleum in the world, import-dependent for gas sources, industrial output affected by electricity shortages, widespread inequality, and poverty of energy.

6.2 Types of Renewable Energy

There are different types of renewable energy. It is a pollution-free type of energy source. In many books, it is referred to as clean energy. The various ill effects of conventional energy, like pollution, etc, can be reduced by the use of renewable energy. The environment we live around can be healthy and free [3]. The various types of nonconventional sources are solar, wind, biomass, geothermal, and ocean energy. The conventional sources are fossil fuels, nuclear energy, etc. Fossil fuels affect environment during their drilling and extraction process and emits a lot of greenhouse gases to the atmosphere [4]. Nuclear energy causes radioactive emission and also affects environment.

6.2.1 Solar Energy

Solar energy is a clean and pure form of energy. It is one of the oldest and primary source. Solar energy is mostly utilized in two forms i.e. solar thermal and solar photovoltaic. There are three types of instruments used to measure the solar radiations like pyranometer, pyrheliometer and sunshine recorder. The various applications of solar thermal are like solar water heating, solar homes, solar dryer, solar distillation, solar cookers. In photovoltaic solar application, solar energy is converted to electrical energy, which can be used in standalone or grid connected form. Even solar collectors can be used to heat the water and this hot water is used as steam in thermal power plant for production of electricity. Solar modules are clustered in series or parallel connection to make one solar panel. Thus, large solar farms are solar arrays, which are combinations of solar panels. Even though energy production from solar has no environmental impact, yet it has many disadvantages like: it is an intermittent source; large surface area is required, which causes wastage of land [5].

6.2.2 Wind Energy

Wind energy is another renewable source, which produces zero emissions, zero fuel price volatility and reduces fossil fuel dependence. The main causes of wind are due to uneven heating between land and sea and uneven heating between equator and pole. Depending upon this there are different types of wind like local winds, seasonal, permanent, which uses flowing air to rotate a wind turbine, which then coupled with a generator produces wind energy. There are basically two types of wind turbine: horizontal axis wind turbine and vertical axis wind turbine. In the HAWT, the nacelle is place at the top along with rotor and the gear box. But in the case of VAWT, the generator assembly is at the foundation. Different types of rotor blades are used for these wind turbines. The power generation from wind depends upon its location of turbine and consistency of wind so careful planning is required while sitting any wind turbine. Even spinning turbine can be a threat to bird life. Wind Energy is a variable source, and also, it cannot be switched on as per our requirement; hence, it can be used as a backup source to reduce fuel demand.

6.2.3 Biomass

Biomass is the energy derived from woods, crop residue, animal waste, municipal solid waste, garbage. Biomass energy is originally obtained from sun. By the process of photosynthesis, carbon dioxide is converted to molecules, which contain carbon like starch, cellulose and sugars in plants, or is stored as chemical energy in animals and even their waste is called as bioenergy. When these bio fuels are heated the chemical energy stored in it are released as heat energy. Now this heat energy can be used as thermal energy like providing hot water to homes and industries or this heat can be converted into steam and used in the turbine for production of electricity. During the process of burning the biomass energy is released in the form of heat energy. Not only this, there are various forms of use of biomass like formation of methane gas and various transportation fuels like biodiesel. Biogas is nothing but methane or land fill gas giving a very rotten garbage smell. Crops like corn and sugar cane can be fermented to produce the transportation fuel, ethanol. The remaining portion of vegetable oils and animal fats can be used to make biodiesel for transportation purpose. The different methods of biomass extraction can be broadly be classified as: (i) physical method, (ii) incineration (direct combustion), (iii) thermochemical method, and (iv) biochemical method.

6.2.4 Geothermal Energy

Geothermal energy is the energy extracted from the interior of the earth. Earth is made of various layers like the inner most is the core, above it like the mantle and at the top lies the crust. Due to plate tectonic movements, the interior molten matter comes to the surface level. Even at some place due to weakness in surface structure, hot water comes out in form of springs. The interior of the earth is at high temperature; thus, this heat is used in the form of hot water or steam to produce electricity. Geothermal energy can be directly used in different forms like cooking, bathing and heating. In different countries like US geothermal energy can be utilized for greenhouse system for planting vegetables, fruits, flowers, fish farming, fun and recreation like spas and resorts, district heating of workplaces. The hot water obtained from earth can also be used for various industrial process like crops drying, leaching of metal ores and different materials used for building. There are various exploration process like digging deep well, drilling, pumping heated water into hot dry rocks. Geothermal energy causes very less damage to atmosphere with very less carbon dioxide emission. It is an ecofriendly source.

6.2.5 Ocean Energy

Ocean energy is used in three forms like wave energy, tidal energy and ocean thermal. When wind moves across the water bodies it transfers some of its energy to the water bodies as wave. Depending upon the speed, time and distance of the wind the strength of the wave energy can be determines. Wave energy is irregular and oscillating low frequency energy, which is available 24 hours a day. In ocean thermal, the temperature difference between deep underground water and upper surface temperature is considered for thermal energy purpose. There are basically two types of ocean thermal energy conversion cycles like closed cycle and open cycle. In closed cycle, the thermal energy of the ocean water is passed to another liquid in the heat exchanger mostly ammonia or any hydrocarbon liquids are used.

6.3 Energy Management

The definition of energy management has been defined by many people in many ways. The ultimate goal is generation of goods and services at least cost and damage to environment [6]. In other words, we can say energy management is defined as achievement and maintenance of optimum energy procurement and its utilization throughout the organization. Bench marking is an important aspect in energy management in which energy consumption are monitored both internally and externally [7]. Hence it acts like a powerful tool for performance assessment and logical evaluation of any scope for improvement. Well-documented data, trend analysis of energy consumption, production, cost helps to analyse the effects of capacity utilization of energy use. Energy management also involves determination of PEP i.e. plant energy performance [8]. It is actually energy usage indicator. It measures the amount of energy used by a plant for manufacturing its product and compares it with the past energy consumption. It helps the top management to finally make out whether the energy consumption has increased or decreased in comparison to the previous year and also helps them to find possible solution to check any leakage.

The main objectives of energy management are: conservation of resources, protection of environment, cost savings.

Figure 6.2 shows the cyclic process of energy management. It is a five step process starting with planning organisation, implementation, verification, company commitment and energy use.

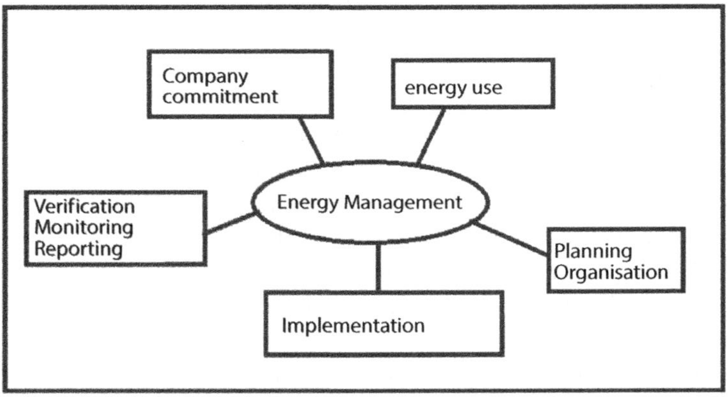

Figure 6.2 Various steps of energy management.

6.3.1 Types of Energy Management

Energy management is basically of two types like demand side management (DSM) and supply side management (SSM). The DSM aims to meet the electricity demand with efficiency and reliability there by maintaining a balance between the demand and supply on the electricity grid. Production factor is another important factor which determines the amount of energy that is need to produce current year's output referring to the plant operation in the similar way as compared to the reference year [9]. It is the ratio of production in the current year to that in the reference year.

6.3.1.1 Demand Side Management

In plain terms DSM means modification of consumer demand pattern for energy requirement. This can be done either by financial incentives or by proper education. The main focus of demand side management is to make consumers consume less energy at peak hours and make a slight shift in their energy consumption pattern to off peak times. Effective demand side management improves load factor and economy [10]. Demand side management has benefitted different sectors like consumers, utility group, society. Some of it are like electric demand by consumers are satisfied, Lowers electricity bill, Improves productivity, service charge are also greatly reduced, improves efficiency and flexibility of the operating systems, enhances costumer services, conservation of energy, environmental protection. Similarly, DSM also faces various hurdles like monopoly by power market, less competition leads to inefficient tariff structure, unaware consumers about energy usage, real demand and supply ends in wide gap, improper incentive schemes, problem in power system reliability and stability.

6.3.1.2 Implementation of DSM

There are different methods for implementation of DSM. Broadly ten such methods of DSM are discussed here.

- Demand forecasting
  Forecasting is one of the method used for analysis of demand and supply. Consumer demand forecasting is one of the effective method for effective DSM planning. These forecast can be either for short duration or long term .It is one of the common method adopted by many electrical companies to predict future energy requirements [11].
- Market Research
  Market research is one of the step stone to understand electricity usage and find out the different hurdles which keep customers unaware of the use of efficient technology. Market research involves identification of end use patterns. Load research is also carried out to find out the different types of load curves like daily basis, monthly, yearly by using sub-meters readings, bills and surveys. Market research includes research in different sector like commercial, domestic, industrial, public utilities etc.
- Identification of end-users and measures.
  One of the most important role of DSM is to find out the target sector and determine the load curve pattern. After identification of end-users it becomes effective for the energy management by the electricity producing companies.

Different sectors have different load consumption pattern. Therefore proper selection of DSM measures will highly impact peak demand of electricity.

- Identification of source of finance
  In DSM management system finance is required for different projects taken by the participants. Administration cost, maintenance, utility monitoring requires finance. Thus it becomes important for DSM to plan out the source of finance and the effective use of it as per the requirement.
- Partnership and viability
  Each partner in a DSM may be participant or utility or government must equally share the cost in order to maximize the profit [12]. The difference between tariffs and investment should be as low as possible.
- Design and selection of programme
  The most important work of DSM is identification of specific programme depending upon the customer class. Then lies execution with emphasis on maximization of profit. A brief report has to be prepared corresponding to that programme. Care to be taken to make the reporting in a uniform manner.
- Cost Benefit Analysis
  It is based on measurements and case study and also financial calculation. Real time cost involved in each and every component of the system is taken into consideration and the benefit and loss are taken into account . Clearly an analysis is carried out how much the system is benefitted with minimal cost incurred.
- Environment and social-economic impact
  Environment is the major concern of every activity. DSM should always make a check and balance to manage the environmental issues and provide economy to the people. It creates social awareness and social equality also.

6.3.1.3 Supply Side Management

Supply side management balances between generation, transmission and distribution of energy in an effective way. The electricity industries have been deregulated to make efficient use of available energy resources. More care should be taken to use high-efficient equipments more than the least efficient equipments. The main objectives of SSM are ensuring sustained availability of reliable energy, meeting increasing electricity demand, reducing environment impact, maintaining socioeconomic balance. On the other hand the various challenges of SSM are like requirement of complete comprehensive information, is required, well balance should be between consumer and supplier, short-term basis approach, lack of technical expertise, ineffective cost decision making.

Supply side management is a managerial procedure for adopting the most cost beneficial method to purchase and distribute energy [13]. The various implementation methods of SSM are as follows: integration of nonconventional source into prevailing organization energy system, power suppliers must be deregulated, load analysis and load management, scope for cogeneration, renewal in procurement strategies by adopting purchase from suppliers other then local bodies, saving opportunities during peak hours.

6.3.2 Ways to Improve Energy Management

There are several ways to improve energy management. A proper follow of these steps can help energy industry to achieve its goal and bring maximum profit.

- Bench marking is the initial step for improvement of energy management.
- Goal setting helps to fix a particular target as to such how much energy can be saved.
- Fix responsibility to the concerned people.
- Internal and external communication must be clear.
- An effective governance structure must be established.
- Investment in power saving technology.
- Exploration of alternative energy.
- Use innovative practices.
- Don't go it alone.

6.4 Energy Audit

Every industry has three major operating expenses, such as energy, materials and manpower. Among the three energy is the top ranker which includes both thermal and electrical [14]. Hence Energy Audit helps in understanding the different ways of energy usage in an organization, identifies areas of possibility of waste occurrence and finds solution for improvement in terms of energy conservation. It gives positive orientation to preventive maintenance, various quality control programmes, energy cost reduction. In other words we can say energy audit is realization of energy conservation methods in practical field. It sets bench mark for managing and planning how to use energy in an organization. Energy audit is basically classified into two types i.e. preliminary audit and detailed audit. It is also classified by function and type of industry, depth to which audit is required and potential and magnitude of cost reduction required.

6.4.1 Types of Energy Audit

- Preliminary audit
- Detailed energy audit

6.4.2 Preliminary Energy Audit

Preliminary energy audit mainly focuses on certain task like:

- Finds out area of energy consumption in an organization.
- Estimates savings from different sector.
- Identification of prime area for attention.
- Identification of improvement or saving opportunities.
- Fixation of reference point.
- Finds out the area where detailed study is required.
- Preliminary audit uses existing audit data to make any suggestion and calculation.

6.4.3 Detailed Energy Audit

Detailed energy audit is a more elaborate process, which evaluates all the major energy using systems. In this case, accurate estimation of energy savings and cost is determined. The major objective of detailed energy audit process is to maintain energy balance [15]. This is based on inventory of systems using energy, balance between current operating conditions effective calculations of energy use, comparison with utility bill charges. Broadly detailed energy audit considers all energy consumption area, all major equipments and all the process consuming energy, detailed calculation, expenditure and cost saving at each level. Detailed energy audit is conducted in three different steps like

Step 1: Preliminary-phase
Step 2: Main phase
Step 3: Post phase.

During detailed energy audit phase different information are required to be collected which are like:

- Consumption of energy: Energy consumption is different for different department, different process, equipment and end-user.
- Data of material balance: Balance of materials information is another factor which is required in the audit process. It includes data like raw materials, bye products, intermediate products, recycles product and final product.
- Energy cost and different types of tariff data: Tariff is the rate at which we buy or sale electricity. Different types of tariff include two part, fixed, blocked rate tariff etc.
- Flow diagram of different working process and materials.
- Various sources of supply of energy: It includes whether electricity is self-generated or from grid.
- Scope for fuel substitution, any process modifications, co-generations.
- Site services generation and distribution processes.
- Effective energy management system adopted and different energy training programmes
- Energy safety measures.

Other than these various data, the energy audit team must have certain basic technical data for carrying out the detailed energy audit process, which includes:

✓ Different types of technology used
✓ Different processes adopted
✓ Details of energy consuming equipments
✓ Quantity and types of input materials
✓ Consumption of water
✓ Amount of fuel consumption
✓ Quantity of steam consumption
✓ Electricity consumption
✓ Different other inputs like cooling water, compressed air.
✓ Different type and amount of waste
✓ Efficiencies and yield.

6.4.4 Data Analysis

Data analysis is an evaluation process of data considering analytical and logical reasoning for examining every component of data collected. Data collection is merely gathering or accumulating different types of data from different sources [16]. Once the data is collected during a detailed energy audit process it has to be properly reviewed and analyzed to draw a conclusion. In Figure 6.3 data analysis is basically divided into three categories descriptive, predictive and prescriptive.

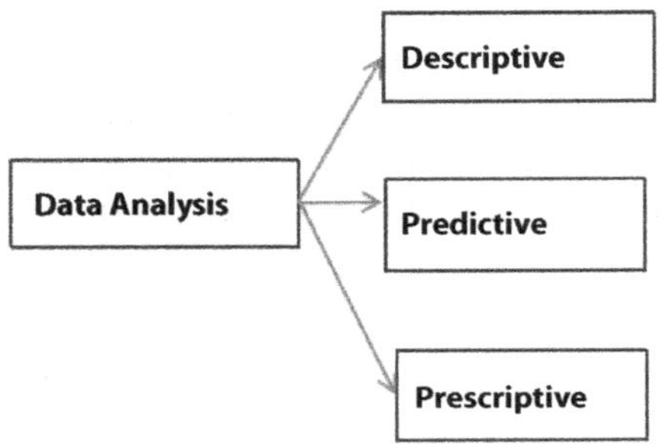

Figure 6.3 Different types of data analysis approach.

There are different types of data analysis methods like

1. Qualitative Data Analysis
   In qualitative data analysis the types of data collected are mostly pictures, words, symbols, observations. It also includes methods and procedures utilized for analysis of data. These data are mostly used for explanation, understanding and interpretation process.
2. Quantitative Data Analysis
   In quantitative data analysis, quantification of data is done for generalization of results which is obtained from a sample. In this case, statistical methods are used to collect raw data and convert it to numerical data.

6.4.5 Detailed Steps in Energy Audit

Step 1: Walk through an audit: It involves resource planning and organization of an energy audit team. Macro level data collection, Informal interview with energy manager, plant and production manager. Organizations of instruments and time frame of work.

Step 2: Conduct brief meeting: This step mainly deal with conducting brief meeting with all divisional heads and organising different awareness programme. It initiates co-operation with different departments.

Step 3: Primary data gathering: It includes collection of different energy and process flow diagram, operation schedule, collection of pre-existing bills.

Step 4: Conduct survey and monitoring: It includes measurements, motor survey, insulation survey, and lighting survey with different types of instruments, and comparison of design data with collected data.

Step 5: Conduction of detailed trials: In this step different power and energy experiments are conducted with guzzlers. For examples load variations monitoring in pumps , blowers and compressor.

Step 6: Analysis of energy use: In this step, detail analysis of energy and material balance, waste and reuse is studied Water analysis is also carried out.

Step 7: Identification and development of energy conservation: Development and redefining of new ideas. Adoption of various energy conservation opportunities. Adoption of new technology.

Step 8: Cost-benefit analysis: Selection of promising projects, prioritization of low, medium and long term measures. Complete analysis of cost structure and benefit to organization.

Step 9: Reporting and presentation to top management: Final documentation and detailed report writing and submission to higher authority.

Step 10: Implementation and follow up: This is the final step followed in the audit process. In this steps whatever improvement option or energy conservation opportunities are implemented followed up for better performance.

6.5 Energy Audit in Solar Plant

Among the various renewable energy sources solar energy is the most dominating one with easy installation and pure form of energy. The solar plant performance has to be regularly monitored for improving its overall efficiency [17]. Hence performance audit of solar plant is undertaken to have a check over its performance and to find out areas where improvement can be done. Solar plants depends on variety of factors like solar insolation, temperature, environmental conditions, circuit design, inverter efficiency, other accessory efficiency, degradation, payback period [18]. In almost all the audit process certain steps are common like field visit, data collection, data analysis, energy calculation, implementation of energy conservation techniques and saving calculation with follow up for certain period. Energy experts prepare energy audit reporting worksheets to collect and analyze site energy information. It gives a detail idea about the recent energy situation, energy producing equipment, and the method of its use, predicted future loads. These sets of data helps engineers to plan PV and generator systems in both off grid and grid connected to develop new specifications and biding documents. It provides a platform for choice between a potential system and configuration.

For carrying out solar audit the following data are required to be collected for the analysis purpose solar irradiance, temperature and climatic conditions, design parameters, different types of losses, efficiency of inverter, modules data, mounting and tracking system, cables, switch gears, transformers, earthing and surge protection, interface unit for grid connection.

The various steps are as follows

- collection of basic information about the site, energy, and various types of fuel cost;
- usage of energy technologies and energy sources applications;
- third, assumptions are considered about cost, efficiencies, and resources of solar. These data can be changed as per different locations and countries;
- data about PV systems;
- sheet containing information about various generators and appliances run by generator;
- details of various types of PV applications, future electric applications as well as nonelectric applications;
- Load profile.

Scheduled inspections of solar panels are an important component of the operational efficiency of a working solar power plant. They are critical to preventing breakdowns and managing warranty claims with equipment suppliers. Quality control of a solar power plant is carried out under normal operating conditions under load and usually does not require shutdown of equipment. It is generally recommended to conduct a scheduled technical inspection of solar power plants twice a year.

Special tools are used to inspect solar panels and pinpoint a large number of problems. The inclusion of thermal imaging checks in the planned maintenance of the solar plants will help the plant to work safer and increase its overall efficiency. Such an analysis is advisable to do in the cold, but sunny seasons (autumn, spring) to obtain more accurate information on the state of the solar modules at your solar power plant.

Finding out the amount of energy required is the first step for installation of solar photovoltaic system. With this peak power, battery size, charge controller and inverter requirement can be decided. The amount of power used is determined by the product of total electrical appliances and time of use or hours of use. During this phase several questions are put up like a) how much energy can be saved b) which appliance to use and which to switch off c) for how many hours to use d) how much power is needed in real time. With answers to all these questions, one can easily make out how much solar panels are necessary. Along with this information like amount of bright sunshine hours and geographical latitude and longitude are also taken into considerations. Next step lies in checking the space for the installation of the panels. There are different types of ways solar panels are installed. But more emphasis is given to the installation with highest exposure to sunlight. Determination of panel orientation is also required for solar PV system. In the audit process the engineers make out that the panels are properly oriented to get better performance. If the location is in the northern hemisphere then the panels must be south facing and if in southern hemisphere the panels must face north.

A major role of audit process is to find out the cost saving in case of grid connected. There are various types of saving opportunities like a) feed in tariff: It is a money back system for surplus electricity back to grid. b) net metering: credit is provided to the costumers for surplus electricity fed to grid. c) reverse metering: in this case, electric bill is reduced by back turning meters. However, for off grid connections, battery bank is highly required which has to be regularly monitored. The audit team makes sure right number of batteries are used for the system, which is decided by taking into consideration the following points like solar panel total power, days of autonomy, efficiency of the system

6.5.1 Technical Inspection Steps of Solar Power Plant

1. Technical condition analysis of various equipments of solar power station like solar modules, inverters, control panels, batteries and transformers.

2. Monitoring of the fasteners, mounting structures and connecting elements of the solar panels.
3. Regular assessment of amount of dust present on the surface of PV modules.
4. Analysis of various types of losses and identification of region of hotspot formation if any.
5. Recommendation for any necessary modifications or solutions in the design of solar plant for reducing losses.

6.6 Energy Conservation

Energy conservation is the process to minimise energy consumption by reducing the energy service or by using energy in a highly efficient manner. Industries take this opportunity to increase the savings by cost cutting the unnecessary energy usage. There are various methods where we can control our energy consumption pattern like avoiding use of light during daytime unnecessary turning on electrical devices when not in use etc. The most important objective of energy conservation is to reduce the energy bill and demand on the natural resources. Energy consumption can be reduced by mere change in our habits and behaviors, which starts at individual level. In smart cities special techniques and methods are used which consumes less energy for same output For examples led bulbs, thermostats, smart strips etc.

Energy conservation has many advantages like reduces energy service requirement,improves quality of environment, improves national security,establishes personal and financial security improves savings, prevents resource depletion.

6.6.1 Energy Conservation Methods

There are various energy conservation methods. Some of the methods are like:

- Put your refrigerator down: Every house hold requires refrigerators for domestic purpose. It accounts for 13.5% of the total energy usage. Thus when we set the refrigerator to 37 to 38 degree fahrenheit energy conservation can be done.
- Use of low energy consuming light bulbs: LED bulbs are more energy saving in comparison to sodium vapour or mercury vapour lamps or tungsten filament bulbs.

- Air filters must be cleaned and replaced: Room heaters and AC consumes much of home energy and often get dirty with deposition of dust for a long period of time. Thus, regular cleaning of air filter is highly recommended.
- Air leaks to be checked: Windows, doors and any openings must be properly sealed to prevent any kind of air leakage. This helps the room to remain hot during winter and cool during summer.
- Air Dry clothes and dishes: during daytime, clothes can be put outside for drying instead of using any dryer option.
- Use of smart strips: Even when the electronic devices are in turned off or standby state some amount of energy is used. Thus, smart strips can be used to prevent this problem. They can be turned off at an assigned time by using remote switches.

6.6.2 Case Study

Kalinga Institute of Industrial Technology, KIIT has around 25 campus at patia, Bhubaneshwar. Energy conservation and energy saving activities are conducted in all the campuses. In this case study, campus 3, (kathojodi campus) is considered for the year 2018 to 2019.

Energy-Saving Projects

- 40-Watt tube light changed to 10 watt led light.
- Installation of solar water heating system in campus 3 girls and boy's hostel.
- Insulation of the wires connecting the PV system.
- Campus street light changed to 8 watt led light.
- Roof top solar installation (standalone system).
- Dedicated feeder from chandaka grid (33KV) providing uninterruptible power supply.
- Installation of capacitor bank for improving power factor and energy saving.
- Old capacitors of ceiling fans are replaced with new capacitor in the laboratories, office rooms, hostel rooms.
- Slogans and banners are displayed in premises to create energy saving awareness.
- Old AC are replaced with new 5 star energy efficient AC.
- Rooftop PV panels were installed supplying electricity during bright sunny days.

Case study 1: 40-Watt tube light changed to 10-Watt led light
It has been studies that tube light consumes more energy than the LED light. And even LED light gives more brighter light and solve the same purpose as using tube light hence unnecessary using higher wattage tube light all the tube lights installed in the campus are replaced with 10 watt LED light. By this way, much of energy and money is also saved.

Case study 2: Campus Street Light 25 Watt CFL Changed to 8 Watt Led Light
Advantages

- Lamp life is higher with high efficiency and long life.
- At ceiling areas provides much brightness
- Least maintenance.
- Eco-friendly and there are no harmful radiations
- Application: Residence, showrooms, cove light, hospital premises, educational institutions, corridors

Case study 3: Installation of Solar water heating system on the top of hostel building
Advantages

- Reduces carbon footprint.
- Requirement of less maintenance.
- Energy payback period is 5–10 years.
- Government also provide rebate on electricity bills as an incentive to encourage more people to use Renewable Energy.
- Cheaper Installation Rate.
- Environment friendly.
- Protecting air quality.
- Protecting water quality.

Case study 4: Roof top solar installation
The rooftop of academic buildings, hostels and other buildings are free from shadow and solar PV panels are installed in first phase of 1.2 MWp (1200 KWp) capacity.

Case study 5: Dedicated feeder from Chandaka grid
Advantages
Provides uninterruptible and quality power supply for campus 3 from grid.

Case study 6: Use of Capacitor bank for improving Power factor

- Capacitor bank is used to provide the reactive power to the system as static VAR compensators.
- It saves energy bill by improving power factor
- It reduces line and cable losses

From Table 6.1, it is seen that total six projects gives a saving of 15.3 lakh kilowatt hour energy and a monetary saving of Rs 1,27,04,744.782/.

Table 6.1 Details of savings from various energy conservation projects.

Case study	**Name**	**KWH Saved**	**LAKH KWH**	**Amount(Rs)**
Case study 1	40-Watt tube light changed to 10-Watt led light	6,68,880	6.68880	39,12,948.00
Case study 2	Campus street light 25-Watt CFL changed to 8-Watt led light.	8665.92	0.0866592	50,695.632
Case study 3	Installation of solar water heating system on the top of hostel building	3,30,000	3.30	19,30,500.00
Case study 4	Roof top solar installation	5,24,939	5.24939	30,70,893.15
Case study 5	Dedicated feeder from chandaka grid (33kv)	-	-	23,17,700.00
Case study 6	Use of capacitor bank for improving power factor	-	-	14,67,808.00
	TOTAL		**15.3248492**	**1,27,04,744.782/**

6.7 Conclusion

Renewable energy sources have greatly reduced the use of fossil fuels. Energy balance can be maintained by the system of cogeneration. In India, technology is developing for the growth in renewable sector by using modern equipments to extract and process energy. Energy management and audit are the pillars of every organization to smoothly manage inflow and outflow of energy. Different types of energy management system is adopted by different organization with a common objective to reduce cost, increase saving, protect environment, and smooth functioning. Energy management mostly uses either clean production technology or pollution control scheme for environment protection. There are various steps in each process and all are co-related with each other. Energy manager has the key role to smoothly co-ordinate all the process and run the organisation energy efficiently. Energy audit shows the detail of energy structure in a system, its source, production, equipments, process-consuming energy, domestic purpose application, etc. In this chapter, energy audit for solar power plant is shown in detail. Similarly, energy audit for other renewable plants can also be carried by following the abovementioned steps for audit process.

References

1. Maamary, H.M.S., Kazem, H.A., Chaichan, M.T., Changing the energy profile of the GCC States: A review. *Int. J. Appl. Eng. Res. (IJAER)*, 11, 3, 1980–1988, 2016.
2. Al Waeely, A.A., Salman, S.D., Abdol-Reza, W.K., Chaichan, M.T., Kazem, H.A.H., Al-Jibori, S.S., Evaluation of the spatial distribution of shared electrical generators and their environmental effects at Al-Sader City-Baghdad-Iraq. *Int. J. Eng. Technol. IJET-IJENS*, 14, 2, 16–23, 2014.
3. Chaichan, M.T., Kazem, H.A., Abid, T.A., The environmental impact of transportation in Baghdad, Iraq, in: *Environment, Development and Sustainability*, 2016, DOI: 10.1007/s10668-016-9900-x.
4. Maamary, H.M.S., Kazem, H.A., Chaichan, M.T., The impact of the oil price fluctuations on common renewable energies in GCC countries. *Renewable Sustainable Energy Rev.*, 75, 989–1007, 2017.
5. Up, J., Leslie, P., Pearce, J., Harrap, R., Daniel, S., The application of smartphone technology to economic and environmental analysis of building energy conservation strategies. *Int. J. Sustainable Energy*, 31, 5, 295–311, 2012.
6. Turner, W.C. and Doty, S., *Energy Management Handbook*, 7th edition, The Fairmont Press Inc, Lilburn, GA 30047, 2009.

7. Ahmed, S.T. and Chaichan, M.T., A study of free convection in a solar chimney sample. *Eng. Technol. J.*, 29, 14, 2986–2997, 2011.
8. Kabir, G., Abubakar, A.I., El-Nafaty, U.A., Energy audit and conservation opportunities for pyroprocessing unit of a typical dry process cement plant. *Energy*, 35, 3, 1237–1243, 2010.
9. International Energy Agency IEA, *Key world energy statistics*, The World Energy Outlook, Statistical Publisher, 2006, Available at: http://www.iea.org/Textbase/nppdf/free/2006/Key2006.pdf [Accessed: 07/ 06/2007].
10. Martinot, E. and Sawin, J.L., *Renewable energy Policy network for 21st century,* REN21 Renewables Global Status Report: 2007 Update, REN21 Secretariat, Paris, 2007.
11. Martinot, E. and Sawin, J.L., *Renewable energy Policy network for 21st century,* REN21 Renewables Global Status Report: 2009 Update, REN21 Secretariat, Paris, 2009.
12. Singal, S. and V. Review of augmentation of energy needs using renewable energy sources in India. *Renewable Sustainable Energy Rev.*, 11, 1607–15, 2007.
13. Subramanian, V.: Renewable energy in India: status and future prospects, in: *Ministry of New and Renewable Energy*, November 2007.
14. GOI, Tenth Five year plan 2002–2007, planning commission, New Delhi, Available at: http://planningcommission.nic.in/aboutus/committee/wrkgrp11/wg11_ renewable.pdf.
15. Pradhan, A. and Panda, B , Performance analysis of Photovoltaic module at changing environmental condition using MATLAB/Simulink. *Int. J. Appl. Eng. Res.*, 12, 13, 3677–3683, 2017.
16. Pradhan, A. and Panda, B., Experimental analysis of factors affecting the power output of the PV module. *Int. J. Electr. Comput. Eng.*, 7, 141–149, 2017.
17. Pradhan, A. and Panda, B., Design and Modelling of simplified Boost converter for Photovoltaic system. *Int. J. Electr. Comput. Eng.*, 8, 1, 132–139, 2018.
18. Panda, B., Pradhan, Panda, B., Pati, N., Jena, C., Design and Implementation of PV Z source Inverter topology. *Int. J. Innov. Technol. Exploring Eng.*, 9, 1, 4391–4395, 2019, DOI: 10.35940/ijitee.A5075.119119.

7

A Smart Energy-Efficient Support System for PV Power Plants

Salwa Ammach[1] and Saeed Mian Qaisar[1,2*]

[1]*Dept. of Electrical and Computer Engineering, Effat University, Jeddah, Saudi Arabia*
[2]*Communication & Signal Processing Lab, Energy & Technology Research Center, Effat University, Jeddah, Saudi Arabia*

Abstract

Despite the revolution acceleration of solar panels, the efficiency of solar power has been a significant concern for many researchers for decades. Besides PVs limited efficiency, metrological factors greatly attenuate PV performance in the field. Three of the main reasons that lead to solar panel's efficiency degradation are the soiling phenomenon that accumulates on the panels' surface, potential hotspots formed on some shaded cells, and limited solar radiation capturing caused by the continuous movement of the sun in the skydome. This paper aims to optimize PV performance by designing and implementing a smart energy-efficient support system for PV panels. The proposed system ensures the optimum generation of electrical power by its triple functionality, hotspots detection, proper orientation of PV panels, and the cleaning process. The latter ensures dust is periodically cleaned of the front surface. Hotspot detection allows the user to take proactive measures before any further complications. The solar tracking mechanism maximizes the total absorbed sun rays by the PV panel. The design criteria are based on resources sharing and minimum cost and power needed to achieve the system's objective. The panels are rotated throughout two motors while causing the unmotorized brushes to slide to the other side by means of gravity, cleaning all the dust on the way. Since PV power plants are usually installed in remote areas, IoT is used to communicate any suspicious hotspots and alert the operator wirelessly. Moreover, the hotspot detection will be carried out using existing electronics and controllers that is already available to serve other tasks. The followed methodology implements a prototype that uses a microcontroller to control the mechanism at

**Corresponding author*: sqaisar@effatuniversity.edu.sa

Suman Lata Tripathi, Souvik Ganguli, Tengiz Magradze, and Abhishek Kumar (eds.) Intelligent Green Technologies for Sustainable Smart Cities, (111–142) © 2022 Scrivener Publishing LLC

minimum cost. The results verify that the system meets the following criteria: low cost, low power, environmental friendly, safe, reliable, and simple

Keywords: Renewable energy, photovoltaic cleaning, light tracking, fault detection, hotspot, machine learning, dust disposition, Internet of things

7.1 Introduction

The rapidly growing demand for electrical energy, the constraint on pollution levels, and the limitation on fossil fuel reserves and supply have led to a rising interest in researching and utilization of renewable energy sources and technologies. At the forefront of reducing greenhouse gas emissions, solar power riveted vast attention in the last several years. This goal paved the way for the proliferation of renewable energy share in global energy production. The use of renewable energy as the primary energy is hailed to be the probable solution for future sustainable energy supply and environmental development. Renewable energy is extracted from sources that are naturally replenishing and virtually inexhaustible in duration. The cleanest and most abundant energy resource is solar power, which riveted great attention in the last several years. Solar energy is a vast, abundant, cost-free, green renewable source of energy.

Solar energy is harnessed and converted into electricity using different technologies in which photovoltaic cells are the most popular and efficient ones. Photovoltaic (PV) technology has taken the world by storm in the energy sector. A photovoltaic is an electric device that converts electromagnetic radiations directly to electricity in the form of direct current by the photovoltaic effect, which is a chemical and physical phenomenon describing the generation of voltage and electric current in a material upon exposure to light.

Despite this, researchers face several hurdles in improving the efficiency of solar cells.

The highest efficiency achieved by a mono-crystalline and multi-junction cell is 26.7% and 46%, respectively, according to a research released in 2019 by Fraunhofer Institute of Solar Energy Systems ISE (*Homepage Fraunhofer-Gesellschaft*, n.d.). Worse still, the metrological factors could result in further degradation of the PV panels on the field. Hence, supporting systems usually accompany the PV power plant's installation to maintain the highest possible energy yield.

Three of the major challenges that attenuated PV performance significantly in the field are the soiling phenomenon, limited captured irradiance, and created hotspots. Many researchers addressed these issues separately and have proposed solutions to overcome these inconveniences effectively.

PV systems must ideally operate automatically and autonomously, but many metrological factors attenuate their performance. The three factors addressed in this project are limited solar irradiance receipt due to the sun's continuous movement on the skydome, dust accumulation over the front surface of the PV panel, and hotspots formation due to shading.

Many electronics required to operate the PV system in an optimal way can result in a complex and costly implementation. On the other hand, all these electronics are mandatory to serve a viable purpose. Therefore, more creative methods must be carried out to combine multiple functionalities in an effective and cost-competitive platform while not sacrificing the performance.

The three functionalities addressed in this context are solar PV tracking mechanisms, panel cleaning, and hot spot detection. This work provides a framework for utility-scale PV power plants support systems to optimize the operation. The proposed system must satisfy the following criteria: low cost, low power, environmentally friendly, safety, practicality, and minimum maintenance.

In short, the questions this chapter is trying to address are: How to improve performance of the PV system with minimum cost? What are possible technical solutions for PV degrading performance on the field? How to optimize the use of electronics resources associated with PV installations?

The PV cell's performance is directly proportional to the sunlight irradiance. PV power plant productivity is boosted when the solar panel is mounted perpendicularly to sun rays. In other words, the more impinging sun rays normal to the PV panel, the more the power output. The reason behind this phenomenon is obvious, the more dust is accumulated on the panel, the more shade is covering the panel, and thus the less irradiance can penetrate to the PV and the fewer photons to motivate electron-hole pairs. This challenge is attributed to either weather conditions like cloudy days, which cannot be controlled, or the sun's continuous movement on the skydome while the panel is immovable.

The azimuth angle, altitude, and geographical coordinates all contribute to the sun's position. The latter is entirely dependent on the coordinates of the solar panel's installation site. This is a constant parameter that never changes. However, the other two parameters are continually changing as defined below.

- Altitude angle: The angle between the line to the sun's center and the horizon, as shown in Figure 7.1, is known as the altitude angle. It begins at 0 degrees at dawn and progressively increases in altitude until it reaches its greatest point at noon. After then, it steadily decreases till sunset.

- Azimuth angle is the angle measured clockwise between the north or south and the projection of the sun's radiations on the horizontal plane as shown in Figure 7.1.

These two angles are varying every season and every instant. To understand the theory behind these angles, the effect of the location of the sun incident on the earth must be understood. Simply, they can be attributed to the rotations of the earth as follows:

- Earth rotation around itself causes the sun to move from east to west, i.e., sunrise to sunset.
- Earth's revolution around the sun caused the sun's actual arch across the sky to change, changing the distance between earth and sun. The sun path is longer in the summer and shorter in the winter. Figure 7.2 is a schematic diagram that shows the sun's position at four different times of the year at the same time of the day. June 21 for summer, September and March 21 for fall and autumn, and December 21 for winter. This revolution effect on the sun's position on the skydome can be moving south toward the tropic of Capricorn or north toward the tropic of cancer.

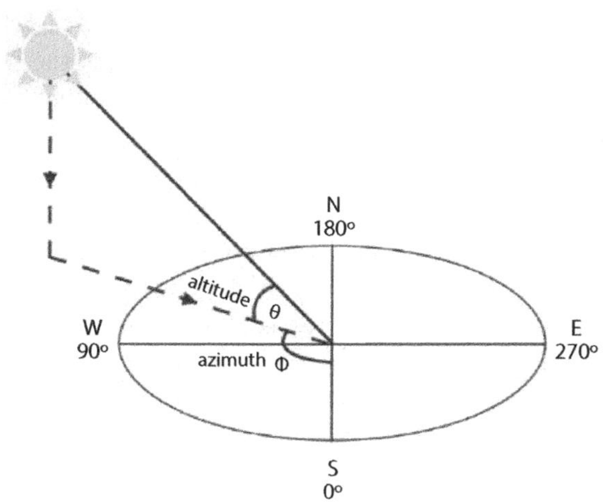

Figure 7.1 Altitude and azimuth angles.

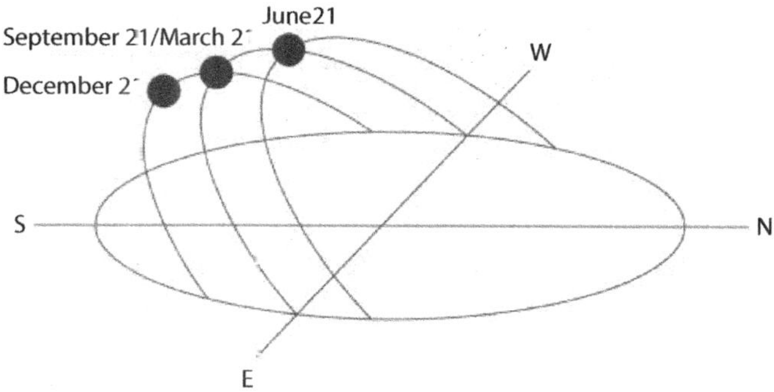

Figure 7.2 Sun path changes over the year.

For example, using coordinates from Jeddah, Saudi Arabia, a chart depicting the solar path for the entire year (*Home ☼ SunEarthTools.Com Solar Tools for Consumers and Designers*, n.d.) may be generated, as shown in Figure 7.3.

These sun changing locations is a challenge that PV system designers have to tackle. Because the sun's location in the sky dome is constantly changing, this orientation can only be achieved by using a moveable tracker that tracks the sun's position. A stationary solar panel, on the other hand, would have a far lower absorption efficiency.

Another factor that negatively impacts the PV performance is the soiling effect that accumulates on the PV panel surface. Solar power plants are

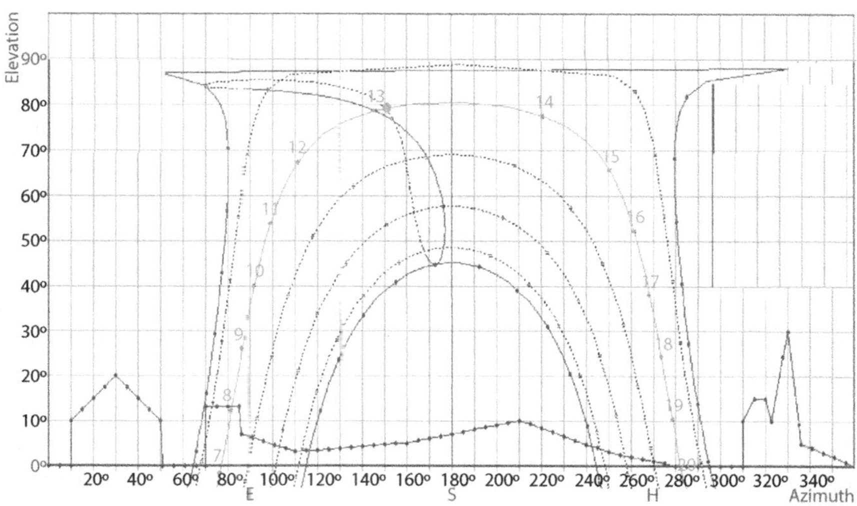

Figure 7.3 Jeddah, Saudi Arabia, annual sun path.

usually installed in arid locations where abundant solar irradiance is available. These lands are characterized by harsh climate conditions and dusty weather that cause degradation to the PV cells and consequently generate less power output. Dust particle size and composition vary according to location. In the middle east and arid areas are higher than in Europe (Jamil *et al.*, 2017), (Haeberlin *et al.*, 1998).

Researchers have investigated the effect of soiling and found that power output is reduced by a range of 2% to 50%, depending on the region (Maghami *et al.*, 2016), (Sarver *et al.*, 2013). On the other hand, a comparison of power output between a static solar panel and a movable one has proved that performance is improved by 40% on average (Rizk & Chaiko, 2008).

Many researchers have studied the characteristics of dust particles and the impact of soiling on power output and have all agreed to one conclusion that dust accumulation is inversely proportional to electric power generation (Jamil *et al.*, 2017), (Haeberlin *et al.*, 1998), (Sayyah *et al.*, 2014), (Caron & Littmann, 2013). The dust blocks the sunrays from reaching the PV panels.

An experiment has been done by Sulaiman *et al.* to study the effect of dust in tropical regions where a comparison has been made between a clean panel against one covered with dirt particles. The result has shown that energy yield degraded by 60% to 74%, 9% to 31%, and 70% to 80% for dust, talcum, and sand, respectively [10].

Hence, continuous activities for cleaning the photovoltaic panels must be exercised in the total absence of water and under severe environmental conditions.

Commercial PV applications connect individual cells in series to obtain the required output voltage. Hotspot development is a localized heating situation that develops within a PV module as a result of solar cell mismatch, partial shadowing, or internal cell faults. Other causes of hotspots include electrical mismatch or cracks during bad fabrication quality or the PV site surroundings. Hot spots in PV modules can have a variety of causes and can point to deeper problems. The cause concerned in this work is related to the environmental condition, which is shading. Other hotspot causes are usually tackled during the fabrication process and reliability testing to produce a robust panel.

The shade can be caused by nearby objects, dust accumulation, any other sources of dirt that stagnated on the surface of the PV, causing blockage to the sun rays reaching the p-n junction of PV cell. The shading of nearby obstacles like buildings, trees, overhead electrical lines, and electrical poles. For example, a rooftop installation, situations in service areas, including satellite dishes, water tanks, air-condition chillers, etc. If such obstacles exist that cause shades directly in the solar panel, then it is advisable to avoid such areas for PV installations. Likewise, for large-scale solar power plants, shadows from transmission lines that may be passing by the PV side can cause shading on the module surface. Also, nearby wind turbines in renewable energy wind and solar farms will cause a shadow on PV arrays.

Another cause of shading is the shade caused by a near module when insufficient space between them or if the tilt angle is large. This is called interrow shading, and it is mitigated by separating their location by a good distance from the module and optimizing the tilt angles, so there is no such shading during the peak hours. This is again limited by the area devoted to the PV installation, and it cannot be extremely high because it will increase the land cost. This is a complex scenario that scientists have been working on. Commercially, if a buyer or investor is planning to install the PV system in a shaded area, the manufacturer will not offer a product and performance warranty.

Shading not only decreases the total electrical energy yielded but also can create thermal hotspots, which eventually destroy the PV panels. The late detection of hotspots can damage the PV cell or module permanently. It speeds up the degradation, ignites a fire hazard in extreme cases, and threatens safety. Many reports can be found about fire accidents that happen in solar PV panels caused by partial shadings. Thus, it is advised that hotspots must be avoided as much as possible as it will increase the cost of maintenance and replacements of PV modules or panels when occurred.

Many efforts are devoted to placing the PV panels in proper locations to avoid shading from nearby objects throughout the whole year (shading direction varies as the sun location varies). However, other factors cannot be 100% avoided. Therefore, hotspots must be monitored closely to avoid PV degradation.

A shaded cell does not only degrade that specific cell characteristic but also causes complications to the whole other cells in the modules even if they were not shaded. In physical terms, a shaded cell in a string of PV cells will operate in reverse bias because there is no photogenerated current in that cell. The reverse bias is created as a result of the negative voltage that appears on the shaded cell. Worse still, the more series cells, the negative voltage across the shaded cell increases, which causes the reverse bias also to increase. Eventually, a breakdown or a saturation region is reached on the reverse bias that causes the hot spotted cell to absorb the power and dissipate it as heat. Since the series electrical connection implies a matching current, the current is limited by the bad cell. That cell is unable to conduct current in forwarding bias, and it is driven by other cells to operate in reverse bias. While all illuminated cells are attempting to produce power, that one bad cell is sinking the power. This condition causes a huge attenuation in the electrical characteristics, such as the I-V curve, of that cell by decreasing the maximum power point (where the PV module should ideally operate at) (Kim *et al.*, 2016; Pannebakker *et al.*, 2017). Of course, designing a PV module and its bypass diode quality are major elements to avoid such problems. To mitigate hot-spotting, Bypass diodes are connected in parallel to the cell to offer a different path for the current. However, a single bypass diode used for every single cell is not an economical or practical option. Instead, a shared bypass diode is used for a group of individual solar cells. Thus, it does not completely prevent the creation of hotspots or mitigate its complications. There are special designs and calculations on the best number of grouped cells sharing the same bypass diode. This is the best practice since it is impractical and costly to devote one bypass diode for each cell. Despite the reduced hotspot caused by bypassing diodes designs, it does not completely solve the hotspots problem (Kaplanis&Kaplani, 2011).

Many techniques have been suggested to mitigate the negative impact of hotspots. They all become more efficient and effective if the hotspot is detected in the early stages and with high accuracy.

In this work, we propose a novel system control for PV panels to track the sun, clean the dust, and detect hotspots at low cost and high efficiency. The methodology implemented is a prototype composed of a control system that combines all these three functions simultaneously. Without additional cost, the same controller can be optimized to embrace all the functionalities together. Usually, a separate controller or system is used for tracking, cleaning, and hotspots detection. This can cost extra resources, electronic equipment, and maintenance costs and efforts. Instead, this system suggests a simple way to utilize shared pieces of equipment more effectively.

To properly integrate the green energy sources, such as PVs in modern grid, the recent technological advancements in signal processing, surveillance and machine learning can also play key roles (Jambi *et al.*, 2018), (Sabo & Qaisar, 2018).

7.2 Literature Review

This section discusses previous work available in the literature about how others addressed the three challenges of the solar cells:limited irradiance due to the sun's changing position, dust accumulation, and hotspots formation. According to the narration mentioned above in the previous section, the tracking system can aid in capturing more sunlight, the cleaning system prevents dust accumulation, and hotspots' accurate detection allows for proactive and fast action to mitigate its impact. These techniques can be responsible for avoiding unfavorable metrological effects on the PV panels.

7.2.1 Solar Tracking System

Researchers analyzed the power output of a static solar panel to that of one with a tracking mechanism and found a 40% boost in performance. As a result, the solar tracking system is used to maximize sunlight reception by automatically adjusting the position of the solar module in response to the sun's location on the planet at any given moment. Solar tracker, either single axis or dual axis, is invented to tackle the limited irradiance inconvenience.

A single-axis mode with slanted PV panel installation is typically the most cost-effective choice, although it might result in inaccuracy and lower electrical production (Abadi, 2014).The absence of the second tracker is a tradeoff between accuracy and an extra motor cost. Others recommended that the tilt angle be manually adjusted annually to compensate for the lack of a second axis tracker (Lubitz, 2011). However, this approach lacks practicality and accuracy because the human positioning to the angle requires a lot of effort to place it in the correct angle. Consequently, dual-axis autonomous tracker is the optimum solution for inaccurate tracking systems.

The most popular technique for tracking principle is using light-detecting resistors (LDR) serving as sensors to the sun irradiance (Petrusev *et al.*, 2016), (Lakeou *et al.*, 2006). Though its regulating approach and/or the number and placement of sensors may change between methods, their theory is based on the same concept, comparing the light detected by these sensors and moving the tracker to the side exposed to the most irradiance. The sensor's alignment can be half-circle (arc shape) or rectangle shape with equal distance between them.

Apart from the primary solar panel used to generate energy, another option employs small-scale photovoltaic cells to function as sensors. The PV cells serving as sensors are compared with respect to their current output. The PV that generates the highest electrical current implies that it is exposed to the higher irradiance, and the motors move accordingly (Rahayani & Gunawan, 2019). The methods listed above will identify the sun's location quite precisely and quickly.

Dual-axis tracking topology is more efficient where both azimuth and altitude angles of the sun are being tracked. The typical mechanical

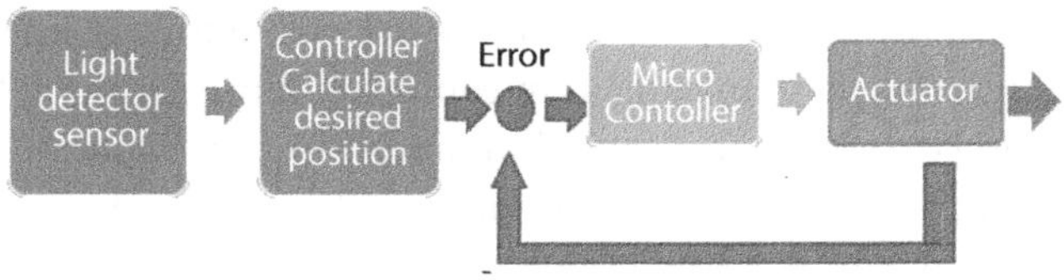

Figure 7.4 Feedback control block diagram.

structure to imitate azimuth and altitude angles is by using two motors. One motor moves horizontally while the other moves vertically.

This closed-loop system alters the panels' angle based on the desired position. The desired position is where the light is equally distributed to all sensors. Once the sun moves, the system will be able to sense it based on changes in light exposure and send it to the controller as an input signal. The microcontroller process it and compare it to the desired location. The controller compares the current position to the desired position and calculates the feedback error. The error is the same angle magnitude in which the system has to move to get zero error and accurate position. The microcontroller generates a signal to actuators, which are motors to rotate the module accordingly. The process block diagram is displayed in Figure 7.4.

The case when systems with higher precision are required, further sophistications can be incorporated by employing the machine learning, sensing and signal processing based viable approaches (S. M. Qaisar & Subasi, 2020), (S. Qaisar *et al.*, 2008), (Abualsaud *et al.*, 2016), (S. Qaisar *et al.*, 2015), (S. M. Qaisar, 2014), (S. M. Qaisar & Subasi, 2018).

7.2.2 Solar Cleaning Mechanisms

Different cleaning techniques have been proposed through the literature, with each has its benefits and drawbacks to mitigate the PV dust accumulation problem.

Manual cleaning is the easiest technique, but it is costly, impractical, and resource-intensive and could cause health issues to the workers, especially for large PV powerplants. Some PV power plants scale over 10 km of land, and, in such cases, it becomes even more difficult to clean all this huge area manually.

Some suggested using limited water quantities to spray the PV surface while wiping the dust and dirt off. This way seems to provide effective cleaning. A robot was developed to spray some water and navigate across a rooftop PV system (Riawan *et al.*, 2018). However, washing the surface with water or special liquids using sprinklers like the ones used in irrigation creates another problem of wasting water resources that are scarce in our world and must be utilized more efficiently for drinking or irrigation purposes. This solution must also be eliminated and even prohibited in dry areas.

In Morocco, the humid weather naturally forms dew droplets on the front glass of the PV panels, which was used and tested in effectively cleaning the dust (Belihi *et al.*, 2019). The dew formation can worsen the dust and make it difficult to clean it because it increases the stickiness (or what is called the surface tension). On the other hand, in few cases, dew can enhance the cleanliness of the solar panel if present in limited quantities. The preliminary results reveal that the impact of dew on tilted panels can be thought of as a positive impact with respect to the flat horizontal panel. However, this conclusion is not finalized and must be supported with stronger shreds of evidence of measuring the PV performance (Belihi *et al.*, 2019).

Blowing air also proved to be efficient in reducing dust accumulation. However, it is not very fruitful when implemented alone. Instead, it is usually compensated with spraying some liquids and wiping brushes. Internet of things (IoT) is employed in (Kumar *et al.*, 2020) to decrease human intervention and move towards more automated systems. IoT's role is to remotely control the cleaning robot by switching them on/off when needed. A PV panel powers the robots to keep renewable and sustainable operation (Kumar *et al.*, 2020).

The most common technique used nowadays in cleaning is using motorized brushes. The detached, stand-alone robots (not connected to the edges of the panel) faces many challenges during implementation on tilted or tracked PV panel such as stability against falling. The attached cleaning mechanism depends on electric motors to slide the brushes across the dirty panels. In case the motors are powered by a battery or any other conventional power source, then they must be periodically maintained and recharged to ensure uninterrupted operation. Some claim that powering the cleaning robots with an extra solar panel saves energy. Nevertheless, this hypothesis is not valid because these solar panels need to be cleaned again to provide sufficient energy to the motors.

In (Gheitasi *et al.*, 2015), the work focuses on the frequency of cleaning. Setting a fixed frequency might not be optimal because the weather condition is rapidly fluctuating, and it should not be taken for granted that dust is formed on a regular basis. For example, every day or every two days. Therefore, current sensors were implemented to check the photogenerated current against a predetermined reference. If the current output were less than this threshold for more than 1 hour, an alarm would be raised to the controller, which will decide whether to trigger the cleaning task or not. The reference is changing with day and night to follow the natural change in solar irradiance. This topology will allow a better resource allocation and usage of cleaning robots more effectively (Gheitasi *et al.*, 2015).

The concept of weighing the benefits and the cost of cleaning the solar project is investigated in (Fan *et al.*, 2020). The cost consideration is the manual cleaning combining manpower and materials. The optimization weights the cost of cleaning versus the cost loss due to dust accumulation. Based on experimental data in China, it was found that the trade-off can be achieved if the cleaning interval is 17 days. Nevertheless, this cannot be generalized to every power plant since some areas are much dustier than others. Besides, other cleaning techniques must be taken into consideration for such investigation to obtain a complete comparative framework since manpower is expected to be much more costly than automated methods(Fan *et al.*, 2020).

PVs with tracking systems have better prevention of dust accumulation than immobile ones by the natural falling of dust particles with the panels' movement despite cleaning the system(Maghami *et al.*, 2016). A decrease of accumulated dust by 50% was observed when the PV panel is mounted on a solar tracker (Dahlioui *et al.*, 2018). The main challenge is fixing the brushes on a movable tracker. Worse still, the movement of the brush must move in a certain direction of gravity to save power. If the brushes rotated in the reverse direction, it would sink extra power-driven by the extra torque required to lift the cleaning system against gravity. All these constraints must be taken into consideration during the design phase.

A study combined tracking with cleaning due to the beneficial effect of the two functions. Nonetheless, each system works independently. This study aims to ensure the cleaning brush is fixed and not falling with the tracker's movement. At the same time, the cleaning mechanism is achieving high efficiency. The brush rotates across the surface by a dedicated motor and is controlled by an Arduino controller. To increase the levels of freedom in the movement of the brush, the brush arm is designed in a pivot shape with two motors. It is a semi-independent cleaning brush, it is not attached to the edges of the PV, but at the same time, it is not a stand-alone robot. Although this system is efficient autonomous, and stable, it requires dedicated motors and extra power to operate successfully (Dahlioui *et al.*, 2018). It is critical to offer a cleaning approach that is included into the biaxial tracker that is defined by the Tilt and Roll kinetics established by (Dahlioui *et al.*, 2018). In that article, a novel cleaning approach for HeliosLite's biaxial tracker, which was recently deployed in Rabat's Solar Energy Platform, was developed. A functional study of the problem has been performed based on a critical analysis of many available cleaning methods for PV modules as well as the tracker's structure. This approach has effectively achieved the key requirements sought by this work: autonomy, stability, cheap cost, and efficiency, based on the employment of a telescopic arm attached to a spinning brush.

A project done by (Tejwani & Solanki, 2010) combines tracking and cleaning methodologies. However, it has a number of limitations. First, the cleaning brush was designed to slide only when the panel is perpendicular to the ground. Mechanically, this structure will force the brush to slide at a very high speed which causes improper cleaning. Second, only two functionalities are applied, which are cleaning and tracking (Tejwani & Solanki, 2010).

The idea of performing PV cleaning by taking advantage of the moving panels is proposed (*Innovative Low Cost Cleaning Technique for PV Modules on Solar Tracker | IEEE Conference Publication | IEEE Xplore*, n.d.). The same concept applies where gravity is the main driver of the sliding brushes while the trackers are rotating the panels. Two mechanical tools were tested to ensure the brushes are efficiently removing dust from the surface. In addition to the mechanical structure, the brush pins themselves must be highly efficient to wipe off the dust on its way to the other end. Thus, a more sophisticated mechanical knowledge and brush structure is needed to enhance the performance. The mechanical structure must support the slow and naturally rotating brushes. At the same time, the brush's teeth must be efficient for different types of dust and dirt. The study is recommending further investigation on the mechanical structure and brushes to achieve the optimum design. The advantages of such a cleaning system are automatic and autonomous system, low cost provided by eliminating the need of extra motors and their power source. The cleaning takes place twice a day, which is more than the motorized cleaning, maintaining sustainability by avoiding using of water, simplicity in construction.

Others are investigating special coating chemical materials to avoid interaction with the dust named self-cleaning coating. It is important to highlight that the chemical coat and smart cleaning mechanism do not contradict each other. In fact, they can be complementary to further mitigate the soiling effect. However, self-cleaning coatings have their pros and cons that need to be studied independently of this study.

7.2.3 Hotspots Detection

Laboratory and field testing before installation can reduce the risk of design and process problems during cell and module manufacture. Solid operating and maintenance plans must be developed for early detection after the system is implemented in order to reduce safety and revenue concerns. These plans completely depend on the accurate and early detection of hotspots. The detection process can follow visual inspection or analytical analysis. In both cases, the bare eye cannot decide whether a PV module suffers from such faults. More sophisticated, and some are simple, methods are proposed in the literature to solve this problem.

The main hotspot detection technique is taking images for the PV cells using thermal cameras (*National Renewable Energy Laboratory (NREL) Home Page*, n.d.). It is a nondestructive test that uses infrared radiations to capture temperature variations of the taken image device. Any variation in temperature from the normal operating temperature is captured in different colors by this special camera. The interpretation of the taken images can take many forms, like image processing or data-driven methods. Usually, eye interpretation is subjective and could not guarantee valid judgment(Gosumbonggot& Fujita, 2019).

On the other hand, during recent years, machine learning has taken the world by storm. Hotspots detection is a perfect scenario for machine learning applications. In the following paragraphs, some work related to hotspot detection via machine learning tools is reviewed.

A study has developed a hot detection methodology using image processing in MATLAB, where the thermograms served as the input data. The algorithm clusters highly impacted spots and define the temperature and size, which in turn can reveal the I-V characteristics(Salazar &Macabebe, 2016). Besides, it can suggest possible reasons that caused these detected hotspots. This technique is rewarding in terms of abolishing the need to disconnect the solar cells to take certain measurements and interrupt its operation(Salazar &Macabebe, 2016).

The Simple Linear Iterative Clustering (SLIC) Super-pixel method, which searches for color similarities, is offered as a comparable topology for processing pictures collected from thermal cameras (Alsafasfeh *et al.*, 2017).

Multiple parts of the picture are converted into super-pixels or patches, which are then tagged as belonging to the item or the backdrop. By dividing the picture into super-pixels, segmentation algorithms may be made more noise-tolerant, and the image's complexity can be reduced from hundreds of thousands of pixels to only a few hundred super-pixels. It might add some level of filtering to eliminate outliers and anomalies. For validation purposes, two identical panels were tested, and the thermal image was processed using Python 2.7 on the Eclipse IDE platform. All artificially created defects were successfully detected by the SLIC algorithm, which confirms the accuracy and fast response in hotspots detection (Alsafasfeh *et al.*, 2017).

Another hotspot detection technique proposed in the literature is also based on taking thermal images of the PV module. The infrared images are captured with the help of an Unmanned aerial vehicle. Then, the images are sent to a control system located in a different area where the image is processed to reveal any hotspots in real-time and saved on a database. The saved images can be played as a video to maintain the hotspots record of the PV panel. Moreover, a normal camera is attached to the vehicle to identify any visual damage to the PV array (Sha *et al.*, 2019). The main advantage of such a technique is that the same device can serve multiple PV power plants. It features mobility and does not require resource allocation for every project. For instance, the same thermal camera and the vehicle are used to scan multiple locations, while fixed thermal cameras are allocated exclusively for the installed location.

A study suggested an analysis of the PV thermal images by transforming them from RGB (Red, Green, Blue) to HSV (hue, saturation, lightness) scheme, and the classifications take place according to the "hue" level. The solar cell edges are eliminated from the analysis, then the mean and standard deviation of the blue content and the grayscale are computed (Chen *et al.*, 2020). Afterward, machine learning takes the lead in determining whether a hotspot is formed or not. The algorithm used is the support vector machine, where the model is trained to detect the hotspots based on the color intensity analysis of experimental data (Chen *et al.*, 2020).

The fuzzy logic technique was proposed to determine degraded PVs due to hot spotting. The input uses real data about PV power plants in the UK includes infrared photos, open-circuit voltage, short circuit current, and percentage of power loss. The verification was carried out via simulation scenarios of PVs suffering hotspots of multiple severity levels. However, this method lacks the ability to detect hotspots under high shading situations. The algorithm's accuracy exceeded 96% in detecting six types of hotspots depending on the number of hot spots from one to more than five. The sixth type is the most deficient type with 11-26% hotspots and is named one hot potted P-V string (*ResearchGate Link*, n.d.).

Instead of thermal images, the data about the PV is collected using digital and analog sensors that measure the electrical characteristics such as voltage and current output, in addition to their derivative (Ben-Menahem & Yang, 2012). The electrical string data are the only input required to accomplish such a task; no other temperature sensors are needed. The derivative can be either executed through a low pass filter or simply by applying numerical mathematics code. The analog to digital signal conditioning is performed via a low-power microcontroller integrated circuit named Cypress Semiconductor. Data should be transferred using controllers equipped with internet features to control the performance of projects installed in remote areas. Cypress semiconductor was selected because it has this feature in addition to many other useful features for such applications. The collected data is transferred to a database and a server. The latter is programmed with a machine learning algorithm (thermal and electrical System ID stages, supported by a Bayesian Maximum A posteriori Probability) that detects hotspots online in real-time and identifies the cause to the best possible estimate. The algorithm was verified using simulated and physical data from a rooftop project. Moreover, the algorithm was programmed to predict potential future hotspots to allow the system owner to be proactive and take fast action accordingly (Ben-Menahem & Yang, 2012).

In the view of machine learning advancements, deep learning using support vector machine is presented in (Wei *et al.*, 2019) as an accurate alternative method not only to detect but also locate hotspots. The algorithm implemented is Faster Region-based Convolutional Neural Networks (RCNN) which is an improved version of the conventional RCNN with respect to a higher speed. This neural network algorithm is popular in object detection, and thus it can perform substantially in detecting hotspots. The infrared input image is analyzed to produce a set of boxes as output. Each box contains an object and its category. Moreover, the study encompasses a conventional image processing method to detect hotspots via Hough Line Detection. It was suggested that the images are collected with the help of an unmanned aerial vehicle. The comparison between digital image processing versus neural networks reveals that the latter is much more accurate but takes more time and requires more storage and processing space. Yet, the time is taken, and the space required is within acceptable limits, which favors machine learning over conventional methods (Wei *et al.*, 2019).

While a thermal image visually displays the high-temperature areas within the PV module, PVs that are suffering from hotspots characteristics also differ analytically. The obstacle that arises here is the confusion between hotspot cell characteristics, normal shaded cell characteristics, and low irradiance solar cell characteristics. If these characteristics were distinguished clearly, the detection algorithm is facilitated. It should be noted that a shaded cell is not necessarily considered a hotspot since the shaded cell at the early stages is not turned into a hotspot yet and can normally operate to a certain time. This imposes an extra challenge on scientists trying to detect hotspots but not shaded cells. Both conditions have similar characteristics, and thus extra efforts must be exerted to differentiate between both. The shaded cell, which has not evolved to a hot spot, does not cause serious performance attenuation.

Moreover, the irradiance intensity is fluctuating throughout the day. In cloudy times, the power output naturally decreases. Again, the discrepancy between hot spot, normal operating shaded cell, and low irradiance must be thoroughly studied in order to develop a robust hotspot detection algorithm. To distinguish between low irradiance and shading condition, analytical difference is studied in (Gosumbonggot & Fujita, 2019). The below two equations are describing the relationship between PV's current and voltage.

$$I_{PV} = (I_{PV,STC} + K_1 \Delta \mathrm{T}) \frac{G}{G_{STC}} \tag{7.1}$$

$$V_{OC} = V_{OC,STC} + K_V(\mathrm{T} + T_{STC}) + \alpha V_T \ln\left(\frac{G}{G_{STC}}\right) \tag{7.2}$$

Where;

- $I_{PV,STC}$ is the module's PV current under standard test conditions (STC, at 1000 W/m^2 and 25°C)
- K_1 is the PV's current temperature coefficient.
- G is irradiation in W/m2
- G_{STC} is irradiation at STC in W/m^2
- $V_{OC,STC}$ *is* open-circuit voltage at STC
- K_V is the temperature coefficient of the PV's voltage
- T is the temperature in °C
- T_{STC} is the temperature at STC in °C
- α is the diode constant
- V_T is the voltage constant

The formulas are used to extract different behaviors of low irradiance and potential hotspots isby assuming $\alpha V_T \ln\left(\frac{G}{G_{STC}}\right)$ from equation 7.2 is a very small value, then the temperature has huge impact on the variation of open circuit voltage. When a PV is suffering from shading, the current and voltage changes radically that it would change with irradiance. This will lead the maximum power point tracker to start fluctuating and giving inaccurate results (Gosumbonggot & Fujita, 2019).

Hot Spot Suppression compares the voltage, current, and temperature of a PV module in partial shadow to that of a module that is artificially irradiated. It can classify the hotspot condition and, if necessary, search for another maximum power point to operate at in a way that decreases the stress on the shaded cell. Hence, it is associated with the maximum power point tracking algorithm. The results show that this method is effective in reducing the temperature of the shaded cell based on experimental and theoretical data analysis (Spanoche *et al.*, 2012).

Since PV projects involve power electronics to assist in photo-generated power extraction and regulation, some of these electronics might be effectively utilized for hotspot detection purposes as well. A dc-dc power converter is used to measure the impedance of a string in a PV panel. The impedance can be interpreted to detect hot spotting at different frequencies (Kim *et al.*, 2016). Since the PV circuit model combines capacitance, resistance, and conductance, each of these components dominates at a certain frequency. The concept is based on a solid understanding of the ideal PV cell impedance. The dc impedance of the equivalent circuit is the sum of parallel and series resistances. Assuming the latter to be constant, the parallel resistance can be found by measuring the total dc impedance, which is obtained from the IV curve. The impedance is defined as the voltage divided by current, and the I-V curve plots the voltage and current characteristics of a PV module. Therefore, it can be concluded that the impedance is the slope of the IV curve, which can be negative in some cases. The steeper the slope indicates a smaller the impedance's magnitude. Due to the assumption that series resistance is constant, the larger the impedance indicates in larger parallel resistance, and the smaller the impedance, the smaller the parallel resistance. Likewise, the parallel capacitance also helps in locating hotspots. There are two types of capacitance in a solar cell: junction and diffusion capacitance. The former is the capacitance resulted from the depletion zone charges. It governs when the voltage magnitude is a small positive number of negative numbers. The diffusion capacitance results from the charges in the p-type and n-type of the semiconductor; it takes the lead when the cell is carrying current in the forward bias. These characteristics of resistance and capacitance allow distinguishing whether the cell is operating in forward or reverse bias or identifying whether a cell in a PV string is impeding the whole string performance compared to the expected outcome. On the other hand, it can be confusingly thought that the decreasing performance is simply due to decreased illumination, for example, on a cloudy day. The experiment done in (Kim *et al.*, 2016) in dark and flashlight indicates that resistance and capacitance are more affected in the bias direction than in illumination at a constant temperature. This discrepancy between the effects is crucial in identifying hotspots. At this stage, an algorithm can successfully develop to detect hotspots based on the AC electrical characteristics and conditions to detect hotspots. An experiment was conducted at high and low frequencies to segregate the measurement of the capacitive region and dc impedance region, respectively. The dc impedance can also be measured in an alternative method according to the I-V curve characteristics. This method depends on the maximum power point tracking algorithm(Kim *et al.*, 2016). The equation to measure the impedance at frequency ω is given by:

$$Z = R_s + \frac{R_p}{(\omega R_p C_p)^2 + 1} + j\left[L_s - \frac{\omega {R_p}^2 C_p}{(\omega R_p C_p)^2 + 1}\right] \tag{7.3}$$

Where Z is the impedance, R_s is the series resistance, R_p is the parallel resistance, C_p is the parallel capacitance, and L_s is the series inductance.

Within the same context, maximum power point tracking is used to detect hotspots in many works. The perturbation and observation (P&O) method is the most widely used method in maximum power point tracking. Its principle is based on gradually increasing or decreasing the voltage and compares the output with the previous cycle. Eventually, the algorithm will stabilize the voltage at the highest power output. The current magnitude variations at constant voltage can reveal information about hotspots in the light of the below formula (Hossam & Itako, 2016).

$$\text{Current change rate} = (I_{V1} - I_{V2}) / \text{ short circuit current } I_o \tag{7.4}$$

Where is the current at certain point (at voltage 1), I_{V2} is the current at certain point (at voltage 2), I_o is the short circuit current when there is no shadowis applied to the panel.

A power conditioning system not only identifies the shading effect but also tracks the maximum Power Point. It is important to recall that the variation in the current can be caused by the limited rays impinging the PV cell, like on a cloudy day. Thus, the algorithm must be able to distinguish among these both causes, and the experiment must be conducted at multiple irradiances (Hossam & Itako, 2016). Furthermore, multiple shadowing schemes such as one shaded cell, two shaded inconsecutive cells, and a big shadow area are needed to affirm the hotspots are correctly detected, and the common factor is distinguishable from low irradiance. The I-V characteristics curve was obtained using I-V 400 PV panel analyzer to measure the output current and voltage of the panel. An analog to digital converter is connected to a dc-dc converter to periodically measure the current and voltage output. The dc converter employs a scanning method to measure the voltage and current, identify the maximum power point, and boost the dc signal. Based on the rate of change of current within 10 seconds period, the code determines whether to flag a hotspot or not (Hossam & Itako, 2016).

An algorithm was developed to track the maximum power point at the shading condition. In this context, a study utilized this topology to distinguish it from hotspots (Gosumbonggot & Fujita, 2019). Since the reverse bias operation is common between normal shading and hotspots, the study emphasizes on the importance of distinguishing between the two topologies. The manufacturer's specifications (open circuit voltage, short circuit current, maximum power, and the number of series and parallel cells) are inputted into the system serving as a benchmark for comparison purposes. The algorithm is divided into three parts, the MPPT, the hotspots detection code, and irradiance change detection embedded with the dc-dc converter. The starting point is the periodic monitoring of the maximum power point. A variation of 10% will initiate the next part of the program to check the irradiance intensity. It was found that the rate of change of the I-V curve varies in the case of shading to that of the hotspot. The slope of the I-V curve at two distinct locations that represent a proportion of the open circuit voltage is used to calculate the rate of change. The method begins with the first reference voltage, 52% of open circuit voltage V_1 [1]. The current is measured as I_1 [1]. The current I_2 [1] is measured when the voltage is adjusted by 0.5V to the following point V_2 [1]. The steps are repeated for m times. Now, the slope is given by the below formula.

$$\text{Slope} = \frac{I_1[m] - I_2[m]}{V_1[m] - V_2[m]} \tag{7.5}$$

Then, the average of these slopes is calculated and compared to a slope threshold that a PV is expected to have when subjected to low irradiance. If the average slope is greater, then a hotspot flag is on, and the PV is said to be suffering from a hotspot. The system was simulated using MATLAB/Simulink software for 2, 3, and 4 faulted panels to ensure the system is valid for all the cases (Gosumbonggot & Fujita, 2019).

Unfavorable hotspots' consequences not only motivate methods to detect and mitigate them but also lead the engineers to develop new ways to design the PV modules. Besides bypass diode, other methods can be a promising candidate for solving this problem. In (Yanli Liu *et al.*, 2010), currents sensors can be connected to the array in a certain setup that allows the identification of the exact location of the hotspot. The connection is called a complex-total-cross-tied array. The sensors then send data to the controller, which is programmed to run a code that compares the photogenerated current from different sensors and outputs the lowest performance cell location [22]. However, this approach reduces the flexibility in the connection of the cells and enforce a specific connection that aid in hotspots detection but might not be the optimum connection for other factors such as size, manufacturer recommendations, and the required electrical output. On the other hand, it can be suitable for certain applications.

7.3 Proposed Solution

This section describes the materials used to execute the project, the materials selection criteria, and the implementation process. Like the previous sections, this section will discuss each function requirement and implementation separately

7.3.1 Solar Tracking

As the sun is moving both north and south during the year, and east and west during the day, dual-axis solar tracker is implemented to maximize the absorbed sunrays. Electrical components needed are listed below:

- Two motors.
- A controller.
- Four Light Detecting Resistors (LDRs)
- Resistors and connecting wires
- breadboard for testing

The dual-axis solar tracker PV operates on the basis of comparing light intensity on four light sensors to identify the direction of the sun, and then the controller sends a signal to both motors to rotate the panels in that direction.

The system consists of The LDRs in the rows and is responsible for the vertical movement of the motor. The two columns of LDRs are responsible for the horizontal movement and detects the azimuth angle. In contrast, the two rows of LDRs detect altitude and are responsible for vertical movement. Each LDR is connected to the analog panel in the Arduino.

The primary light sensors are LDRs. Four light-detecting resistors (LDRs) were arranged in a rectangle in two rows and two columns with some space between them for light detection. The frame that holds the solar panel is equipped with two servo motors. The mechanical component is the main element in the system to ensure the motors are moving without mechanical obstacles. The program for Arduino is uploaded to the microcontroller.

The working of the project is as follows. LDRs sense the amount of sunlight falling on them. If the light is high, it will have a high resistance value. If the light is low, it will give a low resistance value. There are four LDRs: top, bottom, left, and right. When two top LDRs and two bottom LDRs are compared for east-west tracking, the top set of LDRs receives more light, and the vertical servo moves in that direction. The servo travels in that direction if the bottom LDRs get more light. The analog readings from two left LDRs and two right LDRs are compared for angular deflection of the solar panel. The horizontal servo will move in that direction if the left set of LDRs receives more light than the right set. The servo travels in that direction if the right pair of LDRs receives more light.

When more light hits the LDRs, it will have a high resistance value, and thus high voltage drop, which can be translated into motor movement by the controller.

The code used to calculate the desired position is uploaded into the Arduino. There are four scenarios based on the light received by the inputs. They are illustrated in Table 7.1.

Case 1: Horizontal motor rotates right
Case 2: Horizontal rotates motor left
Case 3: Vertical motor moves to the bottom
Case 4: Vertical motor moves up

Table 7.1 Cases of sunlight detection.

	Top right LDR	Top left LDR	Bottom right LDR	Bottom left LDR
Top right LDR	NA	Case 4	Case 1	NA
Top left LDR	Case 4	NA	NA	Case 2
Bottom right LDR	Case 1	NA	NA	Case 3
Bottom left LDR	NA	Case 2	Case 3	NA

Four LDRs connected to 4 10K ohm resistors in series. The LDRs and series resistors are connected in parallel. One side is connected to 5 volts power supply, and the other side is grounded. The power supply and ground used in this project are embedded with the controller. Likewise, the two motors are also connected to 5 volts and ground in the Arduino. In addition, the servo motors are connected to the signal ports in Arduino so it can perform the required action accordingly.

According to software coding, the Arduino UNO is a popular open-source microcontroller board used to control all hardware components. It has a number of digital and analog input/output pins that may be used to connect to other circuits. The board contains 14 digital connections and 6 analog pins, and it can be programmed using the Arduino IDE (Integrated Development Environment) through USB. It may be fueled by a USB connection or a 5 to 10 volt external battery. The pin 5V receives a controlled 5V from the board's regulator. Pins 3, 5, 6, 9, 10, and 11 are used for PWM (Pulse Width Modulation). With the analogWrite() method, you can get an 8-bit PWM output (*Getting Started with Arduino UNO*, n.d.). The principle is shown in Figure 7.5.

For simplicity, the project is implemented at a smaller scale, and hence 5 volts motors were suitable. The structure design of the mechanical part is an essential element in the sun tracker. If the mechanics did not work as it is supposed to be, the system would not function properly. The motor should be steadily fixed from one side, while the other side should have some flexibility to move. The horizontal motor should be fixed on its base and should only turn the upper part. If the horizontal motor is loosely attached to its base, it might rotate the base rather than rotating the required part of the

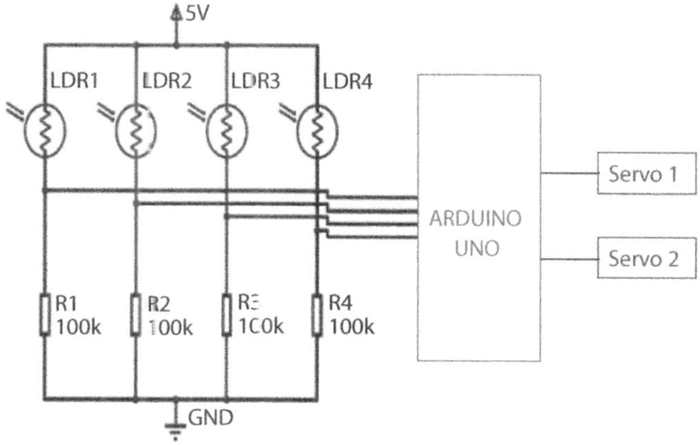

Figure 7.5 Arduino Uno and connecting LDRs circuit diagram.

system. Likewise, the vertical motor can cause undesired rotation if not fixed firmly.

Mechanical structure design comprises of two parts. First, aligning motors and controlling their rotations for suntracking. One motor is aligned vertically to reorient the panel toward the sun's elevation angle, and the other one horizontally to track azimuth angle. Both will be controlled using a single controller. The type of motor has a great influence on the design. A variety of motors available in the market. Each has its advantages and limitations. In essence, the selection of the motor depends on the application and requirements. In our case, two types of motors are popular. Servo and stepper motor. For better selection decisions, their features and functionalities are discussed in the following sections.

- Servo Motor

A brushed DC motor with some sort of positional feedback control attached to the rotor shaft is this type of motor. Closed-loop applications employ DC Servo motors, in which the position of the output motor shaft is transmitted back to the motor control circuit. Resolvers, encoders, and potentiometers are common positional "feedback" devices found in radio control models such as airplanes and boats.

The feedback encoder compares the reference angle with the actual one. It generates a signal to rotate the motor accordingly so that the difference between the reference and the actual angle is zero (Bhargava & Kumar, 2017). Figure 7.6 depicts the block diagram of a closed-loop DC motor.

A servo motor has a speed reduction gearbox built in, as well as a positioning feedback mechanism and some error correction. Because of the gear box and feedback mechanisms connected, the output shaft of a servo motor does not revolve freely like the shafts of DC motors.

The communication between the motor and the controller occurs through the Pulse Width Modulation (PMW) technique. It encodes the messages into pulses signal while decreasing power losses. The supplied

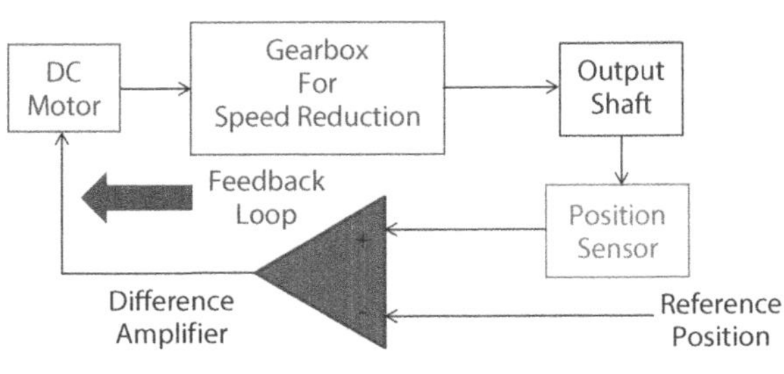

Figure 7.6 DC motor block diagram.

power into the motor is monitored by the high switching frequency of a switch or a transistor. More power is supplied if pulses are on for a longer time relative to of time (Bhargava & Kumar, 2017). Servo motors have three colored wires, the red is Vcc connected to the positive side of any power source, the brown is connected to the negative side to close the circuit, and the orange is the control signal wire that is connected to the controller (Bhargava & Kumar, 2017).

A Stepper motor is an electromechanical device consists of a stator with windings/stages and a rotor just like any motor. Stepper motor is similar to DC motor with a variation in that it operates on electrical pulses rather than continuous current flow. They are engines that move step by step. Each pulse energizes the motor to rotate by a certain angle degree. The more the pulses, the bigger the rotation angle (Lokesh *et al.*, 2018). It receives the input pulses signals and delivers smooth step-by-step motion with controlled positioning and high torque at low speeds (Tatar *et al.*, 2019).

Controlling steps position and speed are done through altering the electrical input of these components within possible limits provided by a specific stepper motor. Each pole of the stator offers a natural stopping point for the motor shaft. The greater number of poles allow higher positioning precision without any positional feedback. In other words, it works in an open-loop control system. Therotor has teeth that fit in the stator magnets. An electrical pulse signal from the micro controller moves the shaft from one pole to the other and stops aligned with the electromagnetic pole in the stator without feedback to the controller. This is a beneficial feature of the stepper motor. Therefore, moving to an accurate position is obtained by the correct number of pulse commands.

The position of the rotor relative to the stator is controlled by the change of the stator polarity position. By varying the stator windings polarities, the rotor can have different positioning (Tatar *et al.*, 2019).

Transistors are used to control the current passing through the coils to create different poles. Also, the built-in driver controls the direction and the speed by changing the polarities of the stat or windings (Tatar *et al.*, 2019).

Another speed control technique is performing time delay of the pulses to speed up or slow down the motor. This way will allow us to know the last position of the rotor by counting the number of pulses generated (Tatar *et al.*, 2019).

The tracking system has improved the electrical output by more exposure to sunlight during the day. The tracker follows the sun accurately, and the panels are in a perpendicular position to the sun's rays.

7.3.2 Cleaning System

The second design to be engineered is the cleaning system. It consists of two rings attached to a cylindrical brush that will be allowed to slide in two parallel lines of the panel's frame. This part of the project will require mechanical expertise interference to achieve the highest possible cleaning efficiency and sliding speed. Moreover, the selection of the brush teeth might need to be designed to ensure optimum cleaning.

Each time the tracker moves the panel at a certain angle that corresponds to the sun position, it triggers the rings to slide to the other side solely by gravity means, the cylinder will rotate across the panels accordingly, and brushes will wipe off dustparticles on its way to the other end. The cleaning will process will take place a minimum of twice a day in east and west directions. At sunrise were the tracker moves 180 degrees towards the east direction while the brush will be located on the left side, it will slide to the right. Also, at sunset, where the panels are directed towards the west, the brushes slide from right to left. This will ensure the panels 'cleanliness and decomposition of dust and sand particles at optimum settings.

7.3.3 Hotspots

This section will be performed in the next part of this project. The initial idea is to integrate the hotspot detection with the tracking controller to save power and resources. The ideal materials and electronics will be discussed in the next part. Moreover, since PV power plants are installed in remote farms, the hotspots must encompass IoT to alert end users.

7.3.4 Modeling and Simulation

The proposed design doubles the functionality of the controller and tracking motors. In fact, two motors that must be used for trackers will aid in the cleaning. While the controller, also used in the tracker, will aid in hotspots detection. This technique is a cost-effective and energy-efficien mechanism and can work at any geographical location. The Following methodology uses a microcontroller and sensors to track the sun's positionto capture maximum sunlight radiations. The cleaning method uses both panels' tracker rotation and gravity to perform its task without any extra motors installed. It cleans in two directions but single-axis twice per day.

Experiment Prototype and physical measurements are to be carried out on the roof of Effat University. The observation can take two weeks to one month. The longer the testing time, the more confident we are about the results. Since PV panels IoT is used to communicate any suspicious hotspots and alert the user/operator in remote locations.

By comparing the proposed system with three identical PV panels, each has a different setup. The comparison is based on electrical output criteria.

- First: fixed with no tracking, added robots cleaning and hotspot detection.
- Second: PV with tracking and hotspots detection excluding cleaning.
- Third: tracking and cleaning with not hotspots detection.

7.3.5 Limitations

The limitations of this project are summarized in the following points:

- The implementation was deemed difficult in the covid 19 times, where many workshops are not available. Also, mechanical background engineer is preferable to achieve high mechanical structure efficiency.
- The fielded comparison in performance between fixed, dirty, not supported with hotspots detection system to our proposed system might take a long time for the dust to accumulate and the results to harvest.
- The cleaning systems designed can clean dust and sand, but they are unreliable in cleaning the birds' droppings and other heavy particles. Therefore, deep cleaning is still required but at less frequency than no cleaning mechanism.

7.3.6 Hypothesis

Combining both tracking and cleaning in a system not only improves the system performance but also reduced cost and power consumption. The proposed tracker in this paper suggests that the brush will slide at 45-85 degrees for the brush to slide slowly and ensures proper cleaning is done. This natural gravity-based movement will take place twice a day. Furthermore, the sharing resources will contribute to detecting hotspots early to avoid any complications.

The proposed system is expected to be the most cost-effective and generates the maximum power over the three other identical PVs with different setups.

7.4 Conclusion

The rise in energy consumption as a result of technical and demographic advancements has resulted in increased pollution, culminating in global warming and a slew of environmental and natural disasters throughout the world. Solar energy has shown to be a green source of clean electrical power in order to address these issues. The environmental factors are degrading the performance of the solar power plants in the field. Three of the major attenuation environmental conditions are limited solar captured irradiance due to the sun's movement, soiling phenomenon that appears on the PV surface especially in arid and dusty areas, and hotspots formation due to continuous cell's shading.

This chapter concerns addressing these problems through designing autonomous energy-efficient solar cleaning and tracking mechanism. Moreover, the system utilizes existing resources to detect hotspots and notify the operator remotely via IoT to take necessary actions accordingly. The proposed design is to attach cleaning brushes that naturally slides across the panel by gravity with the panel rotations due to tracker. This design cuts cost and complexity caused by high frequency of cleaning process, improve performance by maximizing sun irradiance received, and allows proactive measures to be taken by alerting any potential hotspots.

The system is applicable at any geographical location and it is mainly intended for utility scale power plants to further justify the capital cost. However, a prototype is to be tested in Jeddah, Saudi Arabia, to ensure the proposed system is effective and superior over other identical PV panels with different systems. The electrical characteristics of the PV prototype is periodically measured and recorded. Results have proved that this multi-functionality system, its effectiveness, low cost, simple, reliable, requires minimum maintenance and preserving energy.

Future work can include maximum power point tracking and battery management system combined to build one big and complete setup to support PV operations in the field. The incorporation of event-driven methods can improve the performance of these solutions (S. M. Qaisar *et al.*, 2009), (Mina Qaisar *et al.*, 2018). Investigation of this approach is another prospect.

References

Abadi, I., *DESIGN OF SINGLE AXIS SOLAR TRACKING SYSTEM AT PHOTOVOLTAIC PANEL USING FUZZY LOGIC CONTROLLER*, 2014.

Abualsaud, A.A., Qaisar, S., Ba-Abdullah, S.H., Al-Sheikh, Z.M., Akbar, M., *Design and implementation of a 5-bit flash ADC for education*, pp. 1–4, 2016.

Alsafasfeh, M., Abdel-Qader, I., Bazuin, B., Fault detection in photovoltaic system using SLIC and thermal images. *2017 8th International Conference on Information Technology (ICIT)*, pp. 672–676, 2017, https://doi.org/10.1109/ICITECH.2017.8079925.

Belihi, S., Dahlioui, D., Laarabi, B., Barhdadi, A., On the Use of Dew for Cleaning PV Panels in Morocco: Literature Survey and Experimental Results. *2019 7th International Renewable and Sustainable Energy Conference (IRSEC)*, pp. 1–4, 2019, https://doi.org/10.1109/IRSEC48032.2019.9078178.

Ben-Menahem, S. and Yang, S.C., Online photovoltaic array hot-spot Bayesian diagnostics from streaming string-level electric data. *2012 38th IEEE Photovoltaic Specialists Conference*, pp. 002432–002437, 2012, https://doi.org/10.1109/PVSC.2012.6318087.

Bhargava, A. and Kumar, A., Arduino controlled robotic arm. *2017 International Conference of Electronics, Communication and Aerospace Technology (ICECA)*, vol. 2, pp. 376–380, 2017, https://doi.org/10.1109/ICECA.2017.8212837.

Caron, J.R. and Littmann, B., Direct Monitoring of Energy Lost Due to Soiling on First Solar Modules in California. *IEEE J. Photovoltaics*, *3*, 1, 336–340, 2013, https://doi.org/10.1109/JPHOTOV.2012.2216859.

Chen, J., Li, Y., Ling, Q., Hot-Spot Detection for Thermographic Images of Solar Panels. *2020 Chinese Control And Decision Conference (CCDC)*, pp. 4651–4655, 2020, https://doi.org/10.1109/CCDC49329.2020.9164255.

Dahlioui, D., Ayane, S.E., Alaoui, S.M., Barhdadi, A., Dambrine, G., Menard, E., Boardman, J., Innovative Cleaning Technique for PV Modules on Helioslite Biaxial Solar Trackers. *2018 6th International Renewable and Sustainable Energy Conference (IRSEC)*, pp. 1–4, 2018, https://doi.org/10.1109/IRSEC.2018.8702864.

Fan, S., Yao, X., Cao, S., Zhao, B., An Optimization Method Based on Adaptive Dynamic Programming for Cleaning Photovoltaic Panels. *2020 39th Chinese Control Conference (CCC)*, pp. 1565–1568, 2020, https://doi.org/10.23919/CCC50068.2020.9189437.

Getting Started with Arduino UNO, Retrieved April 25, 2021, from https://www.arduino.cc/en/Guide/ArduinoUno.

Gheitasi, A., Almaliky, A., Albaqawi, N., Development of an automatic cleaning system for photovoltaic plants. *2015 IEEE PES Asia-Pacific Power and Energy Engineering Conference (APPEEC)*, pp. 1–4, 2015, https://doi.org/10.1109/APPEEC.2015.7380938.

Gosumbonggot, J. and Fujita, G., Photovoltaic's Hotspot and Partial Shading Detection Algorithm Using Characteristic Curve's Analysis. *2019 9th International Conference on Power and Energy Systems (ICPES)*, pp. 1–6, 2019, https://doi.org/10.1109/ICPES47639.2019.9105518.

Haeberlin, H., Graf, J.D., Fachhochschule, B., *Gradual Reduction of PV Generator Yield due to Pollution*, p. 4, 1998.

Home ☼ SunEarthTools.com solar tools for consumers and designers, Retrieved April 25, 2021, from https://www.sunearthtools.com/.

Homepage Fraunhofer-Gesellschaft, Retrieved April 25, 2021, from https://www.fraunhofer.de/en.html.

Hossam, B. and Itako, K., Real time hotspot detection using scan-method adopted with P O MPPT for PV generation system. *2016 IEEE 2nd Annual Southern Power Electronics Conference (SPEC)*, pp. 1–5, 2016, https://doi.org/10.1109/SPEC.2016.7846122.

Innovative Low Cost Cleaning Technique for PV Modules on Solar Tracker | IEEE Conference Publication | IEEE Xplore, Retrieved April 18, 2021, from https://ieeexplore.ieee.org/document/8702861.

Jambi, L., Alsubaie, S., Qaisar, S.M., *An event-driven power efficient surveillance and lighting system in the Saudi Arabia perspective*, pp. 423–427, 2018.

Jamil, W.J., Abdul Rahman, H., Shaari, S., Salam, Z., Performance degradation of photovoltaic power system: Review on mitigation methods. *Renewable Sustainable Energy Rev.*, *67*, 876–891, 2017, https://doi.org/10.1016/j.rser.2016.09.072.

Kaplanis, S. and Kaplani, E., Energy performance and degradation over 20years performance of BP c-Si PV modules. *Simul. Modell. Pract. Theory*, *19*, 4, 1201–1211, 2011, https://doi.org/10.1016/j.simpat.2010.07.009.

Kim, K.A., Seo, G.-S., Cho, B.-H., Krein, P.T., Photovoltaic Hot-Spot Detection for Solar Panel Substrings Using AC Parameter Characterization. *IEEE Trans. Power Electron.*, *31*, 2, 1121–1130, 2016, https://doi.org/10.1109/TPEL.2015.2417548.

Kumar, S.S., Shankar, S., Murthy, K., Solar Powered PV Panel Cleaning Robot. *2020 International Conference on Recent Trends on Electronics, Information, Communication Technology (RTEICT)*, pp. 169–172, 2020, https://doi.org/10.1109/RTEICT49044.2020.9315548.

Lakeou, S., Ososanya, E., Latigo, B., Karanja, G., Mahmoud, W., Oshumare, W., *Design Of A Low Cost Solar Tracking Photo Voltaic (Pv) Module And Wind Turbine Combination System*, p. 11.406.11, 2006, https://doi.org/10.18260/1-2–1152.

Lokesh, A., Surahonne, A., Simha, A.N., Reddy, A.C., Solar Tracking System Using Microcontroller. *2018 International Conference on Inventive Research in Computing Applications (ICIRCA)*, pp. 1094–1098, 2018, https://doi.org/10.1109/ICIRCA.2018.8597267.

Lubitz, W.D., Effect of manual tilt adjustments on incident irradiance on fixed and tracking solar panels. *Appl. Energy*, *88*, 5, 1710–1719, 2011, https://doi.org/10.1016/j.apenergy.2010.11.008.

Maghami, M.R., Hizam, H., Gomes, C., Radzi, M.A., Rezadad, M.I., Hajighorbani, S., Power loss due to soiling on solar panel: A review. *Renewable Sustainable Energy Rev.*, *59*, 1307–1316, 2016, https://doi.org/10.1016/j.rser.2016.01.044.

Mina Qaisar, S., Sidiya, D., Akbar, M., Subasi, A., An Event-Driven Multiple Objects Surveillance System. *Int. J. Electr. Comput. Eng. Syst.*, *9*, 1, 35–44, 2018.

National Renewable Energy Laboratory (NREL) Home Page, Retrieved April 25, 2021, from https://www.nrel.gov/index.html.

Pannebakker, B.B., de Waal, A.C., van Sark, W. G. J. H. M., Photovoltaics in the shade: One bypass diode per solar cell revisited. *Prog. Photovoltaics: Res. Appl.*, *25*, 10, 836–849, 2017, https://doi.org/10.1002/pip.2898.

Petrusev, A.S., Rulevskiy, V.M., Sarsikeyev, Ye. Zh., Lyapunov, D. Yu., Solar tracker with active orientation. *2016 2nd International Conference on Industrial Engineering, Applications and Manufacturing (ICIEAM)*, pp. 1–4, 2016, https://doi.org/10.1109/ICIEAM.2016.7911461.

Qaisar, S., Fesquet, L., Renaudin, M., Computationally efficient adaptive resolution short-time Fourier transform. *EURASIP, RLSP*, 2008.

Qaisar, S.M., Event driven filtering an intelligent technique for activity and power consumption reduction. *Int. J. Circuits Syst. Signal Process.*, *8*, 2014.

Qaisar, S.M., Fesquet, L., Renaudin, M., *Effective resolution of an adaptive rate ADC*, 2009, Special-session.

Qaisar, S.M. and Subasi, A., Cloud-based ECG monitoring using event-driven ECG acquisition and machine learning techniques. *Phys. Eng. Sci. Med.*, *43*, 2, 623–634, 2020.

Qaisar, S.M. and Subasi, A., *An adaptive rate ECG acquisition and analysis for efficient diagnosis of the cardiovascular diseases*, pp. 177–181, 2018.

Qaisar, S., Yahiaoui, R., Dominique, D., *A smart power management system monitoring and measurement approach based on a signal driven data acquisition*, pp. 1–4, 2015.

Rahayani, R. and Gunawan, A., A design and implementation of dual axis solar tracker using mini photo voltaic as solar sensor, in: *IEEA '19: Proceedings of the 8th International Conference on Informatics, Environment, Energy and Applications*, p. 114, 2019, https://doi.org/10.1145/3323716.3323746.

ResearchGate Link, Retrieved April 15, 2021, from https://www.researchgate.net/publication/336200025_Photovoltaic_Hot-Spots_Fault_Detection_Algorithm_Using_Fuzzy_Systems.

Riawan, I.P.G., Kumara, I.N.S., Partha, C.G.I., Setiawan, I.N., Santiari, D.A.S., Robot for Cleaning Solar PV Module to Support Rooftop PV Development. *2018 International Conference on Smart Green Technology in Electrical and Information Systems (ICSGTEIS)*, pp. 132–137, 2018, https://doi.org/10.1109/ICSGTEIS.2018.8709138.

Rizk, J. and Chaiko, Y., Solar Tracking System: More Efficient Use of Solar Panels. 31, 2008.

Sabo, A. and Qaisar, S.M., *The event-driven power efficient wireless sensor nodes for monitoring of insects and health of plants*, pp. 478–483, 2018.

Salazar, A. and Macabebe, E.Q., Hotspots Detection in Photovoltaic Modules Using Infrared Thermography. *MATEC Web of Conferences*, vol. 70, p. 10015, 2016, https://doi.org/10.1051/matecconf/20167010015.

Sarver, T., Al-Qaraghuli, A., Kazmerski, L.L., A comprehensive review of the impact of dust on the use of solar energy: History, investigations, results, literature, and mitigation approaches. *Renewable Sustainable Energy Rev.*, *22*, 698–733, 2013, https://doi.org/10.1016/j.rser.2012.12.065.

Sayyah, A., Horenstein, M.N., Mazumder, M.K., Energy yield loss caused by dust deposition on photovoltaic panels. *Sol. Energy*, *107*, 576–604, 2014, https://doi.org/10.1016/j.solener.2014.05.030.

Sha, W., Dai, C., Jiang, L., Design of Patrol Monitoring and Control System for Hot Spot of Solar Photovoltaic Module. *2019 International Conference on Intelligent Computing, Automation and Systems (ICICAS)*, pp. 668–671, 2019, https://doi.org/10.1109/ICICAS48597.2019.00145.

Spanoche, S.A., Stewart, J.D., Hawley, S.L., Opris, I.E., Model-based method for partially shaded PV modules Hot Spot Suppression. *2012 IEEE 38th Photovoltaic Specialists Conference (PVSC) PART 2*, pp. 1–7, 2012, https://doi.org/10.1109/PVSC-Vol2.2012.6656784.

Tatar, G., Bayar, S., Alkan, M., FPGA Based Step Motor Control For Solar Panels. *2019 IEEE 13th International Conference on Application of Information and Communication Technologies (AICT)*, pp. 1–5, 2019, https://doi.org/10.1109/AICT47866.2019.8981722.

Tejwani, R. and Solanki, C.S., 360° sun tracking with automated cleaning system for solar PV modules. *2010 35th IEEE Photovoltaic Specialists Conference*, pp. 002895–002898, 2010, https://doi.org/10.1109/PVSC.2010.5614475.

Wei, S., Li, X., Ding, S., Yang, Q., Yan, W., Hotspots Infrared detection of photovoltaic modules based on Hough line transformation and Faster-RCNN approach. *2019 6th International Conference on Control, Decision and Information Technologies (CoDIT)*, pp. 1266–1271, 2019, https://doi.org/10.1109/CoDIT.2019.8820333.

Liu, Y., Li, B., Cheng., Z., Research on PV module structure based on fault detection. *2010 Chinese Control and Decision Conference*, pp. 3891–3895, 2010, https://doi.org/10.1109/CCDC.2010.5498470.

8

A New Hybrid Proposition Based on a Cuckoo Search Algorithm for Parameter Estimation of Solar Cells

Souvik Ganguli*, Shilpy Goyal and Parag Nijhawan

Department of Electrical and Instrumentation Engineering, Thapar Institute of Engineering and Technology, Patiala, Punjab, India

Abstract
This chapter develops a new hybrid bioinspired technique. Two famous algorithms inspired by nature *viz.* the Cuckoo search (CS) and the firefly algorithm (FA) are combined using all their advantages. FA is used to explore the whole search field, while CS is utilized for local search in a global optimization environment. Not only does the bioinspired method work better than parent approaches, but it also overpowers some recent metaheuristic methods. The effectiveness of our approach is demonstrated through some benchmark test functions. This fact is also confirmed by statistical measures and nonparametric research. Also, the hybrid algorithm is used effectively to find the amended double-diode model parameters. The findings of the two models of solar cells show the efficacy of the technique suggested. The fitness plots with the number of iterations validated that the results are more convergent and reliable. The parameter estimation technique can in the future be considered as multidimensional modeling of the diode with the technique proposed. A multitarget approach can also be a good alternative to modeling parameters for various photovoltaic panels.

***Keywords*:** Firefly algorithm (FA), Cuckoo search (CS), hybrid algorithm, amended double-diode model (ADDM), nonparametric test

**Corresponding author*: souvik.ganguli@thapar.edu

Suman Lata Tripathi, Souvik Ganguli, Tengiz Magradze, and Abhishek Kumar (eds.) Intelligent Green Technologies for Sustainable Smart Cities, (143–164) © 2022 Scrivener Publishing LLC

8.1 Introduction

In recent times, hybridization is considered as an essence of high-performance computing algorithms. Plain algorithms cannot always obtain an optimum solution within a stipulated time frame. Hence, pure algorithms are almost always lag behind hybridizations [1]. Thus, researchers are always in pursuit of developing some new hybrid topologies.

Firefly algorithm (FA) [2] is one of the popular nature-inspired techniques, which is influenced by the flash pattern and communication behavior of the fireflies found mainly in tropical environmental conditions. This metaheuristic has undergone manifold improvements and is being applied by many researchers in diverse domains [3–7]. Hybridization is one of the common choices employed by different researchers to extract better performance from an algorithm. FA has similarly being hybridized with other algorithms. Some algorithms like ACO, GA, PSO, DE have been pretty successful when combined with FA [1, 8–10].

The benefit of FA in integrating with other metaheuristic techniques is that it finds its usage equally well in both global as well as local searches. Though the literature suggests quite a handful number of hybrid algorithms with the firefly technique, yet there is enough opportunity for the formulation of new hybrid algorithms. Another nature-inspired technique, coined as the cuckoo search (CS) algorithm [11], working based on the brood parasitism of some cuckoo birds, along with Levy flights random walks, suffer from the drawback of premature convergence for which a hybrid combination constituting FA and CS is conceived.

Thus, a hybrid algorithm with the combination of FA and CS is presented in this chapter based on Talbi's framework [12]. As per Talbi's nomenclature, the hybrid suggested is a low-level, relay type, and heterogeneous type proposition. Thus, in the global optimization-based taxonomy, FA performs exploration while CS is applied for exploitation abilities. The balance between the two features is maintained accommodating 50% of the iterations to each of these characteristics in the hybrid approach.

A host of standard test functions [13] are considered from the literature to justify the exploratory and exploitation features of the suggested technique. Adequate comparison is carried out with the parent [2, 11], as well as the latest reported heuristics [14–16]. Since heuristic algorithms are stochastic, they do not yield the same solution on each test run. Hence, statistical measures like minimum, maximum, average, and standard deviation are considered for the experiment. Thus, the normalized results of the abovementioned parameters are reported. The rank-sum test of Wilcoxon [17] has also been executed to confirm the test outcomes. Since multiple data sets are involved for comparison, the Kruskal Wallis test [18] is conducted to validate the outcomes. Moreover, the Wilcoxon test with Holm-Bonferroni corrections [19] has been performed to further test the significance of the results. In addition to this, the proposed method is tested for parameter assessment of polycrystalline photovoltaic (PV) systems.

Conventionally, in the experimental approach, root mean square error (RMSE) is taken up as the cost function to optimize the unknown solar cell parameters [20]. This chapter employs the sum of square error (SSE), another well-known objective taken from the literature, for modeling PV cells with the help of datasheet information [21]. The fitness function takes into account the open and the short circuit conditions, as well as the maximum power part, added linearly to minimize SSE. Though obtaining the parameters of various models is popular in the literature [22–26], an amended double-diode model [27] is taken up for the investigation. The hybrid technique achieves better results with respect to the other methods on numerous occasions.

The remaining chapter is organized in the way described as follows. The modelling equations of the amended double-diode model are framed in section 8.2. The suggested method is showcased in section 8.3. Section 8.4 presents the results claimed while section 8.5 outlines the main inferences with some directions for future work.

8.2 Modelling of an Amended Double Diode Model (ADDM) and the Objective Function

The newly developed and less popular model namely the amended double-diode model (ADDM) has been employed in this chapter. This model has been taken up owing to the fact that the resistance in series connection with the second diode considered the influence of the grain boundaries. This is so because the resistivity near to grain boundary region is greater in comparison to that inside the crystallites. Hence this model accommodates a current source with two diodes, two resistors namely series and parallel, and another single series resistance R_{S2} as provided in Figure 8.1. There are eight unknown parameters in the model viz. the I_{PV}, I_{01}, I_{02}, α_1, α_2, R_S, R_p, and R_{S2} which needs to be found out. The relation between voltage and current can be written mathematically by the equation given below.

$$I = I_{PV} - I_{01}\left[exp\left(\frac{q(V+IR_S)}{\alpha_1 KT}\right)-1\right] - I_{02}\left[exp\left(\frac{q(V+IR_S-I_{D2}R_{S2})}{\alpha_2 KT}\right)-1\right]-\frac{V+IR_S}{R_p} \quad (8.1)$$

where I_{PV} is the photocurrent, I_{01} and I_{02} denote the saturation currents (reverse) of the two diodes respectively, R_p and R_S represent the parallel and series equivalent resistances, α_1 and α_2 stand for the ideality factors of the two diodes, R_{S2} expresses the series resistance in association with the second diode, 'q' marks an electronic charge, while 'T' symbolizes the temperature in SI units. Last but not the least, 'K' typifies the constant referred to after the name of Boltzmann.

By solving the equation for the second diode we get equation (8.2) denoted by,

$$I_{D2} = \frac{V(R_S+R_p)}{R_S R_p + (R_S+R_p)R_{S2}} \quad (8.2)$$

For the open-circuit case, the current I turns zero, while the voltage V is represented by V_{OC}; thus eq. (8.1) reduces to

$$0 = I_{PV} - I_{01}\left[exp\left(\frac{qV_{OC}}{\alpha_1 KT}\right)-1\right] - I_{02}\left[exp\left(\frac{qV_{OC}}{\alpha_2 KT}\left(\frac{R_S R_p}{R_S R_p+(R_S+R_p)R_{S2}}\right)\right)-1\right]-\frac{V_{OC}}{R_p} \quad (8.3)$$

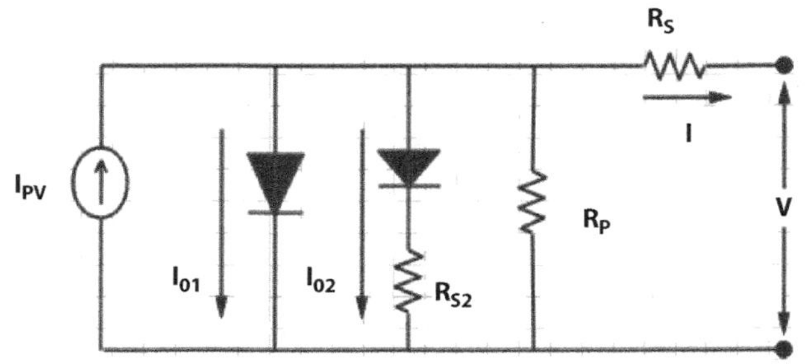

Figure 8.1 An amended double-diode model.

Therefore

$$I_{PV} = I_{01}\left[exp\left(\frac{qV_{OC}}{\alpha_1 KT}\right)-1\right] + I_{02}\left[exp\left(\frac{qV_{OC}}{\alpha_2 KT}\left(\frac{R_S R_p}{R_S R_p + (R_S + R_p)R_{S2}}\right)\right)-1\right] + \frac{V_{OC}}{R_p} \tag{8.4}$$

When the short circuit conditions are imposed on eq. (8.1), the voltage value V drops to zero, whereas I can be equated to I_{SC}; then we get

$$I_{SC} = I_{PV} - I_{01}\left[exp\left(\frac{qI_{SC}R_S}{\alpha_1 KT}\right)-1\right] - I_{02}\left[exp\left(\frac{qI_{SC}R_S}{\alpha_2 KT}\right)-1\right] - \frac{I_{SC}R_S}{R_p} \tag{8.5}$$

Therefore;

$$I_{PV} = I_{SC} + I_{01}\left[exp\left(\frac{qI_{SC}R_S}{\alpha_1 KT}\right)-1\right] + I_{02}\left[exp\left(\frac{qI_{SC}R_S}{\alpha_2 KT}\right)-1\right] + \frac{I_{SC}R_S}{R_p} \tag{8.6}$$

At the instant when maximum power is achieved, $V = V_m$ and $I = I_m$; then eq. (8.1) gives;

$$I_m = I_{PV} - I_{01}\left[exp\left(\frac{q(V_m + I_m R_S)}{\alpha_1 KT}\right)-1\right] - I_{02}\left[exp\left(\frac{q}{\alpha_2 KT}\left(V_m + I_m R_S - \frac{R_{S2}V_m(R_S + R_p)}{R_S R_p + (R_S + R_p)R_{S2}}\right)\right)-1\right] - \frac{V_m + I_m R_S}{R_p} \tag{8.7}$$

The error value for the condition of the open circuit is narrated with the aid of the equation provided below.

$$err_{OC} = I_{01}\left[exp\left(\frac{qV_{OC}}{\alpha_1 KT}\right)-1\right] + I_{02}\left[exp\left(\frac{qV_O}{\alpha_2 KT}\left(\frac{R_S R_p}{R_S R_p + (R_S + R_p)R_{S2}}\right)\right)-1\right] + \frac{V_{OC}}{R_p} - I_{PV} \tag{8.8}$$

On a similar note, the error objective during the occurrence of a short circuit is given with the help of eq. (8.9) provided below as

$$err_{SC} = I_{SC} + I_{01}\left[exp\left(\frac{qI_{SC}R_S}{\alpha_1 KT}\right)-1\right] + I_{02}\left[exp\left(\frac{qI_{SC}R_S}{\alpha_2 KT}\right)-1\right] + \frac{I_{SC}R_S}{R_p} - I_{PV} \tag{8.9}$$

The cost function corresponding to the maximum power point is denoted by

$$err_m = I_{PV} - I_{01}\left[exp\left(\frac{q(V_m + I_m R_S)}{\alpha_1 KT}\right)-1\right] - I_{02}\left[exp\left(\frac{q}{\alpha_2 KT}\left(V_m + I_m R_S - \frac{R_{S2}V_m(R_S + R_p)}{R_S R_p + (R_S + R_p)R_{S2}}\right)\right)-1\right] - \frac{V_m + I_m R_S}{R_p} - I_m \tag{8.10}$$

Consequently, the objective function is written as

$$err = err_{OC}^2 + err_{SC}^2 + err_m^2 \tag{8.11}$$

8.3 Proposed Work

Following Talbi's taxonomy [12], a new technique is proposed with the combination of the firefly algorithm (FA) and cuckoo search (CS) method. FA [2] is inspired by the fireflies found in the environment. This swarm intelligence mimics the social behaviour of the fireflies. Their light intensity governs the communication between any two insects. Further, the fireflies get attracted to each other with varying light intensity. This phenomenon is utilized to encapsulate an objective function, to either minimize or maximize.

On a similar note, the cuckoo search (CS) algorithm [11] is another swarm-based technique that relates to the reproduction behaviour of cuckoo birds. The cuckoo birds usually tend to position their fertilized eggs at the disposal of other cuckoo birds with the hope that they will be being raised by their proxy parents. Quite often, the cuckoo birds can spot out that the eggs in their nests are not of their own. In that case, these eggs are often thrown away by them and even abandon their nests as well. This behaviour of the cuckoo birds is conceptualized to formulate the cuckoo search algorithm.

Both FA and CS approaches have their brilliance and fallacies too. Thus the two algorithms may be gelled together to produce a unique hybrid coined as FACS. In this global optimization method, FA is used to take control of the exploration capabilities while CS takes care of the local search phenomena. The hybrid proposition is evaluated on some high dimensional benchmark problems considered as representative functions to test the exploratory features as well as the exploitation capabilities of the algorithm. For the test problems, the statistical measurements of the results are also mentioned. To confirm the importance of the findings even the Kruskal Wallis test [13] is performed. The convergence characteristics also show that the proposed strategy converges better and more consistently than the comparable methods. In addition, an amended double-diode model is considered for modelling and parameter estimation purposes using the proposed topology. A good number of algorithms are also used for comparison to test the efficacy of the hybrid approach.

8.4 Results and Discussions

The effectiveness of our approach is assessed with the help of several benchmark functions, taken up for the analysis. The test functions are generally graded as unimodal and multimodal respectively. These test functions are so selected to justify the diversification and intensification properties of the proposed methodology. To evaluate the efficiency of our approach, a high dimensional analysis is carried out. In [13] there are already detailed explanations of these test functions. Interested readers may also explore it for appropriate investigation and analysis. Table 8.1 offers some common information about these test functions.

For an ideal optimal value of zero, all test functions 1 to 8 are minimized. A population size of 30 is considered competitive with a maximum number of iterations of 500 for a hundred decision variables with a prescribed search limit as described in Table 8.1. For all of the algorithms considered for comparison, the normal choice of parameters according to literature records was adopted. Since the results of the metaheuristic algorithms are stochastic, 20 independent test runs are performed to produce meaningful statistical conclusions. Therefore, Table 8.2 contains the statistical measurements of their fitness functions. The table shows that the normalized values of the best, worst, average, and standard deviations. The normalized results are found to vary between 0 and 1, 0 being the best value and 1 being the inferior result. With bold letters, the best results are also indicated.

It is found from Table 8.2 that the hybrid algorithm proposed outperforms the parent, as well as some widely cited methods. The rank-sum test of Wilcoxon [17] was conducted to determine the importance of

Table 8.1 Some common test functions and their narratives.

Test function number	**Specific names**	**No. of decision variables**	**Search limits**	**Function type**	**Idealized f_{min}**
1	Sphere	100	(−100, 100)	Unimodal	0
2	Schwefel 2.22	100	(−10, 10)	Unimodal	0
3	Schwefel 2.21	100	(−100, 100)	Unimodal	0
4	Rosenbrock	100	(−30, 30)	Unimodal	0
5	Step	100	(−100, 100)	Unimodal	0
6	Rastrigin	100	(−5.12, 5.12)	Multimodal	0
7	Ackley	100	(−32, 32)	Multimodal	0
8	Griewank	100	(−600, 600)	Multimodal	0

Table 8.2 Statistical results of the different test problems.

Test function number	Methods	Best	Maximum	Average	Standard deviation
1	FACS	**0**	**0**	**0**	**0**
	FA [2]	1.09E–09	2.15E–10	3.39E–10	1.24E–10
	CS [11]	1	2.05E–01	3.51E–01	1.40E–01
	DA [14]	1.65E–01	1	1	1
	SCA [15]	4.31E–01	4.62E–01	3.66E–01	5.26E–01
	SSA [16]	1.77E–01	4.19E–02	6.51E–02	2.88E–02
2	FACS	**0**	**0**	**0**	2.07E–06
	FA [2]	1.28E–13	1.41E–13	1.31E–13	2.65E–06
	CS [11]	1	1	1	**0**
	DA [14]	4.18E–09	2.26E–08	8.59E–09	1.00E+00
	SCA [15]	5.66E–11	1.93E–09	7.45E–10	1.46E–01
	SSA [16]	3.72E–09	5.33E–09	4.45E–09	1.07E–01
3	FACS	**0**	**0**	**0**	**0**
	FA [2]	7.27E–01	1	9.30E–01	1

(Continued)

Table 8.2 Statistical results of the different test problems. (*Continued*)

Test function number	**Methods**	**Best**	**Maximum**	**Average**	**Standard deviation**
	CS [11]	2.88E−01	4.92E−01	3.48E−01	5.59E−01
	DA [14]	5.21E−01	7.13E−01	6.26E−01	7.44E−01
	SCA [15]	1	9.78E−01	1	2.16E−01
	SSA [16]	2.53E−01	3.33E−01	2.99E−01	3.53E−01
4	FACS	2.27E−06	**0**	**0**	**0**
	FA [2]	**0**	1.51E−06	3.48E−08	1.71E−06
	CS [11]	3.94E−02	2.88E−02	2.20E−02	2.56E−02
	DA [14]	7.69E−02	2.75E−01	1.71E−01	2.61E−01
	SCA [15]	1.00E+00	1.00E+00	1.00E+00	1.00E+00
	SSA [16]	2.90E−03	1.97E−03	1.48E−03	1.73E−03
5	FACS	**0**	**0**	**0**	**0**
	FA [2]	8.08E−10	2.40E−10	8.91E−10	1.23E−10
	CS [11]	9.78E−01	2.57E−01	1	1.86E−01
	DA [14]	1	1	2.39E+00	1

(*Continued*)

Table 8.2 Statistical results of the different test problems. *(Continued)*

Test function number	Methods	Best	Maximum	Average	Standard deviation
	SCA [15]	1.92E−01	4.82E−01	1.08E+00	6.32E−01
	SSA [16]	1.38E−01	4.22E−02	1.56E−01	2.64E−02
6	FACS	**0**	**0**	**0**	**0**
	FA [2]	4.22E−01	5.09E−01	4.77E−01	5.54E−01
	CS [11]	7.43E−01	4.85E−01	5.47E−01	1.91E−01
	DA [14]	1	1	1	1
	SCA [15]	1.72E−01	3.78E−01	2.84E−01	5.54E−01
	SSA [16]	3.34E−01	3.10E−01	2.89E−01	2.63E−01
7	FACS	4.14E−02	**0**	**0**	**0**
	FA [2]	**0**	1.03E−01	7.65E−02	1.06E−01
	CS [11]	3.09E−01	6.05E−01	5.44E−01	5.46E−01
	DA [14]	1	7.65E−01	7.16E−01	3.73E−01
	SCA [15]	7.52E−01	1	1	1
	SSA [16]	8.11E−01	5.72E−01	5.04E−01	2.17E−01

(Continued)

Table 8.2 Statistical results of the different test problems. (*Continued*)

Test function number	Methods	Best	Maximum	Average	Standard deviation
8	FACS	6.70E−03	1.00E−03	1.35E−03	4.17E−04
	FA [2]	**0**	**0**	**0**	**0**
	CS [11]	1	2.83E−01	3.33E−01	1.73E−01
	DA [14]	5.68E−01	1	1	1
	SCA [15]	4.55E−01	6.76E−01	5.56E−01	7.04E−01
	SSA [16]	2.37E−01	6.42E−02	7.09E−02	4.47E−02

the findings. In this test, the p-values are determined. Table 8.3 displays the approximate order of the results. Any p-value below 0.05 is deemed unimportant and thus highlighted.

Only a single result is found to be insignificant which is underlined. Rest values are significantly lower than 0.05 as seen from Table 8.3 results. Further, the Kruskal Wallis test [18] is also performed to test whether the data sets produced by the hybrid approach are meaningful or not. Some sample test diagrams (Figures 8.2 to 8.5) of representative test functions are presented for inquisitive readers.

From the test diagrams of Figures 8.2 to 8.5, it can be inferred that nearly 80% of the results are significant. To confirm further, the test of Wilcoxon modified with Holm-Bonferroni corrections [19] is conducted on the obtained results. The corrected p-values are finally presented in Table 8.4. The insignificant test result is underlined.

It is found that mostly 99% of the values are found to be valid confirming the significance of the proposed technique. The proposed method for modelling parameters of amended double-diode versions of some commercial solar cells is used in the testing of a technological application. Table 8.5 provides the details of the datasheet information of the solar cell used.

Table 8.3 The orders of the p-values of the Wilcoxon rank-sum test for checking significance.

Test function number	Hybrid algorithm	FA	CS	DA	SCA	SSA
1	FACS	$\simeq 10^{-8}$	$\simeq 10^{-8}$	$\simeq 10^{-8}$	$\simeq 10^{-8}$	$\simeq 10^{-8}$
2	FACS	$\simeq 10^{-8}$	$\simeq 10^{-8}$	$\simeq 10^{-9}$	$\simeq 10^{-8}$	$\simeq 10^{-8}$
3	FACS	$\simeq 10^{-8}$	$\simeq 10^{-8}$	$\simeq 10^{-8}$	$\simeq 10^{-8}$	$\simeq 10^{-8}$
4	FACS	0.2853	$\simeq 10^{-8}$	$\simeq 10^{-8}$	$\simeq 10^{-8}$	$\simeq 10^{-8}$
5	FACS	$\simeq 10^{-9}$	$\simeq 10^{-9}$	$\simeq 10^{-9}$	$\simeq 10^{-9}$	$\simeq 10^{-9}$
6	FACS	$\simeq 10^{-8}$	$\simeq 10^{-8}$	$\simeq 10^{-8}$	$\simeq 10^{-8}$	$\simeq 10^{-8}$
7	FACS	$\simeq 10^{-6}$	$\simeq 10^{-8}$	$\simeq 10^{-8}$	$\simeq 10^{-8}$	$\simeq 10^{-8}$
8	FACS	$\simeq 10^{-8}$	$\simeq 10^{-8}$	$\simeq 10^{-8}$	$\simeq 10^{-8}$	$\simeq 10^{-8}$

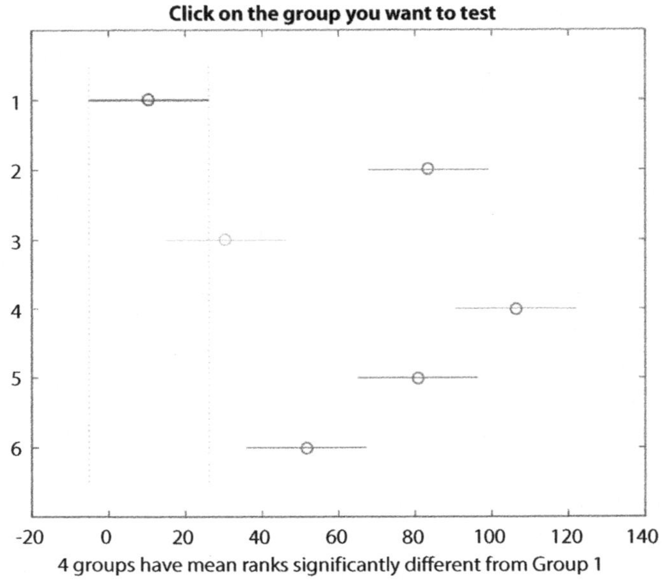

Figure 8.2 Kruskal Wallis test diagram for test function number 1.

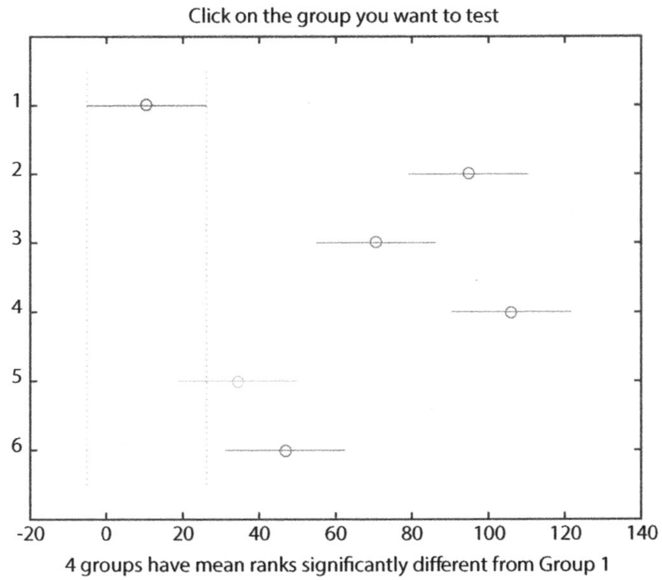

Figure 8.3 The Kruskal Wallis test result for test function number 3.

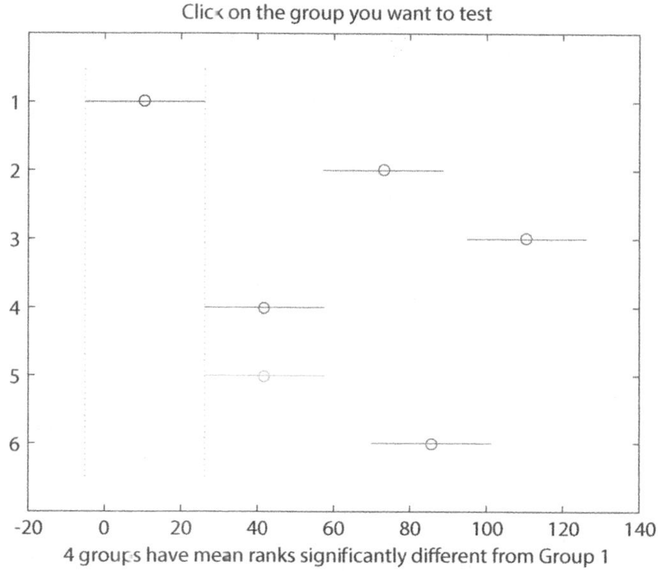

Figure 8.4 Test of significance for Kruskal Wallis test involving test function number 6.

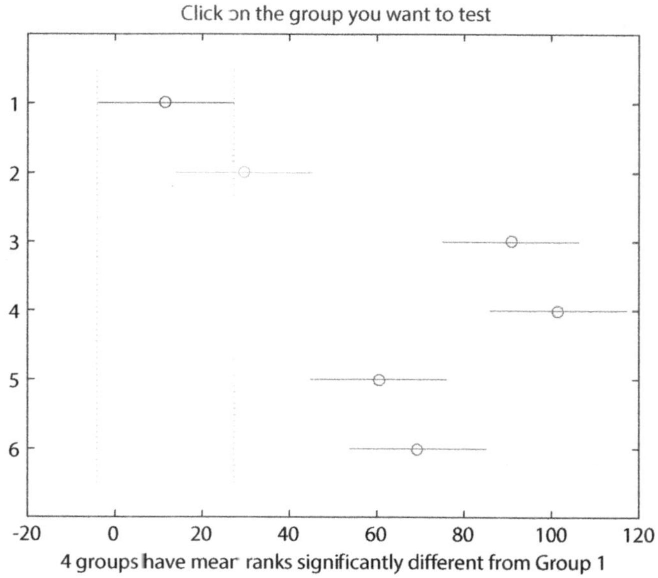

Figure 8.5 The Kruskal Wallis test analysis for test function number 7.

Table 8.4 Rectified p-values using the Wilcoxon test with Holm-Bonferroni corrections.

Test function number	Holm-Bonferroni edited p-values
1	$10^{-6} \times$ [0.33978 0.33978 0.271824 0.203868 0.135912]
2	$10^{-6} \times$ [0.33978 0.33978 0.271824 0.203868 0.135912]
3	$10^{-6} \times$ [0.3336 0.26726 0.3336 0.200445 0.13363]
4	[0.2853 0.00000033978 0.00000033978 0.000000271824 0.000000203868]
5	$10^{-7} \times$ [0.400325 0.400325 0.32026 0.240195 0.16013]
6	$10^{-6} \times$ [0.33978 0.33978 0.271824 0.203868 0.135912]
7	$10^{-5} \times$ [0.11968 0.033978 0.033978 0.0271824 0.0203868]
8	$10^{-6} \times$ [0.220565 0.271824 0.271824 0.203868 0.135912]

Table 8.5 Details of the PV modules [27].

Company	Schutten solar	Solarworld
Model number	STM6 40-36	Pro. SW255
Cell type	Mono-crystalline	Poly-crystalline
Vm [V]	16.98	30.90
Im [A]	1.50	8.32
Voc [V]	21.02	38.00
Isc [A]	1.663	8.88
Ns [cells]	36	60
T [°C]	51	25

Table 8.6 Search bounds for the amended double-diode model parameters.

Parameters of MTDM	Lower limit	Upper limit
I_{PV}(A)	0.001	15
α_1	0.5	7
α_2	0.5	7
$R_S(\Omega)$	0.001	0.5
$R_p(\Omega)$	0.01	500
$I_{01}(\mu A)$	0	1
$I_{02}(\mu A)$	0	1
$R_{S2}(\Omega)$	0	2

Several algorithms are used to solve the eight unidentified parameters (I_{PV}, α_1, α_2, R_S, R_p, I_{01}, I_{02}, R_{S2}) of the amended double-diode model. The boundaries of these parameters are listed in Table 8.6. There are 50 search agents while 1000 is the number of iterations considered for this work.

Table 8.7 provides an assessment of the parameters as well as the error values corresponding to the amended double diode model with different algorithms for both the models described in Table 8.5.

In Table 8.7, it is very clear that when contrast is made with the parent algorithms and some regular heuristic techniques, the suggested approach arrives at the minimum value of error. Besides, the convergence features between the fitness function and the number of iterations are shown in Figures 8.6 and 8.7. The parent techniques are avoided in the figures, while the comparison is made with the GWO, DA, SCA, and SSA.

In the convergence graph, as shown in Figures 8.6 to 8.7, it is clear that FACS converges well and also at the least error value with respect to the other techniques. Thus, compared with the algorithms described, the results of the proposed technique is satisfactory. In the future, the other models of the solar cells can also be explored with the suggested method. For integration with FA for better performance, improvements in the cuckoo search methods can also be adapted. The parameters of the amended double-diode model may also be defined by a multi-objective method integrating opposing goals. Since stochastic methods are used in bioinspired technologies, the outcomes depend, to some extent on parameter setting and terminatory criteria selection.

Table 8.7 Estimated decision variables of the amended double-diode model using different methods for STM6 40-36 and Pro. SW255.

Model	Algorithms	I_{pv}	a_1	a_2	R_S	R_p	I_{01}	I_{02}	R_{S2}	$error_{min}$
STM6 40-36	FACS	1.67	0.50	0.50	0.500	86.53	4.75e-07	3.78e-07	0.903	**2.35e-06**
	FA [2]	1.78	1.45	5.67	0.018	62.89	4.38e-07	3.12e-07	0.124	0.0328
	CS [11]	1.71	1.26	6.14	0.017	34.65	4.33e-07	3.25e-07	0.110	0.0225
	GWO [28]	1.68	0.50	0.53	0.500	89.56	0.00000	3.97e-08	0.623	0.0003
	DA [14]	1.67	1.39	6.09	0.012	158.1	4.72e-07	3.80e-07	0.101	0.0005
	SCA [15]	1.69	0.50	1.43	0.500	59.78	0.00000	6.71e-07	0.000	0.0129
	SSA [16]	1.68	0.50	0.50	0.500	86.46	0.00000	5.04e-07	0.901	0.0005
Pro. SW255	FACS	8.89	1.55	2.81	0.005	348.12	1.03e-06	1.49e-07	1.549	**2.89e-06**
	FA [2]	8.78	0.50	1.49	0.513	234.18	0.00000	3.34e-07	1.223	0.0215
	CS [11]	8.67	0.50	1.43	0.500	231.69	0.00000	3.28e-07	1.555	0.0287
	GWO [28]	8.99	0.50	1.34	0.500	50.85	0.00000	2.66e-07	0.029	0.0013
	DA [14]	8.87	1.54	2.83	0.006	347.1	1.00e-06	1.48e-07	1.537	0.0001
	SCA [15]	8.81	0.50	1.44	0.475	236.6	0.00000	3.60e-07	0.000	0.2515
	SSA [16]	9.01	0.50	0.50	0.500	47.90	0.00000	9.35e-07	1.031	0.0031

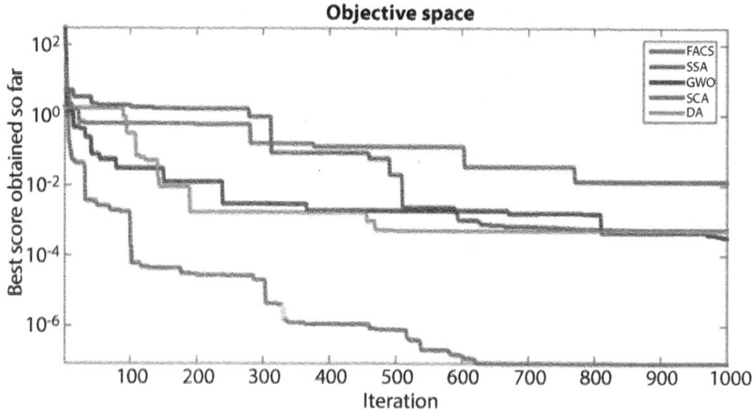

Figure 8.6 Convergence characteristic of the amended double-diode model (Schutten Solar STM6 40-36).

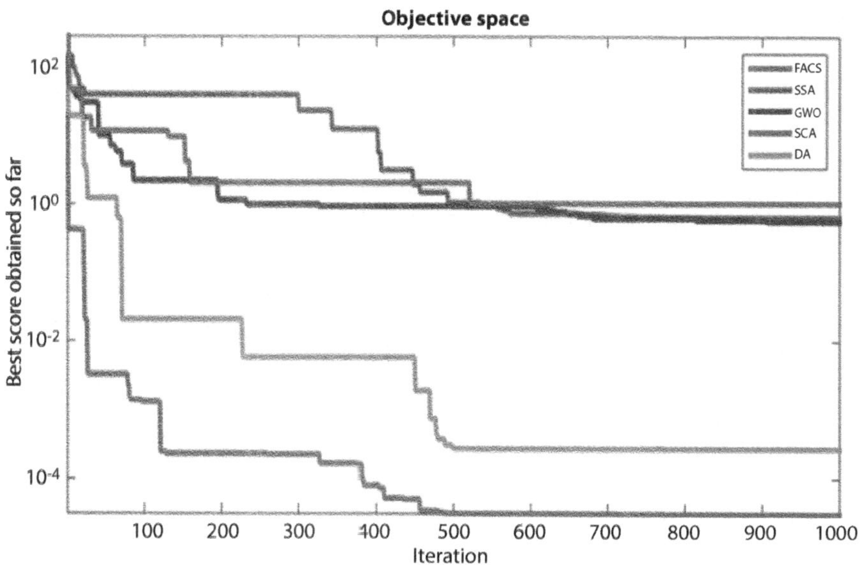

Figure 8.7 Convergence plot for the amended double-diode model (Solarworld Pro. SW255).

8.5 Conclusions

A new bioinspired technique is explored in this chapter. Two popular naturally inspired algorithms are combined to support both of these algorithms, namely the Firefly Algorithm (FA) and the Cuckoo Search (CS) method. FA is utilized to explore while CS performs the local search. In addition to conducting a comparison with the parent approaches, some newly established metaheuristic approaches are also contrasted. With the help of about eight famous test functions, the effectiveness of our approach is proven. Statistical tests and nonparametric tests are also used to confirm findings. This hybrid algorithm was also used to evaluate the amended double-diode model parameters of a target function, namely the sum of square error (SSE). Results from both solar cell models are encouraging. Furthermore, convergence plots ensure that results converge faster and are reliable. To evaluate unknown parameters with the methodology suggested, higher diode modelling can be considered in the future. A multi-objective strategy using two opposing targets may also be an alternate way of defining parameters for various photovoltaic panels as well.

References

1. Rizk-Allah, R.M., Zaki, E.M., El-Sawy, A.A., Hybridizing ant colony optimization with firefly algorithm for unconstrained optimization problems. *Appl. Math. Comput.*, 224, 473–483, 2013.
2. Yang, X.S., Firefly algorithm, stochastic test functions and design optimisation. *Int. J. Bio-Inspir. Com.*, 2, 2, 78–84, 2010.
3. He, L. and Huang, S., Modified firefly algorithm based multilevel thresholding for color image segmentation. *Neurocomputing*, 240, 152–174, 2017.
4. Yelghi, A. and Köse, C., A modified firefly algorithm for global minimum optimization. *Appl. Soft Comput.*, 62, 29–44, 2018.
5. Wang, H., Wang, W., Cui, Z., Zhou, X., Zhao, J., Li, Y., A new dynamic firefly algorithm for demand estimation of water resources. *Inf. Sci.*, 438, 95–106, 2018.
6. Selvakumar, B. and Muneeswaran, K., Firefly algorithm based feature selection for network intrusion detection. *Comput. Secur.*, 81, 148–155, 2019.
7. Wang, W.C., Xu, L., Chau, K.W., Xu, D.M., Yin-Yang firefly algorithm based on dimensionally Cauchy mutation. *Exp. Syst. Appl.*, 150, 113216, 2020.
8. Zhang, L., Liu, L., Yang, X.S., Dai, Y., A novel hybrid firefly algorithm for global optimization. *PloS One*, 11, 9, e0163230, 2016.
9. Xia, X., Gui, L., He, G., Xie, C., Wei, B., Xing, Y., Wu, R., Tang, Y., A hybrid optimizer based on firefly algorithm and particle swarm optimization algorithm. *J. Comput. Sci.*, 26, 488–500, 2018.
10. Senthilkumar, M. and Ilango, P., Energy aware task scheduling using hybrid firefly-GA in big data. *Int. J. Adv. Intell. Paradig.*, 16, 2, 99–112, 2020.

11. Gandomi, A.H., Yang, X.S., Alavi, A.H., Cuckoo search algorithm: a metaheuristic approach to solve structural optimization problems. *Eng. Comput.*, 29, 1, 17–35, 2013.
12. Talbi, E.G., A taxonomy of hybrid metaheuristics. *J. Heuristics*, 8, 5, 541–564, 2002.
13. Jamil, M. and Yang, X.S., A literature survey of benchmark functions for global optimization problems. *J. Math. Model. Numer. Optim.*, 4, 2, 150–194, 2013, arXiv preprint arXiv:1308.4008.
14. Mirjalili, S., Dragonfly algorithm: a new meta-heuristic optimization technique for solving single-objective, discrete, and multi-objective problems. *Neural Comput. Appl.*, 27, 4, 1053–1073, 2016.
15. Mirjalili, S., SCA: a sine cosine algorithm for solving optimization problems. *Knowl.-Based Syst.*, 96, 120–133, 2016.
16. Mirjalili, S., Gandomi, A.H., Mirjalili, S.Z., Saremi, S., Faris, H., Mirjalili, S.M., Salp Swarm Algorithm: A bio-inspired optimizer for engineering design problems. *Adv. Eng. Softw.*, 114, 163–191, 2017.
17. Cuzick, J., A Wilcoxon-type test for trend. *Stat. Med.*, 4, 1, 87–90, 1985.
18. Lou, Y., Yuen, S.Y., Chen, G., Evolving benchmark functions using Kruskal-Wallis test, in: *Proceedings of the Genetic and Evolutionary Computation Conference Companion*, pp. 1337–1341, July 2018.
19. Armstrong, R.A., When to use the Bonferroni correction. *Ophthalmic Physiol. Opt.*, 34, 5, 502–508, 2014.
20. Abd Elaziz, M. and Oliva, D., Parameter estimation of solar cells diode models by an improved opposition-based whale optimization algorithm. *Energy Convers. Manag.*, 171, 1843–1859, 2018.
21. Biswas, P.P., Suganthan, P.N., Wu, G., Amaratunga, G.A., Parameter estimation of solar cells using datasheet information with the application of an adaptive differential evolution algorithm. *Renew. Energy*, 132, 425–438, 2019.
22. Mathew, D., Rani, C., Kumar, M.R., Wang, Y., Binns, R., Busawon, K., Wind-driven optimization technique for estimation of solar photovoltaic parameters. *IEEE J. Photovolt.*, 8, 1, 248–256, 2017.
23. Oliva, D., Abd El Aziz, M., Hassanien, A.E., Parameter estimation of photovoltaic cells using an improved chaotic whale optimization algorithm. *Appl. Energy*, 200, 141–154, 2017.
24. Abd Elaziz, M. and Oliva, D., Parameter estimation of solar cells diode models by an improved opposition-based whale optimization algorithm. *Energy Convers. Manag.*, 171, 1843–1859, 2018.
25. Kumar, C., Raj, T.D., Premkumar, M., Raj, T.D., A new stochastic slime mould optimization algorithm for the estimation of solar photovoltaic cell parameters. *Optik*, 223, 165277, 2020.
26. Liang, J., Qiao, K., Yu, K., Ge, S., Qu, B., Xu, R., Li, K., Parameters estimation of solar photovoltaic models via a self-adaptive ensemble-based differential evolution. *Sol. Energy*, 207, 336–346, 2020.

27. Kassis, A. and Saad, M., Analysis of multi-crystalline silicon solar cells at low illumination levels using a modified two-diode model. *Sol. Energ. Mat. Sol. C.*, 94, 12, 2108–2112, 2010.
28. Mirjalili, S., Mirjalili, S.M., Lewis, A., Grey wolf optimizer. *Adv. Eng. Softw.*, 69, 46–61, 2014.

9

Supervisory Digital Feedback Control System for An Effective PV Management and Battery Integration

Amal E. Abdel Gawad[1], Nehal A. Alyamani[1] and Saeed Mian Qaisar[1,2*]

[1]*Dept. of Electrical and Computer Engineering, Effat University, Jeddah, Saudi Arabia*
[2]*Communication & Signal Processing Lab, Energy & Technology Research Center, Effat University, Jeddah, Saudi Arabia*

Abstract

A limitation of PV-based generation is its weather and daylight dependence. Therefore, researchers motivated to study, design, and implement effective PV batteries-based systems that can viably energize targeted appliances. In the case of unstable PV output power, the switching toward a battery-based power supply should be effectively realized. Thus, the proposed system has the following aims: (1) to ensure that the maximum amount of sunlight will be absorbed by the PV panel during the hours that the sun shines in a particular area; (2) the level of soiling ratio is not above a certain threshold. Otherwise, a cleaning notification is generated and sent to the maintenance center; and (3) in case of any drop of PV output power, PV-to-battery source switching is employed in a viable manner, without causing any major disturbance or damage to the intended load. The solution is realized by developing an intelligent assembly of a digital controller with appropriate sensors and actuators network. The controller will work as a decision maker. Based on sensors' feedback, it will command the actuator to orient PVs to absorb the maximum possible sunlight. Additionally, based on an updated weather parameters database, it can generate and automatically send cleaning notifications with the concerned PV coordinates. Moreover, it manages a stable load supply by effective switching between PV and battery power sources.

Keywords: Renewable energy, negative feedback, linear control, digital system, photovoltage orientation, light tracker

**Corresponding author*: sqaisar@effatuniversity.edu.sa

Suman Lata Tripathi, Souvik Ganguli, Tengiz Magradze, and Abhishek Kumar (eds.) Intelligent Green Technologies for Sustainable Smart Cities, (165–194) © 2022 Scrivener Publishing LLC

9.1 Introduction

The Solar is a primary and renewable energy source. Solar energy is generated by converting the sun's radiation into electricity by exciting silicon electrons. Photovoltaic (PV) modules are majorly used in electricity generation, but more designs and applications have emerged in recent developments. These advancements include flat-plate solar thermal collectors, compound parabolic concentrating collectors, vacuum tube thermal collectors, Fresnel lenses, and parabolic concentrating collectors (Tripanagnostopoulos *et al.*, 2002). The working principle of solar PV and other collectors is different since, in PV modules, electricity is generated by exciting silicon electrons; in contrast, in thermal collectors and concentrators, electricity is generated by converting heat energy into thermal, mechanical, and electrical in the order.

Solar is a renewable energy source considered a cheaper alternative to fossils fuels and coal. After installing PV modules or collectors, the user will generate electricity for the lifetime of the module with little maintenance (Arachchige & Weliwaththage, 2020). Since it is naturally replenished without causing environmental degradation, solar energy can be considered a sustainable source. In energy, sustainability is defined as the ability to produce electricity for an infinitely long time in a safe, reliable, clean, environmentally friendly to meet the current demand without interfering with the ability of future generations to meet their demands (Arachchige & Weliwaththage, 2020). Solar energy can be harnessed anywhere across the world, even though the amount of irradiation varies. When installing a large-scale solar power plant, it is essential to determine the Global Horizontal Irradiance (GHI) to maximize electricity generation by ensuring that the plant is located in areas with the highest GHI. Thus, unlike in other energy sources, we cannot run out of solar, making it a reliable source.

Another motivation for using solar energy is to reduce electricity bills. On a small scale, solar will provide enough or close to enough electricity to run a homestead. Since solar does not require an external fuel source, it will reduce household bills by a large margin. At the industrial level, solar power will provide electricity to run some basic operations such as lighting, heating, cooling, and water pumping. Companies can also install solar energy on a large scale to provide enough energy to run all operations, which will reduce the overall cost of electricity.

Additionally, industries can produce excess electricity and be connected to the grid for export. Grid-tied solar plants will allow companies and individuals to become energy producers and sellers, which increase individual's income level through the Smart Export Guarantee. Therefore, in addition to reducing electricity bills, solar power can become a source of income, especially among large-scale producers.

Users of solar power systems can benefit from the diversity of its application. Using PV modules, an individual can generate electricity from the sun. Similarly, using solar concentrators and collectors, the user can use the sun to heat water for thermal energy generation, space heating, household uses, and cooling. Coupled with low costs of maintenance, households can increase their capacity by increasing areas covered by solar especially using the technology integrated into building materials. Instead of using traditional roofing, solar integrated materials will significantly reduce the cost of installation. Integrating building materials with solar modules has increased the ability and capacity to produce more electricity at small and large scales. The move is a significant step toward generating clean energy and reducing carbon emissions from fossil fuels, natural gasses, and coal.

Advancement in technology has also fostered the development of solar-powered automobiles. Recently, car manufacturers have moved to convert internal combustion engines to electric modules. Electric vehicles are developing at a faster pace, but their development has increased electricity demand exponentially. Utility companies and electricity producers have increased the energy supply from fossils fuels, nuclear, and thermal power plants. That has created unprecedented demand and increased carbon emissions. However, automobile manufacturers are on a speedy development process to design and manufacture solar-powered vehicles (Shaikh *et al.*, 2017). The move will phase off fuel-powered vehicles, increase electric-powered vehicles, and reduce the need for charging by embedding solar modules on modern cars.

Energy sources are identified as valuable energy where energy is extracted directly or utilizing transformation or conversion of this energy such as solid fuels, solar energy, and liquid fuels. A renewable source of energy is identified as sustainable energy (Qazi *et al.*, 2019). The power is alternative to the commonly used nonsustainable sources, such as coal. Traditional electrical energy generation is energy produced from burning fuel to release heat (Ellabban *et al.*, 2014). They are also termed thermal power electrical sources. Renewable power is generated from adequately replenished souse. This paper focuses on comparing renewable sources of energy and traditional electrical energy generation.

Several people confuse renewable energy and traditional electrical generation. But in a real sense, they much different sources of energy. As renewable energy avoids the utilization of fossils such as oil, coal, and natural gases, traditional electrical energy generation utilizes fossil for the generation of power (Ellabban *et al.*, 2014). Renewable energy sources such as solar are termed to be problematic because their nature is uncontrollable by humans. In contrast, traditional electrical power is flexible since they are under the control of human being. The production of energy for both renewable and nonrenewable are different in terms of production. Electricity for most countries is generated from the power plant through electromechanical generators, primarily driven by nuclear fission or by heat engines fueled by combustion (Arachchige & Weliwaththage, 2020). There are other means applicable in power generation, such as the kinetic energy of flowing wind or water. According to Arachchige & Weliwaththage (2020), the comparative of the renewable and nonrenewable energy sources is determined by their origins, with the renewable source marking the prospective of all energy supply. Renewable energy, including PV and wind power, cannot be exhausted. Most industries, homesteads, and other business sectors are looking forward to adopting these energy sources while minimizing nonrenewable energy consumption. Comparing the renewable power sources with traditional electrical energy generation (Heydt *et al.*, 2012), we affirm that nonrenewable takes a considerable amount of time to be naturally replenished while renewable is created quickly. Simultaneously, renewable energy such as solar and wind provide continuous energy with the extended global application. The traditional power generation such as coal gets exhausted with much danger for the countries and industries that depend on their energy supply (Heydt *et al.*, 2012). These differences are the fundamental cause of the alarming shift from consuming nonrenewable energy to renewable sources being the most reliable and sustainable sources.

Solar energy is identified as energy from the sun, which is converted into thermal, electrical energy. It is connected using the evolving solar technologies for a wide range of use such as the provision of light, generation of electricity, including heating water for domestic, industrial, or commercial use (Shaikh *et al.*, 2017). There are several current solar energy projects, such as the Pecan Prairie Solar Project, Commonwealth Solar Project. Solar energy is associated with several applications, such as Solar Water Heating, Solar Distillation, Solar Pumps, Concentrated Solar Power (CSP), and solar PV technology (Malinowski *et al.*, 2017). CSP operates with mirrors and lens as the tracking systems helping to concentrate solar irradiance for heating the solids, liquids, or gases. It directs the concentrated sunlight to a small focal point to produce electricity or heat (Malinowski *et al.*, 2017). On the other hand, PV refers to the system applicable in converting solar irradiance direct to the electric form of energy using PV cells (Malinowski *et al.*, 2017). PV presents the most significant global attention for renewable energy.

In the future, solar technology will ensure solar become cheaper. It will also become an essential source of energy for electric energy generation globally. That will bring a positive impact on the environment, including climate change. These will include production innovations that will come down the pipeline that will reduce the cost of material, such as silver and silicon used in the manufacture of solar cells and two-sided modules, which allow panels to capture solar power. Additionally, solar will reach most of the world since its installation is simple compared to other energy sources. In general, solar installation is expected to increase over the coming years. The world consumption of all forms of energy is expected to increase and has been rising steadily since the 1970s. It also indicates that oil consumption is still the most prominent form of energy used worldwide (*World Energy Outlook 2009 – Analysis*, 2009).

The chart in Figure 9.1 shows the global investment trends by energy source from 2004 through to 2016 (excluding large hydropower). In 2016, solar and wind energy both received 47% of investment (combining to account for 94% of global finance with 226.2 billion United States Dollar (USD)). It is noted that solar and wind energy have been taking an increasing share. In 2006, bioenergy (biomass and liquid biofuels) took a sizable

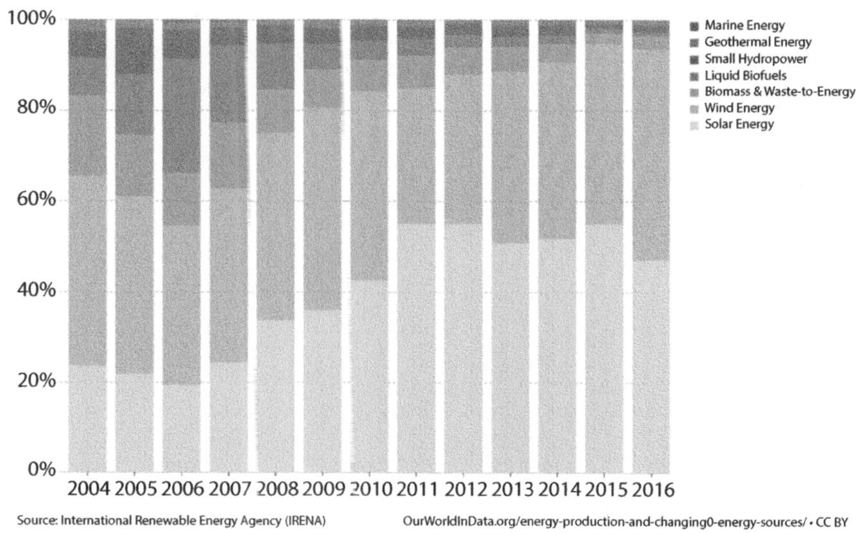

Figure 9.1 Global investment in renewable energy technologies, measured in USD per year.

share of global investment, peaking at 36% with 41.4 billion USD. However, over the last decade, it has dwindled, receiving less than 4% in 2016, with just 9 billion USD. That indicates how much the investors believe in solar and wind energy and see it as the governing renewable future technologies (*Investment in Renewable Energy, by Technology*, 2017).

Given how the average cost of technologies changed from 1976 to 2019, the graph in Figure 9.2 represents that solar PV energy and a lesser degree of solar thermal heat have significantly decreased their cost. This decrease in the cost of solar PV has been accelerating over the past few decades, as shown in the graph. The price of solar PV modules has fallen sharply more than 100 times since 1976. In 1976, the PV module price was 66.1$ per watt-peak ($/Wp). In 2019, the PV module price was 0.38$/Wp (*Solar PV Module Prices*, 2017).

Global horizontal irradiance (GHI) is measured by a pyrometer and is considered the major supplement of the PV supplement. Yang *et al.* (2018) reflect an analysis that guides GHI calculations and their amount across the world. The measurement of all three types of irradiance components, including DNI, GHI, DIF, are achieved at the Baseline Solar Radiation Network (BSRN), being centrally committed under the World Radiation Monitoring Center (WRMC). Solar irradiance measurements

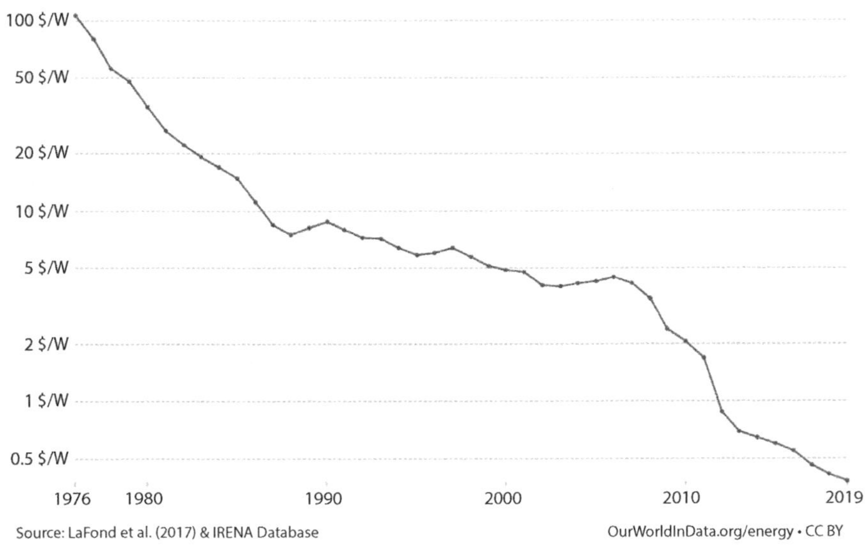

Figure 9.2 Global solar PV module prices (measured in USD per watt-peak), 1976 to 2019.

are considered to revolve around these components. GHI is the total irradiance that is horizontal to the surfaces, and it is regarded as the sum of DNI and DIF.

There are various factors that influence power generation and the efficiency of PV modules, including temperature, angle of incidence of sunray, and distribution of irradiations (Soga & Akasaka, 2003). PV modules produce optimum electricity when the solar radiation hits the surface at an optimal incident angle. Soga & Akasaka (2003) reported that the efficiency of PV modules decreases, depending on the solar incident angle. The efficiency of power generation decreases sharply when the incident angle exceeds 60 degrees and reflectance exceeds 0.2. The decrease in power generation for PV modules prompts design parameters such as oversizing to meet the energy demand. According to Soga & Akasaka (2003), the optimal incident angle is between 50 and 60 degrees. If the solar can maintain this incident angle throughout the day, electricity generation will be optimized. Consequently, solar tracking systems have been developed to ensure maintain the incident angle for PV modules and collectors at this optimal range (Soga & Akasaka, 2003). There are several types of solar tracking systems developed to achieve this goal. Depending on movement and degrees of freedom, there are two types of solar tracking systems: single axis and dual axis. Single-axis solar tracking system has one degree of freedom allowing the PV modules and collectors to follow the sun from east to west, and it is used for utility-scale projects. It is considered to increase electricity generation by between 25% and 35%. In single-axis solar tracking system, the face of the module or collector is oriented at an angle concerning the axis of rotation (Hafez *et al.*, 2018). Single-axis solar tracking system has been applied in a parabolic trough and linear Fresnel solar systems. The primary limitation of using a single-axis tracking system is an inability to orient the module or collector based on yearly changes of the sun's position. The single degree of freedom reduces the efficiency of modules and collectors during cloudy days because of limited rotation.

Solar tracking systems are advantageous because they offer greater energy generation capacity as compared to fixed arrays since they maintain the incident angle at optimum. Solar tracking systems are beneficial, especially in areas with the time of use electricity rates since a user can generate more electricity during peak hours and reduce reliance on expensive grid energy. Their significant limitations include being expensive, which increases the costs of investment. The moving parts will wear quickly, requiring scheduled maintenance to replace broken parts. Further, tracking systems require additional design features and site preparation, which might increase the time of return in investment on a small scale.

Dust removal is a cleaning procedure employed to ensure that PV modules and collectors are in the right conditions to generate electricity. The presence of dust on the surface of the module reduces the overall performance efficiency and the amount of electricity generated. After installation, the cleaning of the solar panel is considered difficult (Syafiq *et al.*, 2018) as dust particles do not allow solar radiation to effectively penetrate the solar pane surface, minimizing the conversion efficiency. Syafiq *et al.* (2018) state that cleaning PV modules and collectors using conventional approaches, including manual wiping or vacuum cleaning, is complex and sometimes not feasible because some typical mechanisms and mounting are ineffective for the operations. Electrostatic Precipitation (ESP) consists of fixed electrodes and is referred to as the best approach to remove dust particles accumulation on the solar panel (Syafiq *et al.*, 2018). The EPS is placed along the sides of the panel, and the PV panel is mounted with a weight sensor used in regular detection of dust thickness in terms of changed weight on the panel for cleaning.

In solar power systems, batteries are used as power tanks and supply it when there is no sunshine. When charging, it is necessary to control the process to prevent overcharge and destruction of cells. One of the techniques of controlling charge in PV is maximum power point tracking (MPPT) which enables fast capture of maximum power point (MPP). The charging controller uses a constant current, constant voltage (CCCV) charging scheme to reduce the time of battering charging. Further, the controller enables access to all system parameters to monitor and administer the process remotely. The implications of the charging technique of battery lifetime were presented with constant current and constant-current voltage with a negative pulse. The equivalent DC resistances of lithium were also considered to consider various health conditions during charging for different temperature conditions and their relationship with the equivalent DC resistance and health condition.

To properly integrate the solar power systems in modern grid, the recent technological advancements in signal processing, surveillance, and machine learning can also play key role (S. M. Qaisar, 2019), (S. M. Qaisar, 2020), (Mian Qaisar & Fawad Hussain, 2020).

The current research purposes of establishing solar power by applying a small panel, ensuring that the energy absorbed by the PV panel during the hour is maximum and effective for the required operations. That will be embarked on several questions that will make the research effective for the current solar power generation. First, the study will try to answer how solar engineers and designers can ensure that the PV panel absorbs the maximum amount of sunlight during the hour of sunshine in an area. The research will also reflect on why the cleaning process is adequate for the PV panel efficiency, considering some effects of soiling on the system surfaces. Finally, the research will explore why switching is performed in a viable manner in the case of PV output drop without causing disturbances or damages to the intended load.

9.2 Literature Review

9.2.1 GHI in the Middle East

Solar radiation reaching the earth's surface is not the same across the world. There are countries and regions with higher solar radiations than others, and when designing a solar park, it is essential to calculate the amount of radiation before deciding on the most suitable location and PV modules. There are several ways of determining and representing the level of radiation reaching the earth's surface. GHI is a mainly used approach used to represent the level of solar radiation; it expresses the number of shortwave radiations received by a surface horizontal to the ground (Zell *et al.*, 2015). Figure 9.3 illustrates the GHI distribution over Saudi Arabia. According to (Zell *et al.*, 2015), the data from thirty stations distributed across Saudi Arabia shows that the annual average daily GHI ranged from 5700 Wh/m^2 to about 6700 Wh/m^2 with a consistently high value within the inland and slightly lower values along the coasts. Saudi Arabia GHI is composing the total solar radiation incident on the horizontal surfaces summing the Direct Normal Irradiance DNI, ground-reflected radiation, and Diffuse Horizontal Irradiance computing the flat-panel PV output.

9.2.2 Types of PV Systems

The work, presented in (Yao & Cai, 2019) reflects three types of PV with their distinct characteristics. The grid-tied type is considered the basic solar installation with a standard grid-tied inverter without any battery storage. It is easy to design and install, with considerable cost having relatively few components. The major disadvantage of this PV type is that it cannot work when the power is off (Yao & Cai, 2019). The second type of PV is the off-grid system which is considered adequate for the customers who cannot connect to the grid system. Off-grid PV is ineffective for the areas with an existing primary electric grid system. Grid-tied with battery

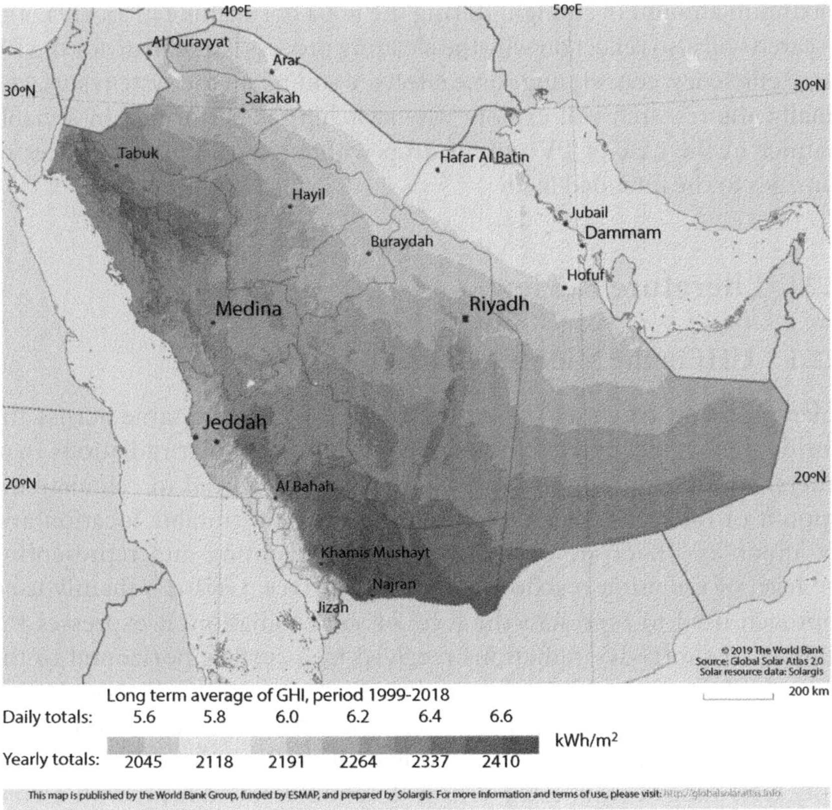

Figure 9.3 Saudi Arabia GHI.

backup or Grid-hybrid system is another type of PV that is advantageous for people along the grid supply requiring a battery backup (Yao & Cai, 2019). The disadvantage of this type of PV is that it is costly and less effective than the other grids.

The solar systems have the capacity to derive energy in two ways, which are through the solar thermal system, where it traps heat and warm water. Also, the solar Photovoltaic (PV) system may be applied to convert sunlight energy to electrical. The electrical energy generated by PV systems is a direct current (DC) that requires an inverter to be converted to alternating current (AC). The conversion of electrical energy into AC help in distributing electricity without affecting the quality. There are various types of PV systems that may assist in converting sunlight energy into electrical energy. The types of PV systems are based on the technology applied.

The grid-connected solar PV system is one of the fastest-growing PV systems. This PV system is characterized by a solar array from residents being connected to the local utility grid. The work of connecting the residents' solar arrays with the local utility grid is facilitated by applying the Service Entrance Panel (SEP) (Abu Eldahab *et al.*, 2014). That will help electrical energy produced by resident solar during the day to be stored in the local utility and transferred to the resident at night. Additionally, the grid-connected solar PV system is required to have an inverter that would ease the work of conversion DC electrical energy to AC. The grid-connected solar PV system is advantageous to users since it enhances the safety of their apparatus. That is achieved when the grid-connected solar PV system enhances the application of inverters that shut themselves automatically when a problem occurs (Abu Eldahab *et al.*, 2014). That limits the amount of energy sent to the resident, a problem in the local utility grid (Spanos *et al.*, 2015). Additionally, the grid-connected solar PV system is identified to be economical to the residents. That is because, during the day, residents are not at home, which makes the PV system maximize the energy produced, which is stored in the local utility grid which they use at night.

One of the disadvantages of the grid-connected solar PV system is having a higher cost of installation. For example, the PV system requires all the residents' solar arrays connected to the local utility grid. Additionally, the grid-connected solar PV system has the drawback of power shortage when many residents are supplied with electrical energy than they send to the local utility grid.

The Off-grid solar PV system is applied to areas that do not have a local utility grid, such as in rural areas. That means that the homeowners need to install solar arrays in their homes without connecting them to the power supply grid (Abu Eldahab *et al.*, 2014). The Off-grid solar PV system is characterized by the employment of batteries that assist in storing electrical energy produced. Additionally, the Off-grid solar PV system may require inverters to convert the DC electricity to AC. That will help the homeowners to meet their power needs. The off-grid solar PV system is advantageous since homeowners have the capacity to calculate the amount of energy they are producing and what they need for their operations. That allows the homeowner to determine the solar array they need to install (Tripanagnostopoulos *et al.*, 2002). Additionally, the Off-grid solar PV system allows users to apply inverters that match their needs. The off-grid solar PV system has the disadvantage of power shortage. For example, when the weather is not favorable, homeowners may experience a power shortage. That would occur since when the power is excess and batteries are fully charged, power may not be transferred to the local utility grid. Additionally, the off-grid solar PV system is identified to be requiring many resources when installing and maintaining.

During the day, sunlight is identified to be varying. For example, the sunlight may not shine at times in the day, which may hinder apparatus that are using the PV system energy not to work effectively. That led to the development of a battery-based solar PV system. The battery-based solar PV systems are characterized by storage facilitates for the electrical energy generated (Abu Eldahab *et al.*, 2014). Batteries that need to be applied in battery-based systems depend on the capacity of the apparatus used to draw energy. The charges to the batteries are required to be regulated through the application of monitors to ensure batteries do not overcharge. That is because monitors are used to determining the level of charge to the battery. Additionally, monitors are applied to detect the current that is flowing into the battery.

The application of a battery-based solar PV system is identified to be advantageous. First, the system helps to apply electrical energy when there is no sunlight, for instance, at night. That would be effective when the users apply a battery bank to connect many batteries to increase storage capacity. Another advantage of the battery-based solar PV system is that users having the ability to determine the time apparatus they are using will be working (Abu Eldahab *et al.*, 2014). That would be achieved by applying a monitor, which helps determine the level of the battery charge.

Although the battery-based solar PV system is advantageous, it is identified to have some drawbacks. First, the battery-based solar PV system installation is identified to be costly. That is because the system requires much equipment such as batteries, inverter, and solar panels (Abu Eldahab *et al.*, 2014). Another disadvantage of the battery-based solar PV system is requiring regular battery replacement. That is because batteries are identified to be having a short life span.

9.2.3 Solar Tracking Systems

The work, presented in (K, 2016) reported on the review of the solar tracking system reflects on the critical components of the solar tracking system. According to (K, 2016), the primary mechanism of solar tracking consists of the control unit, the positioning system, sensing devices, driving mechanism, tracking devices, tracking algorithm. The tracking algorithm within the system determines the effective angle applicable in reflecting the position of the solar tracker (Hafez *et al.*, 2018). The control unit performs the all-tracking algorithm based on the real-time intensity of light reading. The positioning system is applicable in operating tracking devices to face the sun at a concentrated angle (Hafez *et al.*, 2018). This positioning system can be either electrified or hydraulic. Solar tracking systems are categorized into two main types based on degrees of freedom: single-axis and dual-axis solar tracking systems. Based on tracking technology, solar tracking systems are classified into five types: active, passive, semi-passive, manual, and chronological tracking (Hafez *et al.*, 2018). These types of solar tracking systems have several advantages and disadvantages based on their feature and components as well as the operating principles, but their goal is to increase the efficiency of PV modules and collectors.

The active solar tracking system is the most used and studied drive system. Studies on active tracking systems are endeavored to enhance the performance and accuracy of the drive system. For instance, studies have been conducted on signals and sensors, control circuit, and drive system (Hafez *et al.*, 2018). The chronological tracking system uses the same control system like active solar tracking systems but with a different operating principle is regarded as more energy-efficient because its operation depends on collected data to define the accurate tracing position. The chronological tracking system uses less energy, since it does not require real-time calibration (Hafez *et al.*, 2018). Passive, semi-passive, and manual tracking devices are considered simple and lower system movement, making them cheaper to maintain.

The active solar system is identified as a system that determines where the sun path is situated in the sky during the day using the sensor. The sensor is purposely for actuating or triggering the motor drive system toward the sun throughout the day. In case the beam of solar radiation is not perpendicular to the solar tracking system, the light intensity will be different in one sensor than others, making the tracking system and sunlight beam perpendicular to each other. The first tracking system is an algorithm used to calculate the total power in the central receiver in the central receiver of the solar energy system and determine the distribution of the flux density in the system. The error tolerance of the position of the sun is estimated to be between −0.5 to −1 (Hafez *et al.*, 2018). The active tracking system is an arrangement of different control which encompass microprocessor-based control, electrical, optical sensor-based control, date, and time technique, including auxiliary PV cells.

Active tracking system using microprocessors and the electric optical sensor uses at least two PV cells or photoresistors. The output signal of the two variable parameters is conducted where an indication of the difference between the two varying parameters is sent to the drive motor. With respect to the solar tracking system that uses an auxiliary bifacial solar cell, the drive system is triggered by the cells to the destined position in the solar tracking system that utilizes date and time, mathematical algorithm calculated by use of the computer which generates the control signal to the system (Hafez *et al.*, 2018). During the cloudy day, the results from the system are not accurate since the sensor is unable to decide due to low solar irradiance intensity, which is the difference between the sensors.

There are some lucid features in accordance with the auxiliary bifacial solar cell-based, microprocessor, and electric optical-based tracking system. From different scholars, microprocessors and electric optical sensor based are the most used compared with the auxiliary bifacial solar cell because they are widely applied in PV cells that use a tracking system (Hafez *et al.*, 2018). Microprocessor-based and electric optical-based tracking systems are similar since they are used to send a signal to trigger or actuate motors into the proper tracking position based on a measurement from the mounted sensor photoresistors that are placed. Additionally, both types of tracking systems have disadvantages of not efficiently performing during the cloudy day since they are dependent on light that reaches the sensor.

A passive tracker system depends on thermal expansion in the material between the two-point on the sides of the tracker. This material is usually fluid, liquid, or gas. The fluid is inserted into two reservoirs placed opposite to each other in a specific design to vaporize the fluid and change its characteristics in accordance with the change of the path of the sun and time (Hafez *et al.*, 2018). The two tanks are connected so that they make the system move while carrying the condensate fluid from the reservoir with the highest incidence to the reservoir with the minor incidence. The passive solar tracker relies on compressed fluid with a low boiling point to cause the structure to move to an imbalance.

A semipassive tracker system is a technique whereby the solar tracking concentrator tracks the sun by keeping rays perpendicular to the absorber cross-sectional area with minimal mechanical efforts and reduced sun tracking. The system comprises the micro-heliostat array, a Fresnel lens, and s receiver. A semi-passive tracking concentrator is an optical system used to concentrate solar irradiance while tracking the sun with minimum mechanical movement (Hafez *et al.*, 2018). It is also comprised of a Fresnel lens, micro-heliostat array, and a receiver. The micro-heliostat array tracks sun rays and reflects them toward the Fresnel lens. The lens is maintained horizontally to decrease the wind loads in the system. The reiver is stably placed in the lens to release weight which is the power needed for the system to move.

A manual solar tracker is an approach where the system tracks the sun at an angle from one season to another (Hafez *et al.*, 2018). The angle is manually change using the manual gear to ease the system construction and maintenance. The main advantage of a manual tracking system while used to tilt the angel is that it is cheaper than when used in the types mentioned above that implement the second motor.

A chronological solar tracker system is a time-based tracking system where the system collector or the module moves at a fixed angle and fixed rate throughout the day. The actuator or the motor is controlled to rotate at a low speed of 15°/h. This tracking system is typically an open-loop control tracker based on the chronological model of its motion (Hafez *et al.*, 2018). The main advantage of this tracking system is that there is no energy loss due to its low tracking error, thus producing efficient energy.

The case when systems with higher precession are required, further sophistications can be incorporated by employing the machine learning, sensing and signal processing based viable approaches (S. M. Qaisar & Subasi, 2020), (S. Qaisar *et al.*, 2008), (S. M. Qaisar, 2014), (S. M. Qaisar & Subasi, 2018).

9.2.4 Charger Controller

Charger controller is a significant part of the solar systems that will provide overload protection, block reverse currents, and maintain the good health of the batteries. Several parameters affect the controller's design: the charging voltage of the solar panel, solar panel charging current, the amount of power of solar panels, the battery's capacity, the battery voltage, battery temperature, the load current, the amount of power used (*Solar Charge Controller Basics | Northern Arizona Wind & Sun*, n.d.).

9.2.5 Series Regulator

To join or remove the solar array, most engineers use a relay or transistor. Series regulators can be used in both positive and negative ground systems since the relay, or the transistor, can be located in either the positive or negative grounded circuiting. The array will be reconnected; thus, the current will be allowed to flow to the battery again until the battery voltage decreases (SRNE, 2021). Figure 9.4 shows the series regulator circuit diagram.

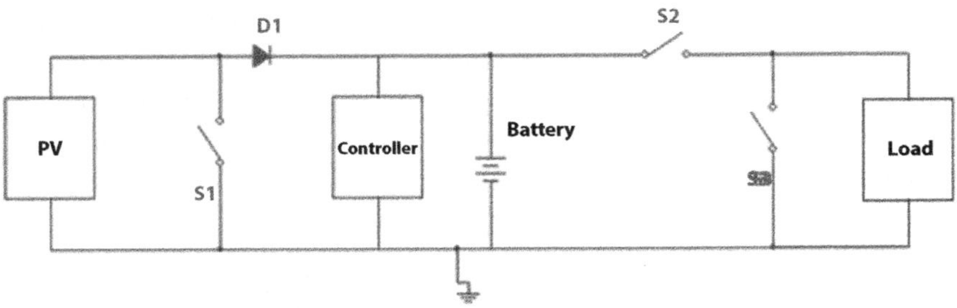

Figure 9.4 Series regulator circuit diagram.

9.2.6 Shunt Regulator

Shunt regulators are considered solid-state devices composed of a blocking diode and a transistor. A transistor will short the solar panels, while the blocking diode will stop the battery from being shorted at almost the same time. Since the blocking diode usually is on the positive side, shunt regulators are only used in negatively grounded systems. When the battery hits a specific voltage, the solar panels are now short-circuited. The array will resume regular operation when the battery voltage decreases, and current will be allowed to flow to the battery once more. The shunt regulators create an amount of heat when the full current from the solar side is being controlled which may affect the system performance. That is the reason behind its limited use to small systems within 20 A or less (SRNE, 2021). Figure 9.5 shows the shunt regulator circuit diagram.

9.2.7 Pulse Width Modulation

The pulse width modulation (PWM) solar charge controller is a second-generation technology device that currently holds the largest market share. The charge conversion efficiency ranges from 75% to 80%. Charge rates are

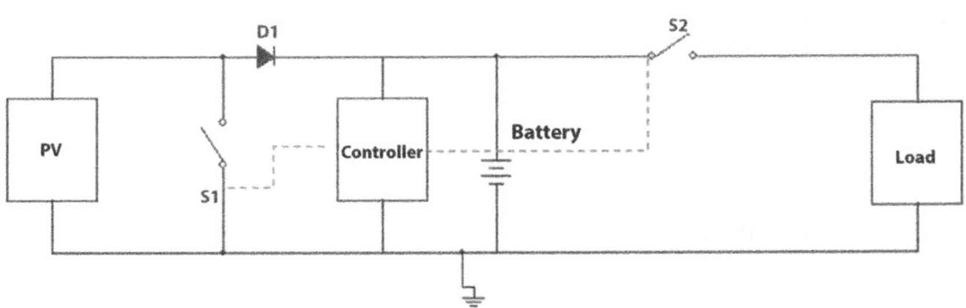

Figure 9.5 Shunt regulator circuit diagram.

adjusted based on battery levels, making control more difficult to achieve optimum levels. PWM controllers improve the efficiency of the three-stage process by using high-frequency electric pulses to alter the amperage provided by changing the pulse lengths periodically. The key benefit of PWM is that it has a shallow power loss in the switching modules. When a transistor is off because there is no current, there is almost no voltage drop across the switch when it is on, and power is transferred to the load. Since power loss is the product of voltage and current, it is identical to zero in both cases. As a result, it is appropriate for small off-grid solar panel systems (SRNE, 2021), (Mohamed *et al.*, 2013).

9.2.8 Maximum Power Point Tracker Charger Controller

The maximum power point tracker (MPPT) solar charge controller, which is a third-generation technology product and the most advanced solar charge controller available. MPPT controllers take the DC input from solar panels, convert it to high-frequency AC, and then adjust the voltage and current to a separate DC voltage and current(*Silicon Chip Online - Build A 12/24V 3-Stage Solar Charge Controller*, n.d.). It can detect the voltage and current of the solar panel in real-time and track the maximum power point in order to charge the storage battery at maximum efficiency. The argument is that the voltage would be specialist team the battery's requirements. They can only be used in negative ground systems because they use the negative line as a reference before switching to the positive line. The system's tracking efficiency is 99%, and its generating efficiency is up to 97%. The battery is well-managed and adequately protected. MPPT charge, constant voltage equalizing charge, and constant voltage floating charge are the three charging modes. MPPT solar charge controllers will undoubtedly replace PWM solar charge controllers as technology advances, and energy debates intensify.

The work, presented in (Abu Eldahab *et al.*, 2014) provided new Maximum Power Point Tracking MPPT approach reflects a significant improvement in the general process of tracking the maximum power point applying a binary search which is further improved than other traditional MPPT approaches. The review demonstrates several strategies, methods, and algorithms, creating an intelligent charging control system as presented by Alvarez *et al.* (2020). In addition, Spanos *et al.* (2015) provided an improved analysis on the life cycle of the three types of batteries, including nickel-zinc, valve-regulated lead-acid batteries, and manganese dioxide. Spanos *et al.* (2015) provides a comprehensive examination with a complete process-based life cycle analysis of batteries. This study further showed significant information for the sizing and selecting the appropriate battery from an optimized hybrid battery and effective energy storage system based on a cooperative game method (Han *et al.*, 2015).

In Abu Eldahab *et al.* (2016), the authors have presented a battery-changer controller showinghowit is usedto control batteryovercharg-ingand, at thesame time, over-dischargingby controlling the flow of currentfromandtothebattery. From the PV module to the battery, the measurements are recorded by the MPPT and the controller. At the start, the system is designed to search the database for the optimum voltage, current, and power. The system helps in applying the battery charging logic at the end.Instead of searching every point within the above PV curve from [0-V_{OC}] linearly, the principle of the new tracking approach considers the application of the binary search technique. Every time, the searching path is divided twice, and the only point at the middle is considered for selecting the half as the new searching path while discarding the other half. The curve demonstrates the search path starting at point 1 to 4, where the optimum power point resides. That curve also shows four checkpoints beginning from the middle point within the intersection of line number 1 and the PV curve. Then, it determines the appropriate direction to move. If the point is positive, the search line goes to the right side, selecting the right half while discarding the other portion. These processes are repeated to the targeted MPP (Abu Eldahab *et al.*, 2016). This instance reflects that only four interactions reach the targeted point when required to check about several points for the traditional linear search method.

The duty cycle for the boosted cycle converter is controlled by the effective pulse width modulation signal, which is generated by a microcontroller. On the other hand, the width of the consequent pulse is determined based on the practical values of the system parameters (Abu Eldahab *et al.*, 2016). These parameters include the panel ambient temperature, panel current, irradiance, battery temperature, and voltage. The program software contains the logic of the MPPT algorithm and the voltage battery charging algorithm.

9.2.9 Reducing the Charging Time

Conversely, the controller utilizes a multistage charging algorithm, considered the safest and practical approach to charging. The principle of operation operates within several stages to reflect optimum results (Abu Eldahab *et al.*, 2016). In the first stage, the battery voltage increases slowly to the optimal voltage level, referred to as the bulk level, using a constant charge current. At the point when the bulk level is voltage is achieved, the second stage of absorption starts. The voltage is effectively maintained at the bulk voltage level at this phase for a specific time while the current is slowly dropping (Abu Eldahab *et al.*, 2016). The float stage begins just after the absorption time, and the voltage is lowered at the float level, and the battery draws little current. The current and voltage regulator circuit is applied to the proposed controller to achieve the specific charging process.

9.2.10 Dust Remover

According to Kazem *et al.* (2020), some weather conditions that adversely affect the operation of the PV cell include an accumulation of pollutants and dust on the panel, acting as the barrier between irradiation and the operation setting of the PV. Kazem *et al.* (2020) give a methodological review of cleaning operation, concentrating on dust problems and implication of solar technology concepts, cleaning methods, and the current proposed cleaning methods. First, the cleaning methodology involves dust concepts, chemical and physical characteristics, and implications on the PV performance. The specific proposed approach is that periodic PV cleaning proved to have a high efficiency and productivity rate. Under the periodic PV cleaning approach, Kazem *et al.* (2020) propose manual cleaning, which depends on the water to remove dust accumulation on the PV surface. The PV surface is covered with tempered borosilicate glass, which is considered easy to clean with water. A tight glass is applicable for the newly installed PV system to help absorb light and the impacts of less reflectivity. The second proposed cleaning methods include semiautomatic roads that require a high-quality large quantity of water and excess labor.

Alvarez *et al.* (2020) reflect many contributions to the optimization of the PV panel cleaning schedule by studying an optimal PV panel cleaning strategy based on both module degradation and radiation profile. They affirm that the PV system yield is significantly impacted due to the soiling on the solar panel surface. These soiling changes within the system operation occur because of environmental conditions. According to Alvarez *et al.* (2020), the panel surface needs to be cleaned to reduce the effects of soiling accumulation. An optimal cleaning is considered the best approach within the text to enhance financial soiling rates, system efficiency, and cleaning cost. The accurate estimation of any return on the investment is considered the best practice within the current planning for the PV project to minimize all financial risks. Most of the integrated renewable generation, including the PV, reflects some challenges during their planning, operation, and maintenance. Alvarez *et al.*, (2020) decisions aimed at creating an enhancement of the PV system's performance are considered during planning and operations. The option reflecting the highest performance efficiency cost relation should be chosen for different solar components under the planning stage. The significant factor under control with the most decisive influence on the performance is cleaning because of the soiling process during the PV system operation. According to Alvarez *et al.* (2020), the yielding loss of the PV panel because of soiling can exceed 1% per day of operation. The location and time are the determinants of the soiling rate estimates being considered to optimize the PV cleaning operation that can implicate in various challenges. In general, the text proposes a comparison between optimal cleaning frequency and optimal cleaning strategy as the practical methodology for the expected annual profile, PV module degradation during the time, cleaning cost, system efficiency of the system component, and the influence of the soiling rate the PV system production.

The analytical economic under soiling of solar PV is crucial when setting up a solar PV plant. The solar PV plant might operate at the very minimum possible costs hence supplying energy at the most considerable price. Chiteka *et al.* (2020) affirms that studying the soiling characteristic while considering each configuration at a locality helps select the most effective criteria that produce reliable energy at a low cost. PV is considered in Chiteka *et al.* (2020) as an avoidable issue that requires concern primarily when the isolation areas are associated with high levels of soiling. There are several approaches with a vast field of studies that embarks PV soiling mitigation, especially asserting the optimization and optimum of the cleaning cycle, depending on the reliable prediction of soiling (Alvarez *et al.*, 2020). Solar energy continues to gain much effectiveness and application in domestic, agricultural, industrial, and commercial areas considering the nonrenewable energy sources' threats. According to Chiteka *et al.* (2020), soiling is one of the commonly overlooked issues within solar energy that can halt the prospect of adopting solar energy as an alternative energy source. A marginal number of resources has been allocated in several nations in addressing the external factors such as soiling influencing the performances of the solar collectors. That cannot work effectively without a comprehensive and practicable approach that will effectively mitigate the soiling issues.

The general understanding of the dust effects on the solar collectors is not effectively understood, and there are lacks of a practical approach to mitigate its adverse impacts. Chiteka *et al.* (2020) considers physical cleaning as the most effective mitigation approach, but it calls for many resources, resulting in a high operation and maintenance cost. As Chiteka *et al.* (2020) depicts, the cleaning cycle optimization is usually characterized by high dust levels with limited water supply (Kazem *et al.*, 2020). Each location has a unique and applicable method determined by the cost of the operations, considering the optimum cleaning schedule.

Al-Masri *et al.*, (2020) embarked on their research on the impacts of different PV models in considering the models on the design of a combined solar array and pumped hydro storage system. They describe several solar models, including single-diode (SD), two-diode (TD), and the ideal single-diode (ISD), as the best evaluation background for the solar array size, reliability, and ecological effects. According to Al-Masri *et al.* (2020), renewable power plants are among the long-term solutions for the oil-importing nations, which creates a background for futures energy supply. The PV renewable energy system is clean and quiet and offers a modular energy resource.

There are several methods for the PV array to choose from when deciding to design the system. Some of the ideal models include ISD, SD, and the TD approaches. The ISD is considered an idealistic model, from its ignorance of leakage and a series of current leakage to the ground losses. According to Al-Masri *et al.* (2020), the SD model is categorized by combining five lumped parameters, including the light generated current, terminal voltage, shunt resistance, series, and leakage saturation or reverse saturation current. The TD model represents two more factors, including the ideal second diode and the second diode's reverse saturation current.

9.3 Proposed Solution

The proposed system is represented in the block diagram as shown in Figure 9.6. This system contains several subsystems or modules; each employs a specific function; all will be presented in this section in detail, then discussed in the next section.

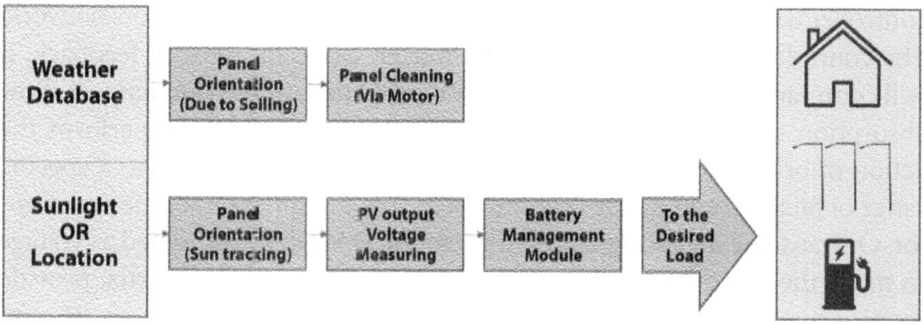

Figure 9.6 The overall block diagram.

The left-hand side of the block diagram presents the parameters that will affect the system as inputs. Including the weather information from the updated database and the amount of sunlight read by the sensors set or the preprogrammed location depends on the tracking mode. However, the designer or the solar engineer can decide the best mode for tracking based on the user or customer requirements, budget, and available tools.

The panel will be orientated using a single-axis 180° servo motor that is a well-known device using in angular positioning applications, toward the east-west direction following the maximum power point criteria either due to LDR sensors or the database information, mainly; GHI and sun location. It is important to remark that the panel(s) will be installed with respect to the calculated tilt angle depends on location in both cases.

Onward, the output voltage of the PV will be measured and compared with other parameters in the PV-to-battery source switching system, also known as a basic battery management system, that will control the power flow from the storage device; the battery, in this case, toward the desired load or from PV to the battery due to charge/discharge process.

In addition, the second possible orientation of the panel(s) can be done due to the soiling ratio and performed by a motor that will move linearly or via a wi-fi device that will send an alarm to the cleaner stuff via the internet of things (IoT) platform.

9.3.1 Single Axis Solar Tracking System

A single-axis solar tracking system is an approach applied in tracking the sun from one side to the others, applying a single point for rotation.

9.3.2 Supervisory Digital Feedback Solar Tracker Control System

This approach is realized by developing an intelligent assembly of a digital controller with light resistance sensors and a servomotor as actuators network. The controller will work as a decision maker, based on sensors' feedback, it will command the motor to orient the PV panel toward the sun that will allow absorption of the maximum possible amount of sunlight. To perform the action of orientation, the ADC will convert the desired voltage to a specific range of bits to compute the optimum step size and direction of the servomotor. Onward, the digital signal will inter to the DAC to be translated as voltage to move the servo motor that has low tolerance moment of inertia, provide more accuracy, and enter back to the loop via the setpoint. The block diagram of the supervisory digital feedback control system is shown in Figure 9.7.

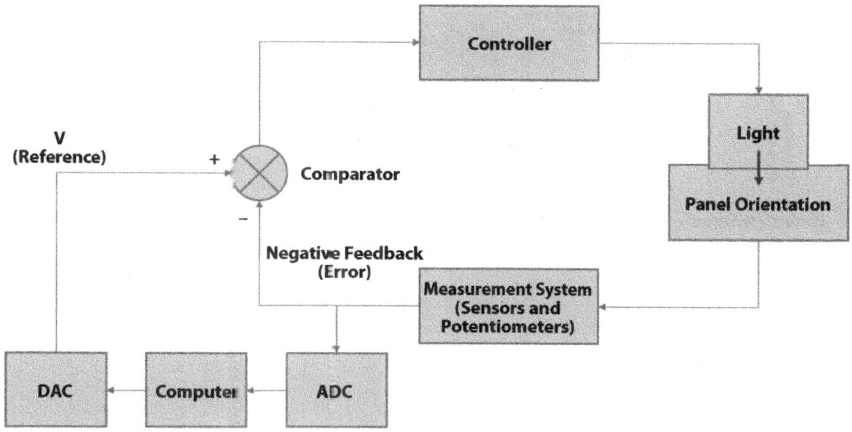

Figure 9.7 Block diagram of the supervisory digital feedback control system.

9.3.3 Database-Based Digital Solar Tracker Control System

This approach is realized by developing an intelligent assembly of a digital controller connected to the updated database from sufficient official association. The controller will fetch the desired data according to the user's input and move it to the computer that will accurately calculate the desired position of the PV depends on its current location and the sun location within the sun path programmed in the database. After that, the DAC will translate this command to voltage signals that can actuate the panel toward the best angle, and the loop will run each 30 days that is the average time to have an effective change in the sun location in most of the middle east regions. The block diagram of the digital control system is shown in Figure 9.8.

9.3.4 Soiling Treatment Module

This module has the same logic as the previous module. The only difference here is the desired time of loop and the panel's orientation; the servo motor may rotate 360° to move the panel to a certain pre-programmed angle that facilitates the manual cleaning process. The block diagram of the soiling treatment module is shown in Figure 9.9.

9.3.5 PV-to-Battery Switching Module

Basically, the proposed system aims to maintain the excellent status of safety and the lifetime of the system while the battery operates charge or

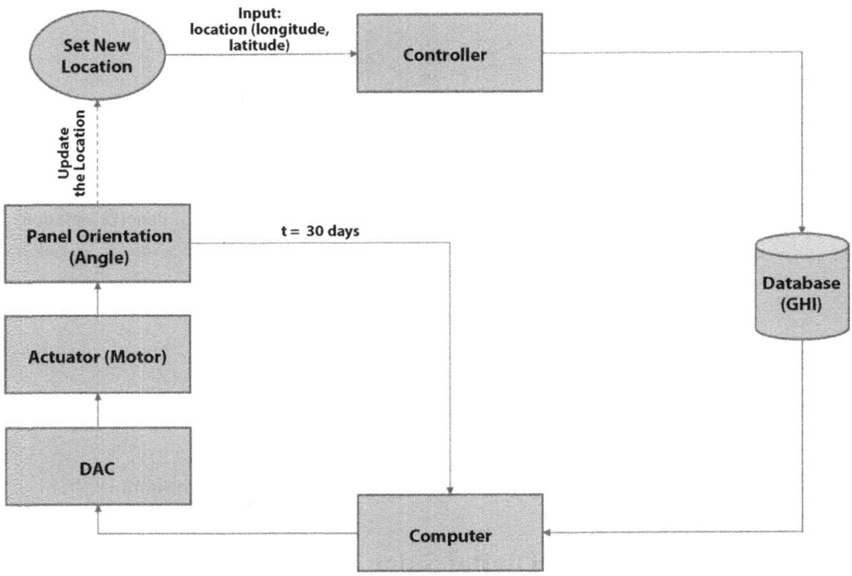

Figure 9.8 Block diagram of the digital control system.

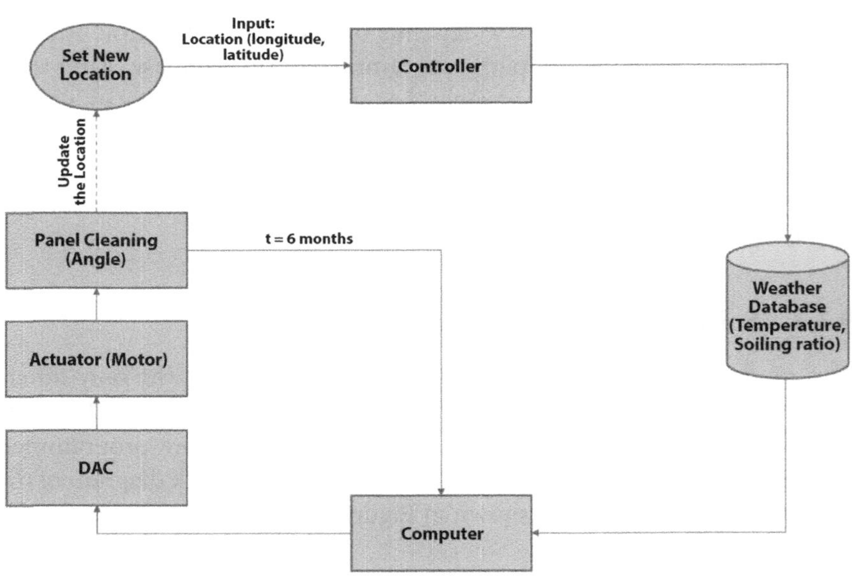

Figure 9.9 Soiling treatment module block diagram.

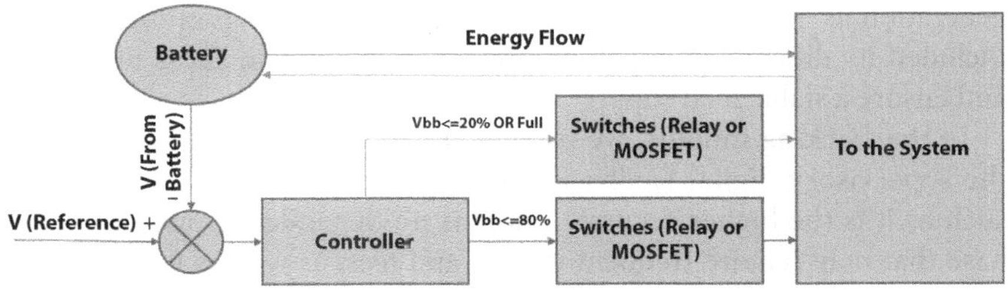

Figure 9.10 PV-to-battery switching module block diagram.

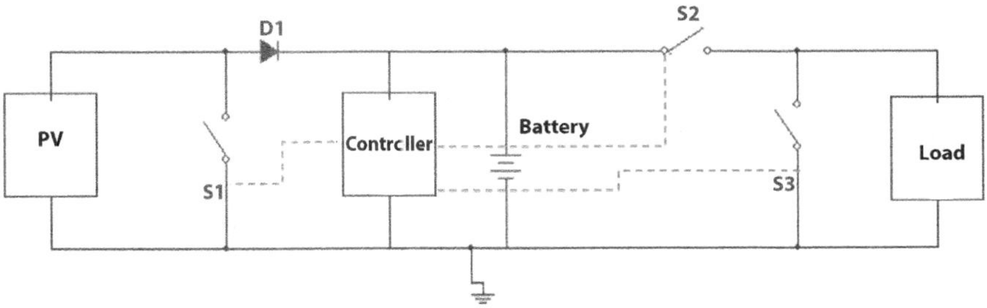

Figure 9.11 PV-to-battery switching module circuit diagram.

discharge processes. It will operate in the same manner as the series controller with minor differences; hence, the charger is connected to the shunt switches as well.

If the battery state of charge is above the threshold of 80%, then the S1 switch will disconnect the PV, while the S2 and S3 switches will connect the load, meaning that the battery is in discharging process. While in case of the battery state of charge is reaching the threshold of 20%, then switch S1 will connect the PV while the S2 and S3 switches will disconnect the load, then the battery will charge through the series charge controller. Note that, at the same time, when the battery charge reaches its maximum that is near 100%, then a shunt controller will short circuit, and the series charge controller will open the PV(s) and block the current to flow toward the battery. The block diagram and circuit diagram of the PV-to-battery switching module is shown in Figure 9.10 and Figure 9.11, respectively.

9.4 Discussion

This considering, the mounted solar photovoltage panels stand-alone or connected to the grid systems, and the major limitation of PV-based generation is weather and daylight dependence. The proposed system included its different modules to provide a solution to these limitations and ensure a stable load supply.

In the tracking module, we presented two different approaches. First, the supervisory digital feedback control system has several advantages, such as it is the higher accurate, there is no dependency upon the database that may require frequent update and fees, as well as network connection and memory utilization. However, the motor may consume more electricity compared with the second approach. It is the best solution for low-cost low power requirements' offline systems. On the other hand, the digital control system connected with the database will be faster in terms of the processor time response as well as the actuator execution that make it compatible with the larger PV solar systems containing critical devices as the load.

In the second module, the level of soiling ratio is measured and should not above a certain threshold. Otherwise, a cleaning notification is generated and sent to the maintenance center due to the IoT platform, or the motor will automatically clean the PV surface.

In the third module, the PV-to-battery Switching system will manage the current flow between different parts of the solar system: from the PV to the battery or from the battery to the desired load based on the battery's state of charge. The benefit of a series controller is its input source consistency. The charging performance is poor due to the fact that the charging current will always pass through two switches. On the other hand, shunt regulators can only be used in photovoltaic systems that produce less than 20 Amperes. This restriction occurs because the S1 switch dissipates a large amount of electricity, which can result in a considerable amount of heat. However, by joining the two systems together, the main feature is to energize the connected devices or loads at any instant of the time, taking advantage of both series and shunt controllers. In addition, the efficiency of the battery will maintain a reasonable at low cost.

The research reflects a considerable technological improvement for the solar system as the most effective renewable power source for most developing nations. The methodology would be effective in advance for the research by considering the impacts of PV modeling. These impacts would effectively evaluate the system on the crucial ecological effects within the area under the study. The limitation is reflected within the analysis based on the recombination losses, temperatures, and series-leakage losses. Our system consists of several components, such as the microprocessor that command the motors, and the available PV panels. There will be some power losses, primarily considering the connection. Generally, the research reflects an endeavor reflection of PV system analysis that is effective for the current study fields and future references.

9.5 Conclusion

The research has reflected on the insights aimed to ensure that the amount of sunlight absorbed by the PV panel and an area's hours is maximum. The study affirms that the best approaches to mitigate soiling and leakages are practical for optimizing the system's operational output. Cleaning is considered an effective way to alleviate the adverse implication caused by soiling within the system. The comprehensive review affirms the methodologies applicable for ensuring maximum operational efficiency and cost-effectiveness of the PV system. Estimating a visual PV panel cleaning approach based on annual module degradation and annual radiation creates the best and most applicable technoeconomic optimization model. Batteries are considered within the research as the influential power tank for the solar power system, playing the role of the power supply when the sun does not shine. A significant part of this research has reflected on the review of the battery changing and its control for extended services of the PV system. The proposed system has a considerable improvement compared to other traditional systems in sunlight tracking and soiling treatment. For the near current development, most of the research needs to extend on effective cleaning, search, and charging controller to complement the effectiveness of the PV panel. There is also a gap in a practical cleaning approach extending to the construction of the ideal PV solar having inbuilt properties to minimize the adversity of cracking, soiling, and low efficiency to the solar surfaces. Generally, the research has made a reference for improving the light tracker's targeted solar control systems. The incorporation of event-driven methods can improve the performance of these solutions (S. M. Qaisar *et al.*, 2009), (Mina Qaisar *et al.*, 2018). Investigation of this approach is another prospect.

References

Abu Eldahab, Y.E., Saad, N.H., Zekry, A., Enhancing the maximum power point tracking techniques for photovoltaic systems. *Renew. Sust. Energ. Rev.*, 40, 505–514, 2014. https://doi.org/10.1016/j.rser.2014.07.202.

Abu Eldahab, Y.E., Saad, N.H., Zekry, A., Enhancing the design of battery charging controllers for photovoltaic systems. *Renew. Sustain. Energ. Rev.*, 58, 646–655, 2014. https://doi.org/10 1016/j.rser.2015.12.061.

Al-Masri, H., Magableh, S., Abuelrub, A., Saadeh, O., Ehsani, M., Impact of Different Photovoltaic Models on the Design of a Combined Solar Array and Pumped Hydro Storage System. *Appl. Sci.*, 10, 3650, 2020. https://doi.org/10.3390/app10103650.

Alvarez, D.L., Al-Sumaiti, A.S., Rivera, S.R., Estimation of an Optimal PV Panel Cleaning Strategy Based on Both Annual Radiation Profile and Module Degradation. *IEEE Access*, 8, 63832–63839, 2020. https://doi.org/10.1109/ACCESS.2020.2983322.

Arachchige, U. and Weliwaththage, S. R. G., Solar Energy Technology, *Journal of Research Technology and Engineering*, 1, 3, 2020.

Chiteka, K., Arora, R., Sridhara, S.N., Enweremadu, C., Cleaning cycle optimisation in non-tracking ground mounted solar PV systems using Particle Swarm Optimisation. *Int. J. Simul. Multidiscip. Des. Optim.*, 11, 9, 2020. https://doi.org/10.1051/smdo/2020004.

Ellabban, O., Abu-Rub, H., Blaabjerg, F., Renewable energy resources: Current status, future prospects and their enabling technology. *Renew. Sust. Energ. Rev.*, 39, 748–764, 2014. https://doi.org/10.1016/j.rser.2014.07.113.

Hafez, A.Z., Yousef, A.M., Harag, N.M., Solar tracking systems: Technologies and trackers drive types – A review. *Renew. Sust. Energ. Rev.*, 91, 754–782, 2018. https://doi.org/10.1016/j.rser.2018.03.094.

Han, X., Ji, T., Zhao, Z., Zhang, H., Economic evaluation of batteries planning in energy storage power stations for load shifting. *Renew. Energy*, 78, 643–647, 2015. https://doi.org/10.1016/j.renene.2015.01.056.

Heydt, G.T., Ayyanar, R., Hedman, K.W., Vittal, V., Electric Power and Energy Engineering: The First Century. *Proc. IEEE*, 100, Special Centennial Issue, 1315–1328, 2012. https://doi.org/10.1109/JPROC.2012.2187130.

Investment in renewable energy, by technology, 2017, Our World in Data, University of Oxford, Oxford, United Kingdom. https://ourworldindata.org/grapher/investment-in-renewable-energy-by-technology, University of Oxford, Oxford, United Kingdom.

Rajan, K., Solar tracking system- A review. *Int. J. Sustain. Eng.*, 10, 2, 2016. https://doi.org/10.1080/19397038.2016.1267816.

Kazem, H.A., Chaichan, M.T., Al-Waeli, A.H.A., Sopian, K., A review of dust accumulation and cleaning methods for solar photovoltaic systems. *J. Clean. Prod.*, 276, 123187, 2020. https://doi.org/10.1016/j.jclepro.2020.123187.

Malinowski, M., Leon, J., II, Abu-Rub, H., Solar Photovoltaic and Thermal Energy Systems: Current Technology and Future Trends. *Proc. IEEE*, 105, 11, 2132–2146, 2017. https://doi.org/10.1109/JPROC.2017.2690343.

Mian Qaisar, S. and Fawad Hussain, S., Arrhythmia diagnosis by using level-crossing ECG sampling and sub-bands features extraction for mobile healthcare. *Sensors*, 20, 8, 2252, 2020.

Mina Qaisar, S., Sidiya, D., Akbar, M., Subasi, A., An Event-Driven Multiple Objects Surveillance System. *Int. J. Electr. Comput. Eng. Syst.*, 9, 1, 35–44, 2018.

Mohamed, M.A.A., Elmahalawy, A.M., Harb, H.M., Developing the pulse width modulation tool (PWMT) for two timer mechanism technique in microcontrollers. *2013 Second International Japan-Egypt Conference on Electronics, Communications and Computers (JEC-ECC)*, pp. 148–153, 2013, https://doi.org/10.1109/JEC-ECC.2013.6766403.

Qaisar, S., Fesquet, L., Renaudin, M., Computationally efficient adaptive resolution short-time Fourier transform. *J. Electr. Comput. Eng.*, Hindwi, 2008. https://doi.org/10.1155/2008/932068.

Qaisar, S.M., Event driven filtering an intelligent technique for activity and power consumption reduction. *Int. J. Circuits Syst. Signal Process.*, 8, 321–329, 2014.

Qaisar, S.M., A two stage interpolator and multi threshold discriminator for the Brain-PET scanner timestamp calculation. *Nucl. Instrum. Methods Phys. Res. A*, 922, 364–372, 2019.

Qaisar, S.M., Baseline wander and power-line interference elimination of ECG signals using efficient signal-piloted filtering. *Healthc. Technol. Lett.*, 7, 4, 114–118, 2020.

Qaisar, S.M., Fesquet, L., Renaudin, M., *Effective resolution of an adaptive rate ADC*, Special-session. 2009.

Qaisar, S.M. and Subasi, A., Cloud-based ECG monitoring using event-driven ECG acquisition and machine learning techniques. *Phys. Eng. Sci. Med.*, 43, 2, 623–634, 2020.

Qaisar, S.M. and Subasi, A., *An adaptive rate ECG acquisition and analysis for efficient diagnosis of the cardiovascular diseases*, pp. 177–181, 2018.

Qazi, A., Hussain, F., Rahim, N.A.B.D., Hardaker, G., Alghazzawi, D., Shaban, K., Haruna, K., Towards Sustainable Energy: A Systematic Review of Renewable Energy Sources, Technologies, and Public Opinions. *IEEE Access*, 7, 63837–63851, 2019. https://doi.org/10.1109/ACCESS.2019.2906402.

Shaikh, M.R., Shaikh, S., Waghmare, S., Labade, S., Tekale, A., A Review Paper on Electricity Generation from Solar Energy. *Int. J. Res. Appl. Sci. Eng. Technol.* 5, IX, 2017. https://doi.org/10.22214/ijraset.2017.9272.

Silicon Chip Online, Build A 12/24V 3-Stage Solar Charge Controller. (n.d.). Retrieved April 28, 2021, from http://archive.siliconchip.com.au/cms/A_112335/article.html.

Soga, K. and Akasaka, H., Influences of Solar Incident Angle on Power Generation Efficiency of PV Modules under Field Conditions. *J. Asian Archit. Build. Eng.*, 2, 2, 43–48, 2003. https://doi.org/10.3130/jaabe.2.b43.

Northern Arizona Wind & Sun. (n.d.), Solar Charge Controller Basics. Retrieved April 10, 2021, from https://www.solar-electric.com/learning-center/solar-charge-controller-basics.html/.

Our World in Data, Solar PV module prices, 2017. https://ourworldindata.org/grapher/solar-pv-prices.

Spanos, C., Turney, D.E., Fthenakis, V., Life-cycle analysis of flow-assisted nickel zinc-, manganese dioxide-, and valve-regulated lead-acid batteries designed for demand-charge reduction. *Renew. Sust. Energ. Rev.*, 43, 478–494, 2015. https://doi.org/10.1016/j.rser.2014.10.072.

SRNE, Different Types of Solar Charge Controllers. SRNE Solar, April 10 2021. https://www.srnesolar.com/blog/power-inverter-from-dc-to-acdcto-ac-converter.

Syafiq, A., Pandey, A.K., Adzman, N.N., Rahim, N.A., Advances in approaches and methods for self-cleaning of solar photovoltaic panels. *Sol. Energy*, 162, 597–619, 2018. https://doi.org/10.1016/j.solener.2017.12.023.

Tripanagnostopoulos, Y., Nousia, T., Souliotis, M., Yianoulis, P., Hybrid photovoltaic/thermal solar systems. *Sol. Energy*, 72, 3, 217–234, 2002. https://doi.org/10.1016/S0038-092X(01)00096-2.

The International Energy Agency, *World Energy Outlook 2009 – Analysis*, 2009, https://www.iea.org/reports/world-energy-outlook-2009.

Yang, D., Yagli, G.M., Quan, H., Quality Control for Solar Irradiance Data. *2018 IEEE Innovative Smart Grid Technologies - Asia (ISGT Asia)*, pp. 208–213, 2018, https://doi.org/10.1109/ISGT-Asia.2018.8467892.

Yao, M. and Cai, X., An Overview of the Photovoltaic Industry Status and Perspective in China. *IEEE Access*, 7, 181051–181060, 2019. https://doi.org/10.1109/ACCESS.2019.2959309.

Zell, E., Gasim, S., Wilcox, S., Katamoura, S., Stoffel, T., Shibli, H., Engel-Cox, J., Subie, M.A., Assessment of solar radiation resources in Saudi Arabia. *Sol. Energy*, 119, 422–438, 2015. https://doi.org/10.1016/j.solener.2015.06.031.

10

Performance Analysis of Tunnel Field Effect Transistor for Low-Power Applications

Deepak Kumar[1]*, Shiromani Balmukund Rahi[2] and Neha Paras[3]

[1]Department of Electrical and Electronics Engineering, University of Petroleum and Energy Studies Dehradun, India
[2]Mahamaya College of Agriculture Engineering and Technology Akabarpur Ambedkar Nagar, Uttar Pradesh, India
[3]Department of Electronics and Communication Engineering, Dr. B R Ambedkar National Institute of Technology Jalandhar, India

Abstract

Tunnel FET is a field-effect transistor, conceived from a P-I-N diode structure. It consists of a lower subthreshold slope than conventional MOSFET due to band-to-band tunneling. Due to the lower subthreshold slope, it has been advocated as a prospective low-power device. This book chapter presents double gate tunnel FET (DGTFETs), drain-voltage (I-V), and capacitance-voltage (C-V) characteristics analysis using technology computer-aided design simulations (TCAD). The digital performance of DGTFETs has been further investigated by incorporating a purdue emerging technology evaluator (PETE) and nanohub simulation tool. It has been observed that the charge carriers' injection in the channel region is due to the band-to-band tunneling phenomenon, which is responsible for the effective coupling of the gate charge with th e drain region, hence may increase the miller capacitance of DGTFET. In addition, to overcome the lower limit of on-sate current, hetero channel DGTFET consisting of $Si_{1-x}Ge_x$ source(x = 40%) has been investigated. The investigation of the performance of $Si_{0.6}Ge_{0.4}$/Si/Si has been observed for HfO_2 gate dielectric material. The average SS (*Avg.*-SS) ~ 37.51 mV/decade at 300 K, while the OFF-state current (I_{OFF}) of ~10^{-18} A/μm drive ON-state current (I_{ON}) of ≈ 4.50×10^{-5} A/μm is obtained for HfO_2 (ε_r ~ 21) as the double gate material.

Keywords: MOSFET, subthreshold swing (SS), power supply, Tunnel FET, High-k, DC performance, analog, PETE, ITRS

**Corresponding author*: d.kumar@ddn.upes.ac.in

Suman Lata Tripathi, Souvik Ganguli, Tengiz Magradze, and Abhishek Kumar (eds.) Intelligent Green Technologies for Sustainable Smart Cities, (195–226) © 2022 Scrivener Publishing LLC

10.1 Introduction

The modern VLSI (very large scale integrated) industry widely uses integrated circuits (ICs) based on conventional MOSFET technology for the development of electronic circuits and systems. The International Technology Roadmap for Semiconductors (ITRS) suggests that scaling down (~up to certain limits) the device dimension of MOSFET increases the transistor's density without affecting the circuit performance [1–6]. The aggressive demand for longer battery life and better performance particularly for mobile devices, circuits, and systems has been rapidly increased in the last few decades. To scale down the conventional MOSFET (while maintaining the gate control), the channel length and the oxide thickness of the device require scaling down by 1/K while the substrate doping is scaled up by K, where K is a scalar constant. The aforementioned dimensional scaling enables the applied voltage scaling down by 1/K. Such scaling is referred to as the Dennard scaling [7]. Scaling down the supply voltages (V_{DD}) can be considered an effective approach to reduce the dynamic power consumption followed by $P_d \sim C_L \times V_{DD}^2$ in conventional VLSI technology to implement ultra-low-power applications (ULPA) for circuits and systems. The aggressive power scaling is becoming, one of the biggest challenges for modern MOSFETs technology associated with ultralow-power applications (ULPA) such as portable electronics, Internet of Things (IoT) technologies, and wearable electronics [8–10].

The trend of the supply voltage (V_{DD}) and threshold voltage (V_{TH}) scaling for various technology nodes is shown in Figure 10.1(a). It is easily depicted that the power supply V_{DD} decreases gradually with device dimensions, but V_{TH} shows insignificant variations.

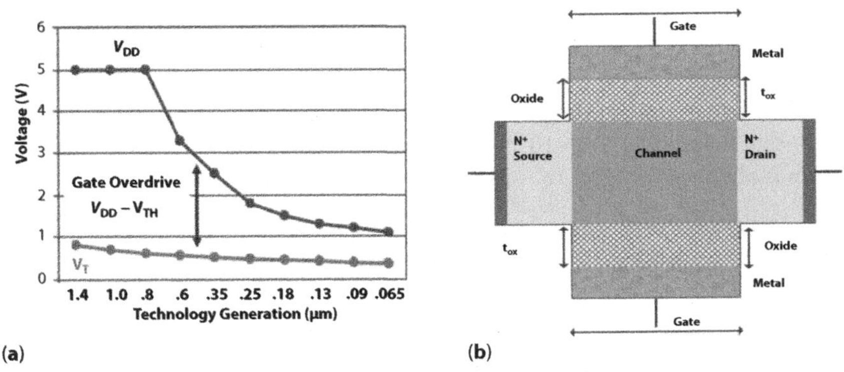

Figure 10.1 (a) Supply and threshold voltage scaling versus technology generation trend [7]. (b) N- channel double-gate MOSFET.

It has been observed from Figure 10.1(a) that as the channel length reduces the linear dependencies between V_{DD} and V_{TH} are vanishing. Figure 10.1(b) shows the schematic diagram for N-channel double-gate MOSFET. As shown in Figure 10.1(a), the V_{DD} is reduced by 80 % of its initial value whereas V_{TH} shows a reduction only by 50%. The most important consequence of scaling is that the gate overdrive voltage ($V_{DD} - V_{TH} \sim V_{GS} - V_{TH}$) does not vary much at advanced technology nodes. When $V_{DD} - V_{TH}$ decreases, the I_{ON} may also reduce, which negatively affects device performances such as a rise in I_{OFF} and dynamic speed ($C \times V_{DD}/I_{ON}$). The low I_{ON} can be increased for advanced technology nodes by (i) increasing V_{DD}; (ii) reducing V_{TH}. Increasing the power supply V_{DD} may lead to enhance dynamic power ($\sim P_{dyamic} \sim C_L \times V_{DD}^2$) during switching operations and leakage power ($P_{leakage} \sim I_{OFF} \times V_{DD}$) consumption when MOSFET is in an off state respectively. It is clear from the stated expressions that both static and dynamic power consumption of the devices increases with V_{DD} [11–20]. The V_{TH} is needed to be scaled down proportionally as the V_{DD} to achieve optimum I_{ON} to maintain the satisfactory circuit performance of CMOS devices. The transistor density is increasing for advanced technology nodes, which may also lead to an increase in the static leakage power consumption in electronic devices and systems as shown in Figure 10.2. The scaling down V_{TH} further increases I_{OFF} significantly as seen from Figure 10.3. This increases static leakage power dissipation followed by $\sim V_{DD} \times I_{OFF}$ across integrated circuits (ICs).

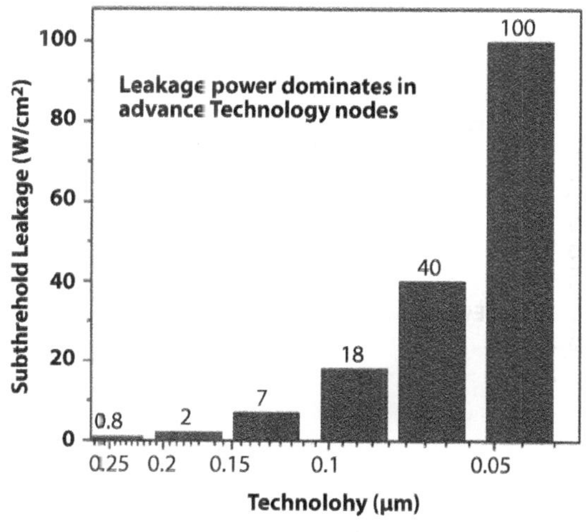

Figure 10.2 Sub-threshold leakage versus technology, showing the leakage trend with scaling [15].

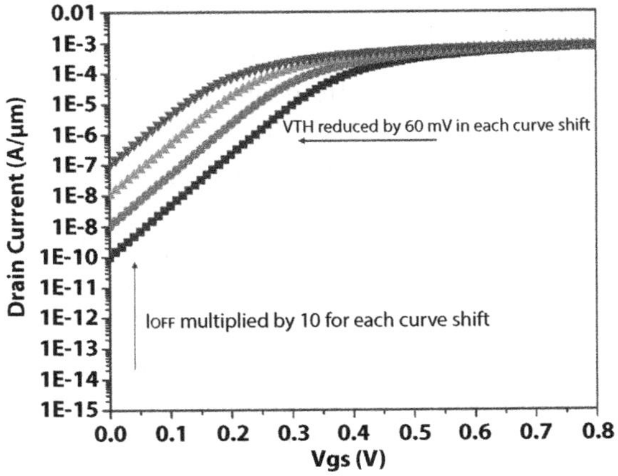

Figure 10.3 A typical $I_d - V_g$ curve for MOSFET showing leakage current is increased by 10 times for every 60 mV decrease in threshold voltage V_{TH}.

The transfer characteristics (I_{ds}–V_{gs}) of N-MOSFET for gate length = 30 nm, t_{ox} (HfO_2) = 1.0 nm, work function (WF) = 4.2 eV is shown in Figure 10.3. For an average 60 mV reduction in V_{TH}, the off current (I_{OFF}) i.e. sub-threshold current increases by 10 times its initial value which also results in increasing the leakage power consumption in electronic devices. The simulation results depicting the energy levels across the body can be seen from Figure 10.4. During the OFF state (i.e. V_{gs} = 0.0 V, V_{ds} = 0.75 V), the source-channel junction barrier is preventing the thermionic emission of

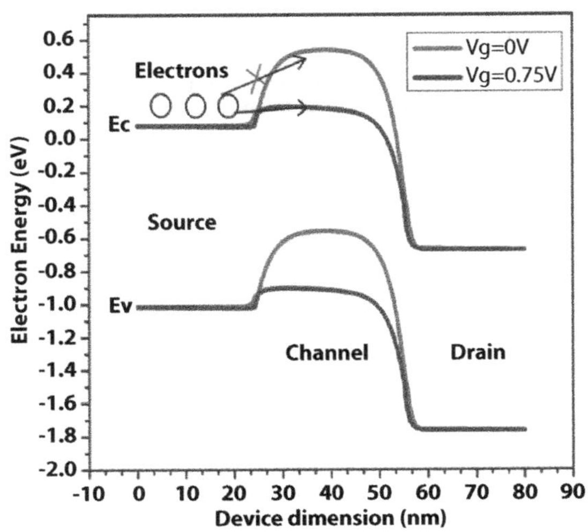

Figure 10.4 Energy band diagram of N-MOSFET, WF = 4.2 eV.

electrons. In ON-state ($V_{gs} = V_{ds} = 0.75$ V), reduced source-channel energy barrier width enables the thermionic emission of carriers creating the drain current (I_{ds}). The property of thermionic emission of electrons does not allow the SS to go below 60 mV/decade for MOSFETs as given by Eqn. 10.1.

10.1.1 Limitation of Conventional MOSFET

In Eqn. 10.1, the C_{dep} is defined as depletion capacitance and C_{ox} is the oxide capacitance for the conventional MOSFET. The minimum SS value is restricted to 60 mV/decade for a good gate-controlled MOSFET i.e. for $C_{dep} << C_{ox}$. So, to achieve a larger value of $log_{10}I_{dS}$ (decade change in ON-state current of the device), we need higher values of V_{GS} for the device with an SS of 60 mV/decade. Thus, this inability of MOSFET to achieve SS < 60 mV/decade limits its gate transfer characteristics to improve further. The ideal numerical value for conventional MOSFET at room temperature is 60 mV/decade, which is written as Eqn. 10.1. Then Eqn. 10.1 indicates MOSFETs suffer from high leakage current (I_{OFF}), SS limit, velocity saturation, and short channel effects (SCEs) [11, 12].

$$SS = \left(\frac{d(log_{10} I_{ds})}{dV_g} \right)^{-1} = 2.3 \frac{k_B T}{q} \left(1 + \frac{C_{dep}}{C_{ox}} \right) \tag{10.1}$$

For room temperature Equation 10.1 can, be written as

$$SS = \frac{\partial V_g}{\partial \Psi_s} \frac{\partial \Psi_s}{\partial (\log I_d)} = \left(1 + \frac{C_{dep}}{C_{ox}} \right) \times \frac{kT}{q} \ln 10 \rightarrow 60 \frac{mV}{decade} \,|\, T = 300K \tag{10.2}$$

In Eqn. 10.1, k_B is Boltzmann constant having numerical value ~1.380649×10^{-23} J/K and q is electronic charge having numerical value ~1.6×10^{-19} C [20–30].

10.1.2 Subthreshold Slope Devices

The devices with steep SS are becoming essential to achieving voltage scaling for a small leakage current (I_{OFF}) for maintaining the power dissipation, ultralow-power uses in IoTs. Kinds of literature for development in low-power electronics reported some steep slope devices summarized in Figure 10.5.

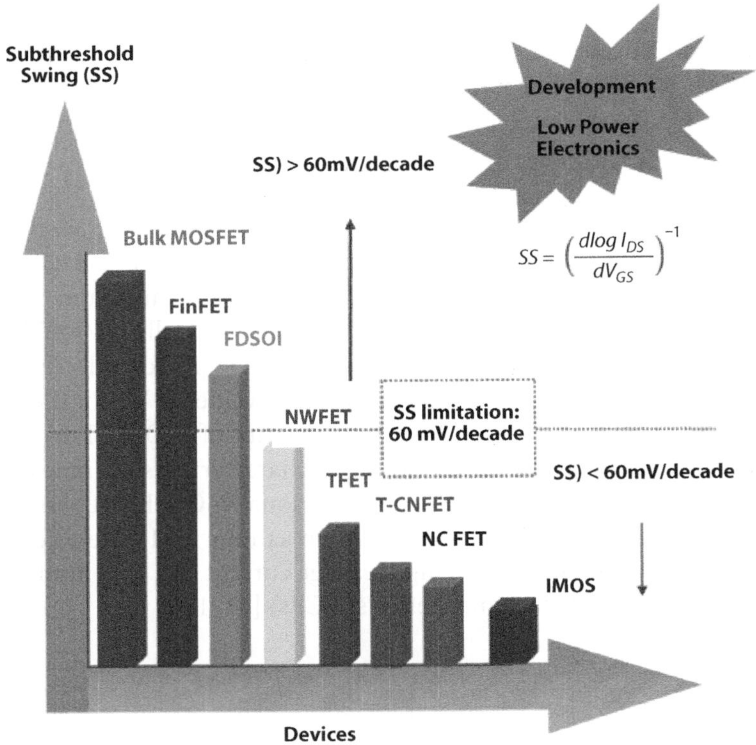

Figure 10.5 SS improvement for various FET devices.

It has been observed that conventional field-effect transistors (i.e., Bulk, Fin FET, and Fully-depleted Silicon-on-insulator (FDSOI)) have SS > 60 mV/decade, while nanowire field-effect transistor (NWFET), carbon nanotube-based TFET (T-CNTFET) impact ionization MOS-based transistor (I-MOS), negative capacitance (NC) FETs have SS < 60 mV/decade as trend shown in Figure 10.5. Among the listed steep slope devices, Tunnel FET (i.e. TFET) and Negative Capacitance FET (i.e. NCFET) are most popular due to compactable process technology. The advantages of Tunnel FET are low I_{OFF} which leads to low leakage per device and its high I_{ON}/I_{OFF} ratio can be suitable for ultralow-power and memory applications [25–30]. In the present book chapter, authors have dedicated Tunnel FET and a brief application window.

Eqn. 10.2 can be written as

$$SS = \frac{\partial V_g}{\partial \Psi_s} \frac{\partial \Psi_s}{\partial(\log I_d)} = m \times n \tag{10.3}$$

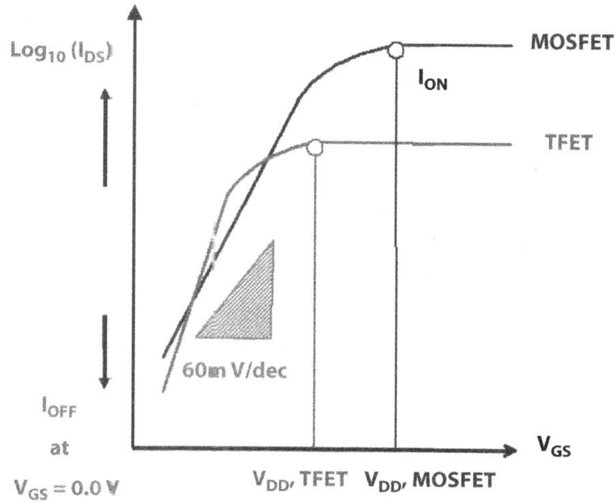

Figure 10.6 Transfer characteristics comparison of TFET with Bulk MOSFET.

In Eqn. 10.3, the terms m and n are known as body factor and transport factor. In tunnel FET, SS is controlled by transport factor 'n', known as tunneling. While in the case of negative capacitance FET, SS is controlled by body factor "m" is less than 1, followed by Eqn. 10.4.

$$m = \frac{\partial V_g}{\partial \Psi_s} = 1 + \frac{c_s}{c_{ins}} > 1 \tag{10.4}$$

Figure 10.6 shows the comparison of current-voltage characteristics between conventional MOSFET and Tunnel FET. From Figure 10.6, it has been observed that at the same operating condition, tunnel FET shows lower SS and off-sate current and conventional MOSFET i.e., $I_{OFF\text{-}TFET} < I_{OFF\text{-}MOSFET}$ and steep SS i.e., $SS_{TFET} < SS_{MOSFET}$. This indicates lower power dissipation than conventional FET structures. This is an indication for the replacement of conventional FET devices, circuits, and systems with lower SS devices for smart low-power applications in IoTs and memory design [31–50].

10.2 TFET Structure and Simulation Setup

The TFET has a P-I-N type structure analogous to conventional MOS transistors, comprising of an intrinsic channel that separates the source and drain regions. A metal gate on top of the channel separated by a thin dielectric layer controls the channel potential. The major differences between these two FET devices that bring the steep change in their I-V/C-V characteristics are charge transport and working principle. TFETs are based on quantum-mechanical tunneling of charge carriers whereas conventional MOSFETs are based on thermionic drifts diffusion transport of charges. To analyses, the charge transport properties of the TFET, a DGNTFET and DGPTFET structure are considered for the study, shown in Figure 10.7(a–b) with the dimensions of the gate length L_g = 30 nm, body thickness t_{si} = 7.0 nm, gate oxide thickness t_{ox} =1.0 nm; body dielectric ε_{si}= 11.8 ε_o, oxide dielectric ε_{ox} = 21.0 ε_o (HfO_2). For NTFET, the source is uniformly doped with Boron (B) of 1.0×10^{20} cm^{-3}, doping of Phosphorus (P) of 1.0×10^{20}cm^{-3} is done at the drain region and B of 1.0×10^{16}cm^{-3} is doped in the channel region respectively. Similarly, for PTFET, the source is uniformly doped with P of 1.0×10^{20} cm^{-3}, doping of B of 1.0×10^{20} cm^{-3} is done at the drain region and B of 1.0×10^{16} cm^{-3} is doped in the channel region respectively. TCAD simulations are performed to carry out the electrical characteristic (I-V/C-V) by considering dc and ac signal at 5 MHz) analysis. Nonlocal B2BT model, Fermi statistics, SRH recombination, Poisson's, and continuity equations are coded in the simulation script to achieve device simulation. Simulations also considered gate work function and the width of the device of 4.2 eV and 1.0 μm, respectively. A similar simulation model is used for devices listed in Figures 10.14a to 10.14c.

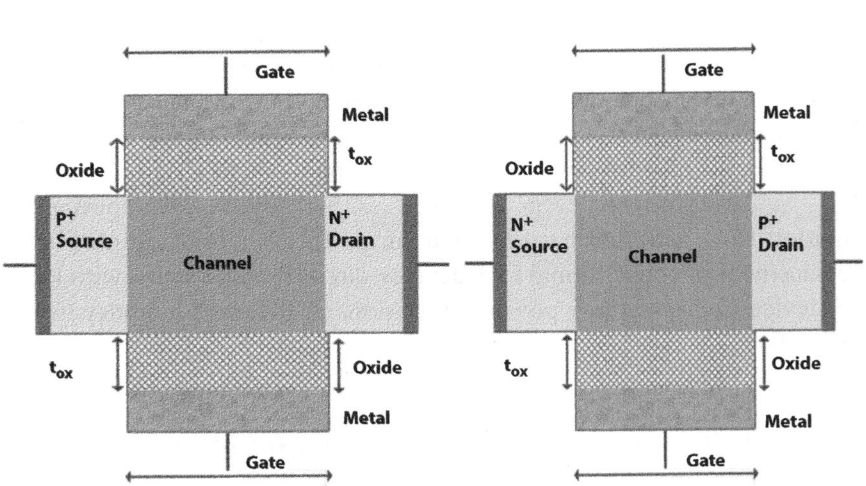

Figure 10.7 (a) Device structure of N-type DGTFET. (b) Device structure of P-type DGTFET.

10.3 TFET Working Principle

For better electrostatics using HfO_2 as a gate dielectric, the strength of the electric field below the gate dielectric is increased, which increases the tunneling probability of charge carriers. The 2D TCAD simulations are performed for the device shown in Figure 10.8 to carry out the energy band diagrams to investigate the tunneling of electrons. At V_{gs} = 0 V and V_{ds} = 0.75 V, the wider source-channel tunnel junction barrier offers low carrier transmission (T_t) resulting very low magnitude of I_{OFF} (~10^{-14} A/μm). For biasing voltage $V_{gs} = V_{ds}$ = 0.75 V, the electron tunneling gets initiated after pushing down the energy bands of the intrinsic region hence enhanced T_t across the valence band of the N+ source region (VB) and the conduction band (CB) of the intrinsic channel region leading to current conduction. The amount of drain current (I_{ds}) is majorly determined by the energy barrier width associated with Band-to-Band (BTB) Tunneling for NTFET and an exponential increase in current is observed when this barrier width gets reduced [16–21, 52] as shown in Figure 10.9.

With increasing V_{gs}, the energy bands in the channel shift downward. The continuous sweeping up of the V_{gs} causes the CB of the channel to move more downward, which further reduces the barrier at the source-channel junction. This initiates the electrons to tunnel an insufficient number from VB of the source to CB of the channel and the drain current starts rising due to tunneling. Here, the V_{TH} can be defined by V_{gs} where the I_{ds} = 100 nA.

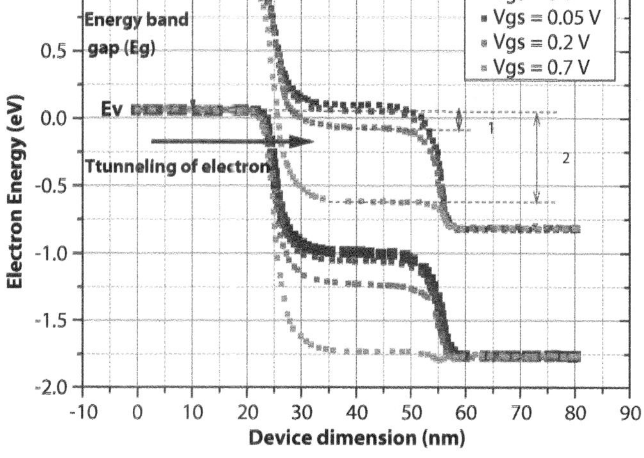

Figure 10.8 Simulated energy band diagram of N-type TFET, work function = 4.2 eV.

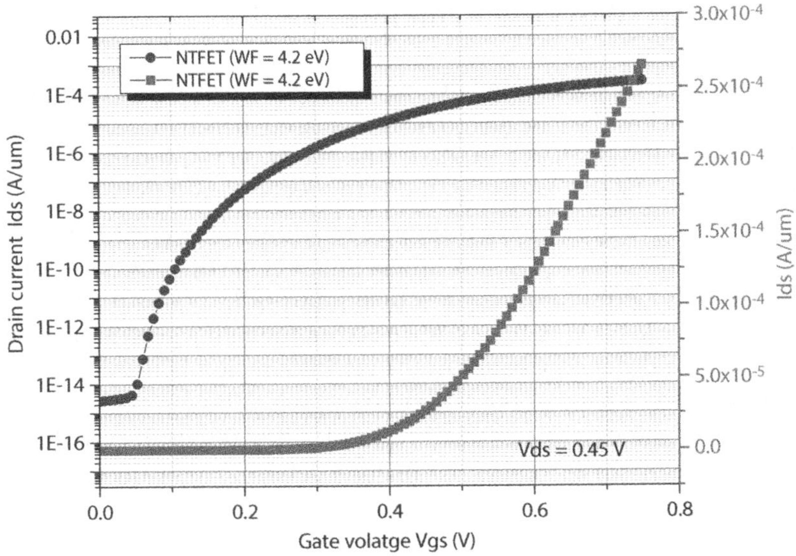

Figure 10.9 Transfer characteristics of double gate TFET.

Figure 10.10 presents the C-V characteristics simulation of N-TFET to analyze the miller capacitance (C_M) in TFET. The gate capacitance (C_{gg}) partitioning in the TFET significantly differs from that of the MOSFET because of the fundamental difference of the charge conduction mechanism. The coupling of gate charge with source and drain is equal in MOSFETs but is

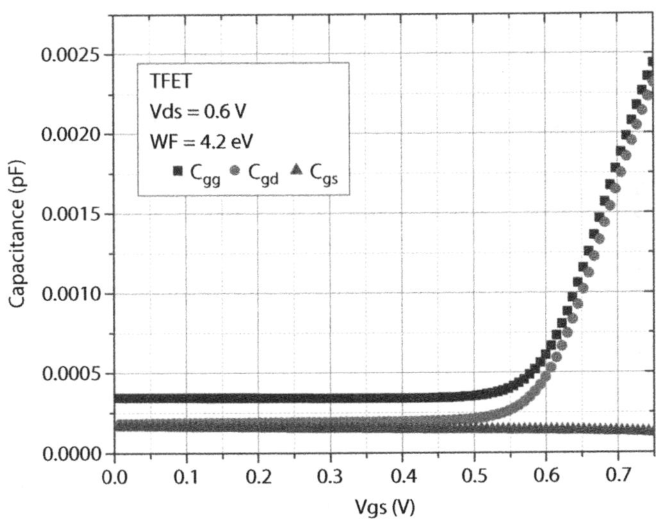

Figure 10.10 Simulated CV plot of Tunnel FET showing C_{gd} follows C_{gg}.

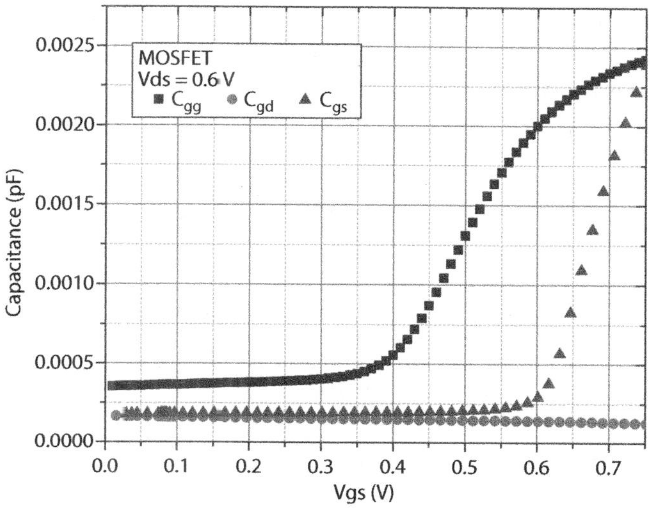

Figure 10.11 C-V characteristics of N-MOSFET.

unequal in TFETs. The gate charge coupling with drain is predominant as compared to source due to the mechanism of BTBT decoupling the gate charge from the source. This results in high gate-drain capacitance C_{gd} [22, 46]. It is seen from the abovementioned figure that C_{gd} is following the C_{gg}. This results in the high miller capacitance in TFETs. It will surely affect the dynamic transient responses like frequency response and high peak overshoots during the transient response of TFET digital circuits. These types of drawbacks can be improved by modulating device geometry and by employing a hetrojunction source-channel interface with high k dielectrics.

Figure 10.11 shows the C-V characteristics of N-MOSFET. There is equal channel charge coupling to the source and drain in MOSFETs and thus, C_{gd} and C_{gs} contribute equal capacitance for C_{gg} ($C_{gg} = C_{gd} + C_{gs}$) in the linear region, i.e. $C_{gd} = C_{gs}$. In the saturation region, $C_{gd} \sim 0$ and C_{gs} become 2/3 C_{gg}. C_{gg}, $C_{gs,}$ and C_{gd} are termed as the gate, gate-source, and gate-drain capacitance. It is also clearly depicted that C_{gd} is not following the C_{gg} which means MOSFET provides the lower magnitude of miller capacitance and does not much affect the dynamic transient responses [23, 47].

10.3.1 Transport Mechanism in TFET

10.3.1.1 Band to Band (BTB) Tunneling Transmission

The expression of BTBT transmission (T_t) of carriers is achieved by Wentzel Kramers Brillouin (WKB) approximation with a consideration that the tunnel barrier is a triangular-shaped potential barrier as shown in Figure 10.12 (a-b). By employing the WKB theory, T_t is given as Eqn. 10.5; where k(x) is the quantum wave vector of the electron inside the triangular barrier (see Eqn. 10.6), h^* is $h/2\pi$; where h is Planck's constant.

$$T_t \sim \exp\left(-2\int_{x_1}^{-x_2} k(x)|\,dx\right) \tag{10.5}$$

$$k(x) = \sqrt{\frac{2m^*}{h^{*2}}(PE - E)} \tag{10.6}$$

Figure 10.12(a) shows the triangular energy barrier for quantum tunneling of charge carriers. A shaded tunneling region is represented in a triangular region for considering the ON-state of DG NTFET as shown in Figure 10.12 (b). Along the device length, the potential energy (PE) of the electrons reduces from energy bandgap (E_g) at tunnel junction to zero. The rate of reduction in PE is represented as (E_g/2-qFx), where E_g is the energy bandgap of the semiconductor material at the tunnel junction, and F is the electric field measured in V/m. The final expression of k(x) is given by Eqn. 10.7 [18].

$$k(x) = \sqrt{\frac{2m^*}{h^{*2}}\left(\frac{E_g}{2} - qFx\right)} \tag{10.7}$$

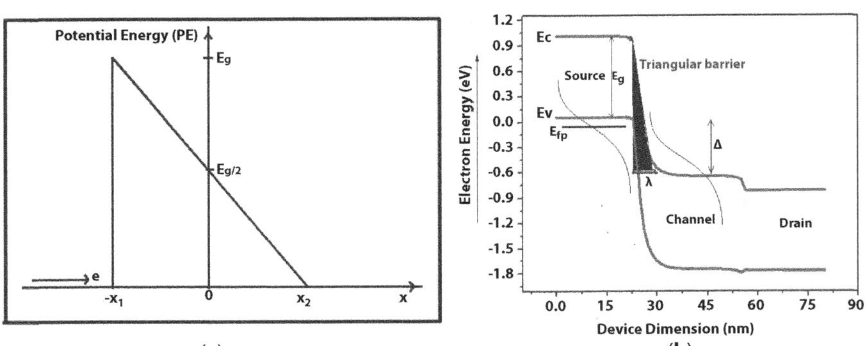

Figure 10.12 (a) Representing triangular-shaped potential barrier as a tunnel barrier. (b) Energy band of NTFET showing the WKB analysis.

Now, Eqn. 10.7 can be used with Eqn. 10.5 to obtain the expression for T_t as given by Eqn. 10.8.

$$T_t \sim exp\left[-2\int_{x_1}^{-x_2}\sqrt{\frac{2m^*}{h^{*2}}\left(\frac{E_g}{2}-qFx\right)}dx\right] \tag{10.8}$$

In the triangular barrier, it is assumed that at $\mathrm{x}=\mathrm{x}_1, E_g = \frac{E_g}{2}-qFx$ and that at $\mathrm{x}_2, \left(\frac{E_g}{2}-qFx\right)=0$, the T_t (eqn. 10.8) is further expanded to Eqn. 10.9.

$$T_t \sim \exp\left[-\frac{4\lambda}{3}.\frac{\sqrt{2m^*}}{hF}.(E_g)^{1.5}\right] \tag{10.9}$$

The λ can be expressed in the terms ε_{ox}, t_{ox}, t_{si}, and gate geometry given by Eqn. 10.

$$\lambda = \sqrt{\frac{\varepsilon_{si}t_{si}t_{ox}}{2\varepsilon_{ox}}} \tag{10.10}$$

In Figure 10.12 (b), the height of the triangle along the y-axis is $\Delta\phi + E_g$, and a screening length λ is referred with the spatial extent of electric field F [19]. The energy range corresponding to tunneling is represented by Δϕ. For the vital conditions for BTB tunneling to occur: (i) availability of states in source region (ii) availability of states in the channel region (iii) sufficiently narrow energy barrier for tunneling, (iv) momentum conservation [17, 19].

From the Figure 10.12.b, the slope of the energy bands provides a specific electric field thus F can be replaced by $\Delta y/\Delta x = (\Delta\phi + E_g)/\lambda$ [20]. By Rearranging the Eqn. 10.6, the following is obtained as Eqn. 10.8 [24].

$$Tp \sim \exp\left[-\frac{4\lambda}{3}.\frac{\sqrt{2m^*}}{h(\Delta\phi+E_g)}.(E_g)^{1.5}\right] \tag{10.11}$$

Moreover, the present study of TFET provides an opportunity to apply and test much of the theory and information. There are two reasons why the observation and exploration of this FET are worthwhile. First, the TFET device can allow further scaling of power supply (V_{DD}) with improved FET characteristics than conventional FET. Second, the process technology of this device follows the conventional CMOS process technology [34–38]. This feature allows for further reduction of the production cost of the TFET based circuit and system for future development.

10.3.1.2 Kane's Model

For tunnel FET, the BTBT current (I_{BTBT}) can be modeled by the following Eqn. 10.12 [45].

$$I_{BTBT} = q \int G \, dv = qWL_{gate} \int G \, dx \tag{10.12}$$

As shown in Eqn. 10.12, W is the width and L_{gate} is gate length, G is the charge generation rate ($cm^{-3}s^{-1}$) of tunneling in TFET, which contributes to current conduction. It is calculated by Kane's Model as given in Eqn. 10.13.

$$G = A\left(\frac{F}{F_0}\right)^P \exp\left(-\frac{B}{F}\right) \tag{10.13}$$

The variables A and B are the material-dependent parameters. $F_0 = 1$ V/cm, P = 2 and 2.5 for direct and indirect BTBT respectively. Prefactor A and exponential factor B for direct and indirect tunneling are defined as follows [28, 29]:

$$A_{dir} = \left[\frac{g\pi m_r^{1/2}(qF_0)^2}{9h^2\left(E_g^\Gamma\right)^2}\right]\left(\frac{9}{\pi^2}\right) \tag{10.14}$$

$$B_{dir} = \frac{\pi^2 m_r^{1/2}\left(E_g^\Gamma\right)^{3/2}}{qh} \tag{10.15}$$

$$A_{ind} = \frac{g(m_c m_v)^{3/2}(1+2N_{TA})D_{TA}^2(qF_0)^{5/2}}{2^{21/4}h^{5/2}m_r^{5/4}\rho\varepsilon_{TA}E_{g'}^{7/4}} \tag{10.16}$$

$$B_{ind} = \frac{2^{7/2}\pi m_r^{1/2}E_{g'}^{3/2}}{3qh} \tag{10.17}$$

Table 10.1 Listed parameters for Kane's Model (Eqn. 10.14-10.17).

Parameters	Name
g	Degeneracy factor ($g = 2g_c g_v$)
m_r	Reduced tunneling mass
E_g^{Γ}	Direct bandgap at the Γ point
m_v	Carrier effective mass in VB
m_c	Carrier effective mass in CB
N_{TA}	Occupied number of the transverse acoustic phonon at temperature , $1/[exp(\varepsilon_{TA}/kT) - 1]$
k	Boltzmann constant
D_{TA}	Deformation potential of transverse acoustic phonons
ρ	Mass density
ε_{TA}	Transverse acoustic phonon energy
$E_{g'}$	Minimum bandgap

The parameters for Kane's model (for Eqn. 10.14–10.17) are mentioned in Table 10.1 for TCAD device simulations. A correction to the prefactor A_{dir} is done by multiplying it by $(9/\pi^2)$ [46]. While using the indirect model, a more physically correct formula that includes the emitted/absorbed phonon effect as mentioned in should be incorporated. For $Ge_{0.6}Si_{0.4}$: $A_{dir} = 1.34 \times 10^{20}$ cm^{-3} S^{-1}, $B_{dir} = 54.05$ MV/cm, $A_{ind} = 2.44 \times 10^{15}$ cm^{-3} S^{-1}, $B_{ind} = 16.8$ MV/cm [31–33].

10.4 Subthreshold Swing (SS) in Tunnel FETs

The motivation to express the equation of subthreshold swing (SS) for TFET is to only understand the different parameter dependencies as compared with SS of MOSFET by considering BTB current for reverse-biased p-n junction as given by Eqn. 10.18 [17, 25].

$$I = aV_{eff}Eexp^{\frac{-b}{E}} \qquad (10.18)$$

$$a = Aq^3 \frac{\sqrt{2m^* / E_g}}{\pi^2 h^{*2}} \tag{10.19}$$

$$b = \frac{4\sqrt{m^*} E_g^{\frac{3}{2}}}{3qh^*} \tag{10.20}$$

where A is the cross-sectional area of the TFET device, V_{eff} is the junction biasing and E is the electric field at the tunnel junction. The parameters *a* and *b* are defined by Eqn. 10.19 and Eqn. 10.20 respectively. Then, from the Eqn. 10.15–10.17 the SS is computed as Eqn. 10.21.

$$SS = -\ln(10)\left[\frac{1}{V_{eff}}\frac{\partial V_{eff}}{\partial V_{gs}} + \frac{E+b}{E^2}\frac{\partial E}{\partial V_{gs}}\right] \tag{10.21}$$

According to the first term in Eqn. 10.21, a design of TFET should be in such a way that V_{gs} may indirectly control the tunnel-junction bias V_{eff}. The V_{eff} can be also varied by TFET geometry with a thin and/or high dielectric gate material and an ultrathin body so that the gate field can effectively modulate the channel. For an equivalent t_{ox} ~ 1 nm, dV_{eff}/dV_{gs} = 1, and the first term of Eqn. 10.21 is approximately inversely related to V_{eff}. This results in much steeper SS at lower V_{gs} and increases with V_{gs}. The second term suggests that dE/dV_{gs} should be maximized to achieve steep SS.

The performance of Si DGTFET is also analyzed by digital simulations for the TFET inverter and benchmark circuits (see Table 10.2 and Table 10.3) mentioned in Purdue Emerging Technology Evaluator (PETE) simulator [39]. Figure 10.13 and Table 10.2 show the VTC performance parameters respectively for N-type DGTFET.

Table 10.2 VTC parameters based simulation in PETE.

High noise margin N_{MH}	Low noise margin N_{ML}	A maximum gain of the inverter	Input high voltage V_{IH}	Input low voltage V_{IL}	Output low voltage V_{OL}	Output high voltage V_{OH}
0.31012 V	0.31012 V	17	0.43988	0.31012	0	0.75

Table 10.3 Performance of DG TFET inverter and benchmark circuits mentioned in PETE simulator.

	Charging delay of inverter (in ps)	**Discharging delay of inverter (in ps)**	**Average delay of inverter (in ps)**	**Energy per cycle (in fJ)**	**Average leakage power (in fW)**	
Trans char of inverter not loaded	2.82	1.65	2.23	0.67	8.90	
Driving constant load		Delay of inverter driving a constant load (in ns)	Switching Energy per cycle of an inverter driving a constant load (in pJ)	Leakage power of Inverter driving constant Load (in fW)		
	SINGLE-STAGE INVERTER	1.27	0.297	8.89		
		Optimal Number of Stages to Drive the Load	The Ratio of Size of One stage to its previous stage	Delay of buffer chain driving a constant load (in ps)	Switching energy per cycle of buffer chain driving a constant load (in pJ)	Leakage power of buffer chain driving a constant load (in pW)

(Continued)

Table 10.3 Performance of DG TFET inverter and benchmark circuits mentioned in PETE simulator. (*Continued*)

	Charging delay of inverter (in ps)	**Discharging delay of inverter (in ps)**	**Average delay of inverter (in ps)**	**Energy per cycle (in fJ)**	**Average leakage power (in fW)**	
	SIZED BUFFER CHAIN	5	2.7601	13.69	0.057	0.77
Tran char of logic gates driving an inverter		Average charging delay (ps)	Average discharging delay (ps)	Average delay (ps)	Average switching energy per cycle (in pJ)	Average Leakage of NAND gate (in pW)
	2 INPUT NAND Gate	2.81	2.53	2.67	0.002	0.014
	2 INPUT NOR Gate	4.29	1.65	2.97	0.0013	0.0134
	2 INPUT XOR Gate	10.24	7.89	9.06	0.0013	0.0178
		Delay (ps)	Dynamic Power (uW)	Leakage Power (ps)		

(*Continued*)

Table 10.3 Performance of DG TFET inverter and benchmark circuits mentioned in PETE simulator. (*Continued*)

	Charging delay of inverter (in ps)	**Discharging delay of inverter (in ps)**	**Average delay of inverter (in ps)**	**Energy per cycle (in fJ)**	**Average leakage power (in fW)**	
The power performance of chain of logic gates	5 STAGE RING OSCILLATOR	22.36	150	0.04		
	10 stage NAND Gate	35.67	124	0.134		
	10 stage NOR Gate	49.588	350	0.134		
	Critical Path Delay of RCA (in ns)	The dynamic power of RCA running at max speed (in uW)	Leakage power of RCA (in pW)	Total power of RCA (in uW)		
Summary of trans analysis 0f 8-bit ripple carry adder.	0.158	97.95	0.604	97.95		

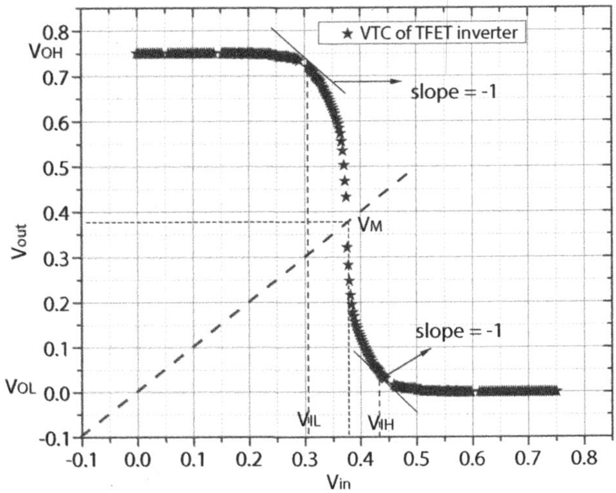

Figure 10.13 Voltage Transfer Characteristics (VTC) based simulation in PETE.

10.5 Performance of Hetrojunction Tunnel FET

10.5.1 Transfer Characteristics Analysis of TFET Devices

In the following section, the impact of high-k gate dielectric materials in hetero channel DGTFET is reported. In heterostructure channel, two semiconductors, $Si_{1-x}Ge_x$ and (Si=1.12eV) materials in source and another material in drain/channel region were used for improving the tunneling current by reducing tunneling region. The reported results in this section were carried out on the Silvaco-ATLAS device simulator tool. The device proposed in this work is an N-type double gate heterodielectric TFET. Figure 10.14 (a–c) also shows the 2D cross-sectional schematics of the DGTFETs.

Table 10.4 contains a summary of the device design parameters which were used in device simulation and analysis. The gate oxide thickness (t_{ox}), which is one of the most crucial device design matrix parameters for TFET device, is taken as 2.0 nm. The thickness of the silicon (T_{Si}) body, which can also be referred to as, the junction depth of the proposed device, is taken as 10.0 nm. The channel length, also referred to as the region from source to drain is 50.0 nm. Uniform doping of 1.0×10^{20} cm^{-3} and 5.0×10^{18} cm^{-3} has been used in drain and source regions respectively. The gate metal work function is taken as 5.2 eV.

The transfer (I_{ds}–V_{gs}) characteristic of the n-channel DG-TFET at a constant drain voltage of 0.5 V is presented in Figure 10.15 clearly shows

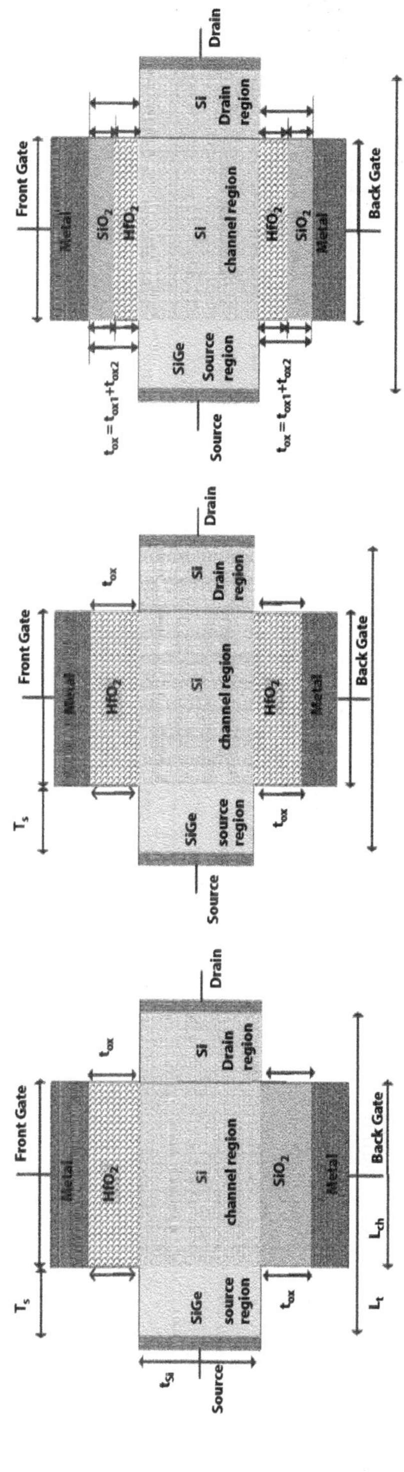

Figure 10.14 (a) S1, (b) S2, and (c) S3 showing the 2D schematics of the DGTFETs.

Table 10.4 Device design matrix parameters for structures (S1, S2, and S3).

Work function (φ_M)	Doping levels for the source (N_S)	Doping level for drain (N_D)	Doping level for the channel (N_C)	Gate oxide material thickness (t_{ox})	The total length of the device (L_t)	Channel length (L_{ch})	Silicon film thickness (t_{Si})	Source and drain lengths (L_s/L_d)
5.2 eV*	1.1×10^{20}-cm^{-3}	5.1×10^{18} cm^{-3}	1.1×10^{15} cm^{-3}	2.0 nm	250.0 nm	50 nm	10 nm	100 nm

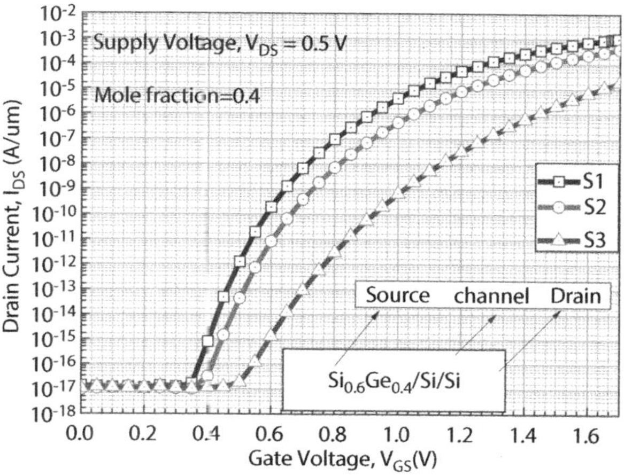

Figure 10.15 The I_{DS}-V_{GS} characteristics for a constant drain to sources voltage (V_{ds}= 0.5 V).

the variation of drain current with change in gate voltage. The electrical parameters such as I_{ON}, I_{OFF}, I_{ON}/I_{OFF}, average SS, and threshold voltage (V_{TH}) are extracted from the three device designs shown in Figure 10.14 (a–c) and tabulated in Table 10.5 for clear comparison among the three designs. The best device performance has been obtained in structure-2 (i.e. front gate (FG): HfO_2 and back gate (BG): HfO_2) at 300 K temperature. From Table 10.4, it has been noticed that for structure 2, the best V_{th} and the average subthreshold values (Avg-SS) increases slightly. It has been observed that the average SS for all three structures remains less than the value of the MOSFETs (60 mv/dec). Also, a high value of I_{ON}/I_{OFF} in the proposed devices is of the order of 10^{13}. Low-power consumption is ensured by the low I_{OFF}(of the order of 10^{-18}) of the device.

The transconductance ($g_m = dV_{gs}/I_{ds}$) for the results listed in Figure 10.15 is shown in Figure 10.16 which is further utilized to compute the unity gain cutoff frequency (f_T) as an analog parameter.

The proposed hetero channel double gate (HC- DG) - TFET exhibits maximum electric field (for V_{DS} = 0.5 V) at the interface of $Si_{1-x}Ge_x$/ Si (source/channel region) are illustrated in Table 10.6 for the TFET device structures. According to the obtained results, it is noticed that a peak electric field of 2.27 MV/μm is visible at the source-channel interface of the device. The largest electric fields have been obtained at the junction in all three devices values of which are mentioned in Table 10.5.

Table 10.5 Impact of gate dielectric materials on device design parameters.

Devices	Structure	I_{ON} (A/μm)	I_{OFF} (A/μm)	I_{ON}/I_{OFF} ratio	V_{th} (V)	Average-SS (mV/dec)
S1	FG: HfO_2 BG: HfO_2	4.50×10^{-5}	2.11×10^{-18}	2.13×10^{13}	0.75	37.51
S2	FG: HfO_2 BG: SiO_2	2.66×10^{-5}	2.06×10^{-18}	1.30×10^{13}	0.75	38.13
S3	FG & BG: HfO_2/SiO_2	1.20×10^{-5}	1.76×10^{-18}	6.81×10^{12}	0.90	38.96

Note : FG : Front Gate, BG : Back Gate, A : Ampère, μm : Micromètre. mV : millivolt

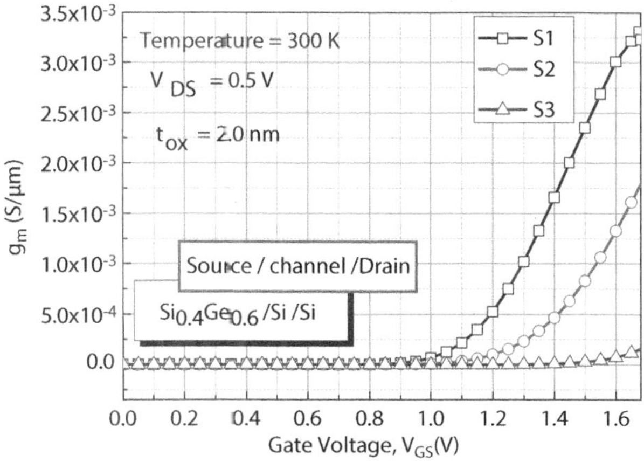

Figure 10.16 Transconductance variation with applied V_{gs} in designed DG-TFET for: (a) S1, (b) S2, and (c) S3.

Table 10.6 Impact of gate dielectric materials on tunneling electric field FG: Front Gate, BG: Back Gate).

Parameter	S1 FG: HfO_2 BG: HfO_2	S2 FG: HfO_2 BG: SiO_2	S3 FG : HfO_2/SiO_2 BG: HfO_2/SiO_2
Maximum E(V/μm)	2.27×10^6	2.13×10^6	1.62×10^6

The maximum electric field decreases from 2.27×10^6 V/μm to 2.13×10^6 V/μm and 2.27×10^6 V/μm to 1.62×10^6 V/μm for S2 and S3 respectively as a reference to device S1 which indicates the effect of the double layer of dielectrics (HfO_2/ HfO_2) at both FG and BG. This type of structure of a double layer of dielectric gate material is more capable to bind the EF at the source-channel junction more strongly [48].

10.5.2 Frequency Analysis of TFET Devices

The C-V characteristics are required in analyzing the frequency and switching properties of the device. Figure 10.17 shows the device capacitance device structures S1-S2-S3. As seen from Figure 10.17, the gate capacitance (C_{gg}) increases as the gate voltage (V_{GS}) increases. The largest capacitance is obtained in the case of structure, S1 where high-*k* gate dielectric HfO_2 is used in both front and back gates. Applying high-*k* material in gate

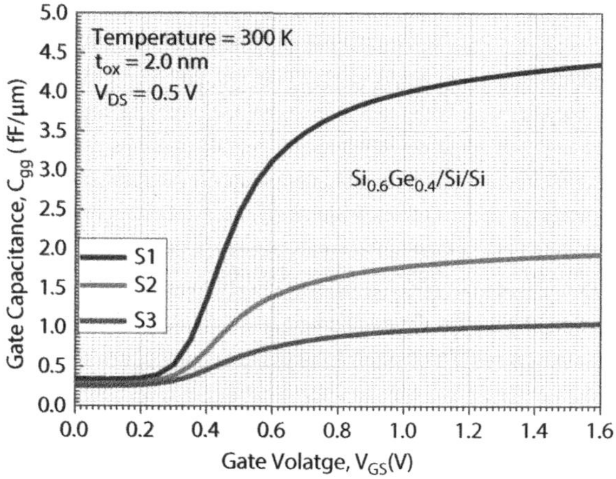

Figure 10.17 The C-V characteristic of DG-TFET variation with V_{gs} in designed-TFET for: (a) S1, (b) S2, and (c) S3.

dielectric increases the capacitance and enhances the gate control over the channel, which is desirable, a parameter for circuit applications.

Modern devices are working in the RF range in the development of an advanced communication system. It becomes essential for device engineers to analyze the RF performance of the newly developed devices [51]. The simplest device, we are using today have the analog and digital components embedded together in the same chip. So, here we are listing the analog parameter (unit gain cutoff frequency, f_T). The cut-off frequency, f_T refers to the frequency at which unity gain is obtained. A large value of f_T is required so that the newly designed device provides significant gain at high frequencies. It is estimated by the ratio of g_m over C_{gg} formulated as, Eqn. 10.22 [40–44]:

$$f_T = \frac{g_m}{2\pi(C_{gs} + C_{gd})} = \frac{g_m}{2\pi C_{gg}} \tag{10.22}$$

As seen from the results g_m (see Figure 10.16) is more sensitive for the high k dielectrics (device S2), a similar trend is also clearly depicted from the Figure 10.18 where f_T is maximized to ~200 GHz, ~150 GHz, and 110 GHz for devices S1, S2, and S3, respectively at ~V_{GS} = 1.7 V. The structure S1 shows better gate controlling ability for the higher gate current thus shows higher f_T among the defined structures S1-S3.

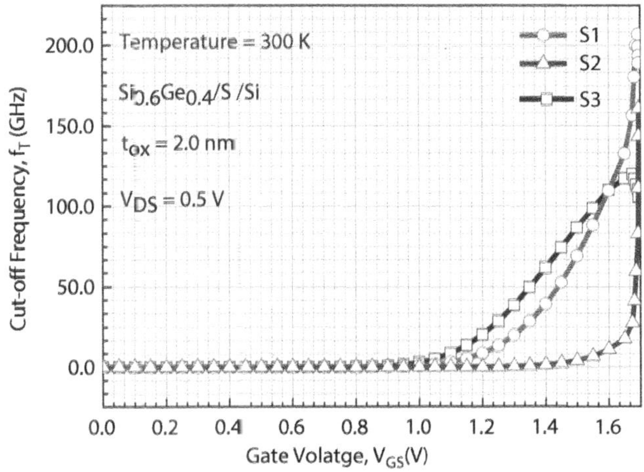

Figure 10.18 Cutoff frequency, f_T variation with input V_{gs} in designed DG-TFET for: (a) S1, (b) S2, and (c) S3.

10.6 Conclusion

Structure S1 has a front gate of high-*k* gate dielectric material HfO_2 (dielectric constant = 25) and the back gate is of low-*k* gate dielectric material SiO_2 (dielectric constant = 3.9), device structure S2 has both front and back gates of high-*k* material HfO_2. In structure S3, the total dielectric thickness is divided into two halves composed of HfO_2 and SiO_2 in both front and back gates. Exhaustive simulations were performed to analyze the device performance in terms of dc and analog parameters can be modulated to be achieving the optimum performance. PETE nano-hub simulator can be considered as one of the fastest and effective ways to evaluate the benchmarked digital circuit performance based on TFET devices. A high ON-state current of 4.50×10^{-5} A/μm is achieved in the proposed device with a very low OFF-state current of 2.11×10^{-18}A/μm at a V_{TH} of 0.75 V along with steeper SS of 37.51 mV/dec are important factors for the suitability of the device as low-power applications. In addition to this, the proposed device has a very high f_T and gain thus TFET can be implemented for high-frequency applications. To improve ON current, enhanced coupling between gate and channel is required by selecting a superior gate dielectric material and by modulating the device geometry. Thus, a lot of research work is required to explore the gate leakage modeling and selecting gate dielectrics other than SiO_2 and Hafnium oxide for ON current improvement respectively. In addition, to investigate the device characteristics of the TFET device, circuit simulations must be employed to analyze the digital and analog applications.

References

1. Gargini P., The international technology roadmap for semiconductors (ITRS): Past, present and future. in: GaAs IC Symposium. *IEEE Gallium Arsenide Integrated Circuits Symposium. 22nd Annual Technical Digest 2000.* (Cat. No.00CH37084), IEEE, pp. 3–5, 2000.
2. Song, J., Yu, B., Yuan, Y., Taur, Y., A review on compact modeling of multiple-gate MOSFETs. *IEEE Trans. Circuits Syst. I Regul. Pap.*, 56, 8, 1858–1869, 2009.
3. Ionescu, A.M. and Riel, H., Tunnel field-effect transistors as energy-efficient electronic switches. *Nature*, 479, 7373, 329–337, 2011.
4. Agarwal, T.K., Nourbakhsh, A., Raghavan, P., Radu, I., De Gendt, S., Heyns, M., Verhelst, M., Thean, A., Bilayer graphene tunneling FET for sub-0.2 V digital CMOS logic applications. *IEEE Electron Device Lett.*, 35, 12, 1308–1310, 2014.
5. Cao, W., Yao, C.J., Jiao, G.F., Huang, D., Yu, H.Y., Li, M.F., Improvement in reliability of tunneling field-effect transistor with pnin structure. *IEEE Trans. Electron Devices*, 58, 7, 2122–2126, 2011.
6. Núñez, J. and Avedillo, M.J., Comparison of TFETs and CMOS using optimal design points for power–speed tradeoffs. *IEEE Trans. Nanotechnol.*, 16, 1, 83–89, 2016.
7. Dennard, R.H., Gaensslen, F.H., Yu, H.N., Rideovt, V.L., Bassous, E., Leblanc, A.R., Design of ion-implanted MOSFET's with very small physical dimensions. *IEEE Solid-State Circuits Soc. Newslett.*, 12, 1, 38–50, 2007.
8. Li, W. and Woo, J.C., Vertical P-TFET with a P-Type SiGe pocket. *IEEE Trans. Electron Devices*, 67, 4, 1480–1484, 2020.
9. Ahmed, S.Z., Truesdell, D.S., Tan, Y., Calhoun, B.H., Ghosh, A.W., A comprehensive analysis of Auger generation impacted planar Tunnel FETs. *Solid-State Electron.*, 169, 107782, 2020.
10. Xie, Q., Xu, J., Taur, Y., Review and critique of analytic models of MOSFET short-channel effects in subthreshold. *IEEE Trans. Electron Devices*, 59, 6, 1569–1579, 2012.
11. Datta, E., Chattopadhyay, A., Mallik, A., Omura, Y., Temperature dependence of analog performance, linearity, and harmonic distortion for a ge-source tunnel FET. *IEEE Trans. Electron Devices*, 67, 3, 810–815, 2020.
12. Rahi, S.B. and Ghosh, B., High-k double gate junctionless tunnel FET with a tunable bandgap. *RSC Adv.*, 5, 67, 54544–54550, 2015.
13. Schmidt, M., Schäfer, A., Minamisawa, R.A., Buca, D., Trellenkamp, S., Hartmann, J.M., Zhao, Q.-T., Mantl, S., Line and point tunneling in scaled

Si/SiGe heterostructure TFETs. *IEEE Electron Device Lett.*, 35, 7, 699–701, 2014.

14. Boucart, K. and Ionescu, A.M., Double-gate tunnel FET with high-$\kappa $ gate dielectric. *IEEE Trans. Electron Devices*, 54, 7, 1725–1733, 2007.
15. Ionescu, A.M. and Riel, H., Tunnel field-effect transistors as energy-efficient electronic switches. *Nature*, 479, 7373, 329–337, 2011.
16. Boucart, K. and Ionescu, A.M., Threshold voltage in tunnel FETs: Physical definition, extraction, scaling and impact on IC design, in: *ESSDERC 2007-37th European Solid State Device Research Conference*, IEEE, pp. 299–302, September 2007.
17. Sze, S.M., Li, Y., Ng, K.K., *Physics of Semiconductor Devices*, John Wiley & Sons, 2021.
18. Knoch, J. and Appenzeller, J., A novel concept for field-effect transistors-the tunneling carbon nanotube FET, in: *63rd Device Research Conference Digest, 2005. DRC'05*, vol. 1, IEEE, pp. 153–156, June 2005.
19. Streetman, B.G. and Banerjee, S., *Solid State Electronic Devices*, vol. 10, Pearson/Prentice Hall, Upper Saddle River, NJ, 2006.
20. Colinge, J.P., Multiple-gate soi mosfets. *Solid-State Electron.*, 48, 6, 897–905, 2004.
21. Graef, M., Schwarz, M., Holtij, T., Hain, F., Kloes, A., Iñíguez, B., Two-dimensional bias dependent model for the screening length in double-gate Tunnel-FETs, in: *2013 14th International Conference on Ultimate Integration on Silicon (ULIS)*, IEEE, pp. 29–32, March 2013.
22. Gnani, E., Baravelli, E., Gnudi, A., Reggiani, S., Baccarani, G., Capacitance estimation for InAs Tunnel FETs by means of full-quantum k· p simulation. *Solid-State Electron.*, 108, 104–109, 2015.
23. Khatami, Y. and Banerjee, K., Steep subthreshold slope n-and p-type tunnel-FET devices for low-power and energy-efficient digital circuits. *IEEE Trans. Electron Devices*, 56, 11, 2752–2761, 2009.
24. Zhang, Q., Zhao, W., Seabaugh, A., Low-subthreshold-swing tunnel transistors. *IEEE Electron Device Lett.*, 27, 4, 297–300, 2006.
25. Taur, Y. and Ning, T.H., *Fundamentals of Modern VLSI Devices*, Cambridge University Press, 2013.
26. Mookerjea, S., Krishnan, R., Datta, S., Narayanan, V., Effective capacitance and drive current for tunnel FET (TFET) CV/I estimation. *IEEE Trans. Electron Devices*, 56, 9, 2092–2098, 2009.
27. Garg, S., Saurabh, S., Exploiting within-channel tunneling in a nanoscale tunnel field-effect transistor. *IEEE Open J. Nanotechnol.*, 16, 1, 100–8, 2020 Oct.
28. Kao, K.H., Verhulst, A.S., Vandenberghe, W.G., Soree, B., Magnus, W., Leonelli, D., De Meyer, K., Optimization of gate-on-source-only tunnel FETs with counter-doped pockets. *IEEE Trans. Electron Devices*, 59, 8, 2070–2077, 2012.
29. Choi, W.Y. and Lee, W., Hetero-gate-dielectric tunneling field-effect transistors. *IEEE Trans. Electron Devices*, 57, 9, 2317–2319, 2010.

30. Rahi, S.B., Asthana, P., Gupta, S., Heterogate junctionless tunnel field-effect transistor: Future of low-power devices. *J. Comput. Electron.*, 16, 1, 30–38, 2017.
31. Thoti, N., Li, Y., Kola, S.R., Samukawa, S., Optimal inter-gate separation and overlapped source of multi-channel line tunnel FETs. *IEEE Open J. Nanotechnol.*, 1, 38–46, 2020.
32. Kao, K.H., Verhulst, A.S., Vandenberghe, W.G., Soree, B., Groeseneken, G., De Meyer, K., Direct and indirect band-to-band tunneling in germanium-based TFETs. *IEEE Trans. Electron Devices*, 59, 2, 292–301, 2011.
33. Blaeser, S., Glass, S., Schulte-Braucks, C., Narimani, K., von den Driesch, N., Wirths, S., Zhao, Q.T., Line tunneling dominating charge transport in SiGe/Si heterostructure TFETs. *IEEE Trans. Electron Devices*, 63, 11, 4173–4178, 2016.
34. Bordallo, C.C., Martino, J.A., Agopian, P.G., Rooyackers, R., Vandooren, A., Thean, A., Claeys, C., Analysis of analog parameters in NW-TFETs with Si and SiGe source composition at high temperatures, in: *2015 30th symposium on microelectronics technology and devices (SBMicro)*, IEEE, pp. 1–4, 2015.
35. Yang, S., Lv, H., Lu, B., Yan, S., Zhang, Y., A novel planar architecture for Heterojunction TFETs with improved performance and its digital application as an inverter. *IEEE Access*, 8, 23559–23567, 2020.
36. Lee, J.C., Ahn, T.J., Yu, Y.S., Work-function engineering of source-overlapped dual-gate tunnel field-effect transistor. *J. Nanosci. Nanotechnol.*, 18, 9, 5925–5931, 2018.
37. Jiang, Z., Zhuang, Y., Li, C., Wang, P., Liu, Y., Drive current enhancement in TFET by dual source region. *J. Electr. Comput. Eng.*, 1, 2015, 33–43, 2015.
38. Narang, R., Saxena, M., Gupta, R.S., Gupta, M., Device and circuit level performance comparison of tunnel FET architectures and impact of heterogeneous gate dielectric. *JSTS: J. Semicond. Technol. Sci.*, 13, 3, 224–236, 2013.
39. Yang, S., Lv, H., Lu, B., Yan, S., Zhang, Y., A novel planar architecture for Heterojunction TFETs with improved performance and its digital application as an inverter. *IEEE Access*, 8, 23559–23567, 2020.
40. Mallik, A. and Chattopadhyay, A., Tunnel field-effect transistors for analog/mixed-signal system-on-chip applications. *IEEE Trans. Electron Devices*, 59, 4, 888–894, 2012.
41. Ghosh, P., Haldar, S., Gupta, R.S., Gupta, M., An investigation of linearity performance and intermodulation distortion of GME CGT MOSFET for RFIC design. *IEEE Trans. Electron Devices*, 59, 12, 3263–3268, 2012.
42. Gupta, A.K., Raman, A., Kumar, N., Design and investigation of a novel charge plasma-based core-shell ring-TFET: analog and linearity analysis. *IEEE Trans. Electron Devices*, 66, 8, 3506–3512, 2019.
43. Kaya, S. and Ma, W., Optimization of RF linearity in DG-MOSFETs. *IEEE Electron Device Lett.*, 25, 5, 308–310, 2004.
44. Chung, H.T., Chen, C.T., Li, Y.S., Liu, Y.W., Liao, C.Y., Huang, W.H., Cheng, H.C., Effect of Crystallinity on the Electrical Characteristics of Poly-Si

Tunneling FETs via Green Nanosecond Laser Crystallization. *IEEE Electron Device Lett.*, 42, 2, 164–167, 2021.

45. Prabhat, V. and Dutta, A K., Analytical surface potential and drain current models of dual-metal-gate double-gate tunnel-FETs. *IEEE Trans. Electron Devices*, 63, 5, 2190–2196, 2016.
46. Kumar, D., Performance evaluation of double gate tunnel FET based chain of inverters and 6-T SRAM cell. *Eng. Res. Express*, 1, 2, 025055, 2019.
47. Naima, G. and Rahi, S.B., Lower power circuit and system design hierarchy and thermal reliability of tunnel field effect transistor. *Silicon*, 14, 3233–3243, 2022.
48. Paras, N. and Chauhan, S.S., Insights into the DC, RF/Analog and linearity performance of vertical tunneling based TFET for low-power applications. *Microelectron. Eng.*, 216, 111043, 2019.
49. Paras, N. and Chauhan, S.S., Temperature sensitivity analysis of vertical tunneling based dual metal Gate TFET on analog/RF FOMs. *Appl. Phys. A*, 125, 5, 1-9, 2019.
50. Kim, H.W. and Kwon, D., Low-power vertical tunnel field-effect transistor ternary inverter. *IEEE J. Electron Devices Soc*, 9, 286–294, 2021.
51. Tayal, S., Vibhu, G., Meena, S., *et al.*, Optimization of device dimensions of high-k gate dielectric based DG-TFET for improved analog/RF performance. *Silicon*, 14, 3515–3521, 2022.
52. Kumar, D. and Jain, P., Performance of Dual Metal-Double Gate Tunnel Field Effect Transistor with Different Dielectrics, in: *Proceeding of International Conference on Intelligent Communication, Control and Devices*, Springer, Singapore, pp. 927–933, 2017.

11

Low-Power Integrated Circuit Smart Device Design

Shasanka Sekhar Rout[1]*, Salony Mahapatro[2], Gaurav Jayaswal[3] and Manish Hooda[3]

[1]*Dept. of ECE, GIET University, Gunupur, India*
[2]*Asiczen Technologies, Bhubaneswar, India*
[3]*Technology Development Division, Semi-Conductor Laboratory, Department of Space, Punjab, India*

Abstract
We are leading the way in the most significant human advancement of all time. One such area is wireless electronics systems. In recent wireless systems, high level integration using low power integrated circuit and CMOS process plays an essential task for low power operation. For handset wireless systems, the low power design is vital due to the battery capacity constraint. So, the volume and weight of the transceivers has to be reduced by the help of low voltage and low power techniques. The mounting demand for computing with low cost, has led to quicker, lesser, more reliable ICs that need low power. The technology associated with these advances can be integrated into the advance electronics systems to construct them extra functionality along with diminish power requirements.

This chapter presents the design of basic circuits and design aids for low power consumption. The basics of power dissipation in ICs along practical circuits are offered to help the readers for familiar with the low power design. This chapter also provides different power optimization approaches at the system, algorithm, architecture, logic, and device levels. Here main point of focus includes, VLSI circuit design, random access memory circuits (RAMs), VLSI design methodology for low power IC design. This chapter also focuses on low power flood sensor, voltage controlled oscillator (VCO) and Gilbert mixer design.

***Keywords*:** CMOS, IC, mixer, power consumption, RAM, transceiver, VCO, VLSI

**Corresponding author*: ssrout1988@gmail.com

Suman Lata Tripathi, Souvik Ganguli, Tengiz Magradze, and Abhishek Kumar (eds.) Intelligent Green Technologies for Sustainable Smart Cities, (227–246) © 2022 Scrivener Publishing LLC

11.1 Introduction

In recent wireless systems, high-level integration using low-power integrated circuit (IC) and CMOS process plays an essential task for low-power operation. For portable wireless systems, the low-power design is vital owing to the battery capacity constraint. So, the weight and capacity of the transceivers can be optimized with the help of low-power techniques. Nowadays, each circuit faces power consumption issues, in which portable devices aim for longer battery life and high-end circuits aim for reliability. There is no prediction of power consumption depending upon the transistors based on Moore's law [1]. Commonly, high performance and low area are two contradictory constraints in IC design. The IC designer action is going on to solve the trading-off parameters. The power dissipation problem is not a design condition but an addendum. Also, power consideration is the crucial design criteria in individual portable devices like mobile phones and pacemakers for a long-term basis. Therefore, low power is required for maximum battery life in our communication system. A standard communication system consists of transmitter, channel, and receiver [2] as presented in Figure 11.1. The whole system is implemented in regular life's devices, like mobiles, wireless sensors, laptops, etc. Therefore, all the subblocks should be operated in low power to make a balance for easy communication.

This chapter presents the design of basic circuits and the design aids for low power consumption. Here, the fundamentals of a very large-scale integration (VLSI) system are focused on from the process technology to the architecture level. The basics of power dissipation in ICs along with the practical circuits help, the readers become familiar with the low-power design. This chapter also provides different power optimization approaches at the system, algorithm, architecture, logic, and device levels. The main

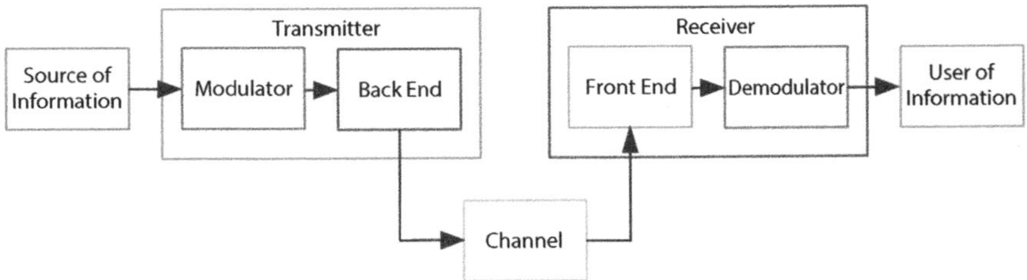

Figure 11.1 Representation of regular communication system [2].

point of focus include, VLSI circuit design, random access memory circuits (RAMs), and VLSI design methodology for low-power operation. This chapter also focuses on low-power estimation techniques, low-power flood sensor, voltage controlled oscillator (VCO), and Gilbert mixer design.

11.2 Need of Low Power

Numerous reasons motivate the rising of the power dissipation issues. Amongst, battery systems like laptops, palmtops, and mobiles arise from the required external battery capacity. A notebook consumes 10 Watt powers and uses 1.5 battery pounds with 3 hours of recharging operational time. But, advanced battery technology like Ni-MH provides a huge energy density of 30 Watt-hour per pound with a low battery life [3]. The low-power operation is vital for portable devices and to optimize the power of many ICs. The high clock frequency is required for high operation and large integration density in ICs. But, the power dissipation is too much at higher frequencies in IC design. Reliability is another problem in high power dissipation. So, many ideas are implemented to decrease the systems on chip (SoC) power consumption [4]. The power dissipation problem can be solved, depending on the scaling technology. As a result, the operation depends on low operating voltage in which power saving can be done. Not only the power is diminished but also the weight and cost will reduce in battery operating devices. In VLSI technology, low-power designs are demanding constraints, which impact various applications like different portable devices, microprocessors, LCD displays, hard disk, pagers, etc. Low power is essential to make longer the battery life of these systems. Table 11.1 represents the analysis of the battery in a cellular system [5].

Table 11.1 Analysis of battery in a cellular system.

Cellular system	**Motorola Microtac**
RF power	600 mW
Battery	NiMH 600 mAh
Weight	167 gm
Battery life	8 hours standby
Total power load	3900 mW

11.3 Design Techniques of Low Power

The low-power approaches should be implemented to diminish the power dissipation of the whole system during the design process while maintaining the performance. It is mandatory to observe the power distribution in a processor during the optimization process. Therefore, the individual blocks, which consume more power, should be optimized to save power. So, the power reduction should be done by system integration, algorithm, architecture design, circuit level, and process technology [6, 7]. The various design levels of an IC system for power reduction are is illustrated in Figure 11.2.

11.3.1 Power Optimization by IC System

The system integration is vital for the total process of power minimization. Few approaches are as follows:

i. Implement a high level of system integration, i.e., integration of memory elements and other ICs, including analog and digital designs
ii. Use of low system clock for power reduction

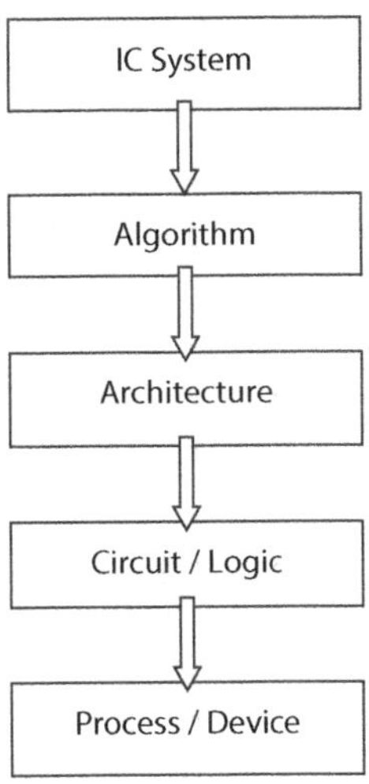

Figure 11.2 Design levels of IC system for power reduction [6].

11.3.2 Power Optimization by Algorithm Section

The algorithm section also helps to reduce the power by:

i. Data coding for lowest switching operation
ii. Reducing the steps of operation, which results less hardware resources implementation

11.3.3 Power Optimization by Architecture Design

Here, many techniques can be implemented for low power in different architectural designs, which are as follows:

i. Minimization of the global busses
ii. Parallelism and pipelining concept for low-power architecture
iii. Reduction of instruction sets for execution
iv. Silent mode of unused blocks by different power management approaches

11.3.4 Power Optimization by Circuit Level

Different concepts can be implemented to optimize the power at logic or circuit level. They are listed as follows:

i. Implement multi threshold logic circuit
ii. Minimize the switching operation by logic reduction
iii. Repeat encoding of sequential logic circuit
iv. Implement static design over dynamic design
v. Minimize buses and clocks
vi. Optimize supply voltage and transistor size
vii. Optimizing internal swings and device counts
viii. Improvement of power by custom design

11.3.5 Power Optimization by Process Technology

In actual circuits, the power dissipation is optimized by the reduction of the supply voltage. But, there is an enhancement in delay due to the approach of supply voltage to a threshold voltage. Therefore, the devices should be scaled to sort out this issue. The following merits are listed due to the scaling for low-power design.

i. Power savings due to the combination of the entire structure into a particular chip
ii. Reduction of parasitic effects due to the small geometry
iii. Better device characteristics due to current flowing capability improvement
iv. Standby power trade-off due to the multiple threshold devices
v. High level of density integration for low-power operation

11.4 VLSI Circuit Design for Low Power

The simple model of VLSI circuit design is used to calculate the power dissipation and delay. Nowadays different issues in advanced CMOS VLSI circuit designs exist as follows:

i. Clock sharing in VLSI systems
ii. Power dissipation in input/output circuits
iii. Clock skew problem
iv. Power dissipation elements in a CMOS logic gate
v. Ground and supply voltage bouncing
vi. Low-power design techniques
vii. Switching activity approach

11.4.1 Power Dissipation of CMOS Inverter

One needs to identify the different power elements to optimize VLSI circuit power consumption. Power dissipation is divided into two types, which include maximum power and average power dissipation. The average power dissipation is more vital as per battery lifetime implementation. CMOS inverter includes mainly two power dissipation components, static and dynamic [7].

11.4.1.1 Static Power

Static power dissipation is obtained by the leakage current and input voltage current drawn. The leakage current is owing to the parasitic diodes in the inverter. It is directly proportional to supply voltage (V_{DD}) and temperature. Total static power due to the leakage current (I_d) is given in equation (11.1) and it is neglected [8].

$$P_{s1} = I_d V_{DD} \tag{11.1}$$

Total static power due to input voltage-current (I_{Dmean}) is given in equation (11.2) and it increases significantly with temperature. This value is not allowed for battery-used devices as it is in standby mode.

$$P_{s2} = I_{Dmean} V_{DD} \tag{11.2}$$

Therefore, total static power is represented as in equation (11.3)

$$P_s = P_{s1} + P_{s2} \tag{11.3}$$

11.4.1.2 Dynamic Power

Here, the power dissipation is owing to the current used in charging and discharging the total output load capacitance (C_L). It is affected by the worst supply voltage (V_{DD}) and frequency (f), not by temperature. The dynamic power dissipation is given in equation (11.4).

$$P_d = C_L V_{DD}^2 f \tag{11.4}$$

11.4.1.3 Short Circuit Power Dissipation

The inverter still dissipates the switching energy in the absence of parasitic and load capacitance. There is overloaded power dissipation on behalf of the short-circuit current, depending on the slow changes in input voltage. It also depends on the temperature and supply voltage [9]. The short circuit power dissipation is avoided by considering $V_{DD} <= (V_{Tn} + V_{Tp})$ [10], which is represented in equation (11.5).

Where V_{Tn} and V_{Tp} are the threshold voltage of nMOS and pMOS, respectively.

$$P_{sc} = I_{Dmean} V_{DD} \tag{11.5}$$

11.4.1.4 Other Power Issue

The entire power dissipation (P_T) of an inverter circuit is represented in equation (11.6).

$$P_T = P_s + P_d + P_{sc} \tag{11.6}$$

Other power dissipations include temperature effect and worst power estimation. It occurs due to the junction temperature, fast process, and maximum power supply.

11.4.2 Capacitance Estimation of CMOS Logic Gate

The power dissipation and speed of a CMOS gate depend on the output load capacitance (C_L). The total load capacitance is the sum of input capacitance (C_{in}), parasitic capacitance (C_p), and wiring capacitance (C_w).

$$C_L = C_{in} + C_p + C_w \tag{11.7}$$

11.5 Circuit Techniques for Low Power

The total power dissipation mainly depends on static power and dynamic power. But, some circuit techniques can be implemented to achieve low power by maintaining the total performance [11].

11.5.1 Static Power Technique

The low threshold voltage (V_T) is one vital source for static power dissipation. The V_T should be less than one-fifth of the supply voltage to preserve the performance at a low power supply. The subthreshold current increases by lowering V_T due to the exponential relationship in between V_T and current. The total standby current can be more for VLSI IC design, but it is not accepted for low-power design. Therefore, there are many techniques that are used to lower this current with the help of low V_T by reverse biasing of the gate to source voltage (V_{GS}).

11.5.1.1 Self-Reverse Biasing

It is used to diminish the static power dissipation in the stand-by form of a memory driver [12]. The circuit diagram is shown in Figure 11.3, which reduces the subthreshold current in the driver circuit. A pMOS transistor (Mp) is inserted in between the supply voltage and common source node X.

The driver pMOS transistors (Mp_1.... Mp_n) are the same size transistors. Here, the threshold voltage of Mp transistor is just higher than the

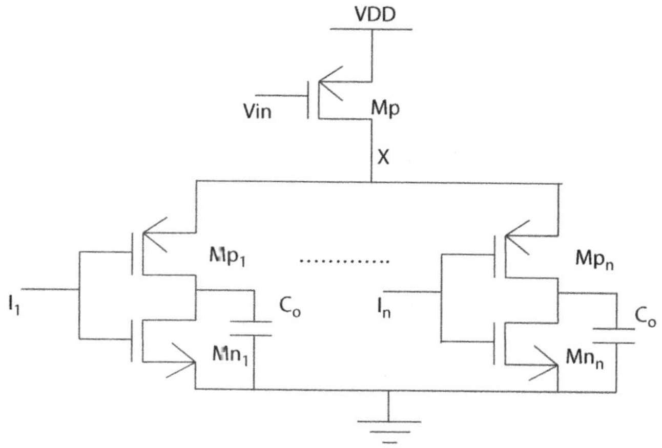

Figure 11.3 Self-reverse biasing circuit [12].

threshold voltage of driver transistors. The size of Mp is made larger than the driver transistors to not affect the drive current. When Vin is high, Mp is off and the total sub#threshold current should be n times each driver current, which results in very high current. Therefore Mp reduces and limits the subthreshold current. All pMOS driver transistors work on self-reverse biasing V_{GS} voltage due to the reduced common source node voltage at X, which hugely lowers the subthreshold current.

11.5.1.2 Multithreshold Voltage Technique

It is applied to any CMOS logic to reduce the subthreshold current. Here both high and low threshold voltage devices are used, so called multi threshold voltage technique [13]. In the standby mode, the power dissipation reduces due to the limited subthreshold current by the high threshold voltage device. Other circuit design rules can be implemented to lower the static power dissipation as follows:

i. By avoiding pseudo nMOS circuits in the specific design
ii. By avoiding low threshold voltage devices in I/O buffers
iii. By avoiding low dc current level converters

11.5.2 Dynamic Power Technique

At high frequency, the power dissipation of CMOS circuit is increased severely and effects on the circuit performance. So, low-power circuit approaches are used to diminish the dynamic power of the CMOS circuit [14–16]. The dynamic power dissipation referring to the equation (11.4) can be reduced by:

i. By reducing the output capacitance C_L
ii. By reducing the unused block clock frequency
iii. By reducing the supply voltage in the circuit operation

Some rules can be followed to design the low dynamic circuits as follows:

i. By using low input capacitance
ii. By avoiding dynamic logic style
iii. By selecting low capacitance-based technology
iv. By reducing the switching activity
v. By using minimum device size in non-critical path

11.6 Random Access Memory (RAM) Circuits for Low Power

In recent time, all computers demand high density and high speed static RAM (SRAM). Similarly, laptops, memory cards, notebooks, etc, need low-power and high-density SRAM. Both low supply voltage and low-power circuit technique are used for low-power dissipation in SRAMs, in which memory cell plays an essential role in the design [17]. Low-power dynamic RAM (DRAM) does the battery backup operation. SRAM is preferred over DRAM due to the no refresh operation and high speed.

11.6.1 Low-Power Techniques for SRAM

The total power dissipation in SRAM is due to both active and standby power. The active power includes the components like decoders, memory arrays, sense amplifiers, buffer, etc. Except the memory array, all other component power dissipations depend on the operating frequency, total capacitances, and voltage swing. The standby power consumption is done by lowering the operating voltage [18]. The active power consumption can be reduced by different ways as follows:

i. Reducing the power supply
ii. Lowering the word line and all cells capacitances

iii. Applying multi-step static CMOS decoding
iv. Lowering the dc current with the help of the pulse operation concept

11.6.2 Low-Power Techniques for DRAM

The total power dissipation in DRAM is due to both active and data retention power. The active power includes the components like decoders, sense amplifiers, memory array, refresh circuit, etc. [19, 20]. The active power consumption can be reduced by different ways as follows:

i. Lowering the dc power
ii. Reducing the internal supply voltage
iii. Decreasing all capacitance value

The data maintenance power in DRAM is mostly due to the dc power and refreshes operation by the peripheral circuits. By reducing the circuit frequency and dc current, this power can be diminished [21].

11.7 VLSI Design Methodologies for Low Power

The design methodologies for low-power operation include different levels of abstraction like physical, gate, architectural, and algorithm [22]. Here, all techniques for power reduction are connected with dynamic power dissipation.

11.7.1 Low-Power Physical Design

Several techniques are used to diminish the power of physical design. Here, floor planning, and placement & routing concepts are used for low-power physical design. Floor planning allows the space allocation in a chip for a particular position of modules. It selects a particular implementation for every module to decrease the total power per area of the chip [23]. The placement and routing depend on functional blocks, gate arrays, standard cells, etc. The products of switching activity capacitances are used for low-power placement design.

11.7.2 Low-Power Gate Level Design

In low-power gate-level design, power is traded for area and speed. Here, many concepts are focused to optimize the internal and switching activity capacitances in both technology-dependent and independent phases.

11.7.2.1 Technology Mapping and Logic Minimization

The technology mapping transfers a logic function to a technology dependant circuit for low-power operation. Here, every logic function is divided into two input gates to reduce total power dissipation by lowering the whole switching activity. This mapping is an essential step for gate arrays and standard cells designs [24]. It is following the constraints lie area, power, and delay [25]. The Boolean function is optimized by using logic minimization, which results in power optimization. The signal switching activity is optimized by a restructured logic circuit in the technology independent phase [26]. The delay is the main constraint for power minimization. The power dissipation is enhanced by the incremental area with no delay.

11.7.2.2 Reduction of Spurious Transitions

There is a consumption of power dissipation (30%) due to the unnecessary switching transitions [27, 28]. So, the path delay is balanced by varying the logic structure to optimize the power consumption. Also, the optimized size of logic gates is used to steadiness the delay of the paths. The self-timing technique is used to lower the logic depth and glitch power [28].

11.7.2.3 Power Reduction by Precomputation

Precomputation is synthesized to optimize the logic circuit. It is applied to multiple functions. The selective precomputation adds a duplicate combinational logic and registers to reduce the power savings.

11.7.3 Low-Power Architecture Level Design

The architecture uses many components, like multiplier, adder, memory element, etc. The effect of low-power architecture level is more important than the low-power logic level design. The different concepts are implemented for low power, such as parallelisms, power management, distributed processing, and pipeline. Parallelism is implemented to lower the power dissipation at the cost of the area by maintaining the throughput [27]. In the pipeline, the power dissipation is lowered by increasing the switching capacitances. Though the registers of pipeline structure reduce the logic depth, glitch power is also reduced. The integration of both parallelism and pipeline will reduce power dissipation, due to the reduced supply voltage. The several local hardware resources are optimized than the worldwide hardware in the distributed performance. The power management concepts have a vital role to avoid exhausting power in usual and standby operational mode [29]. Both the static and dynamic power managements are used for low-power architectural design.

Static one allows the power saving in standby mode, whereas dynamic one allows shut down of several blocks to operate a particular application.

11.8 Power Reduction by Algorithmic Level

The algorithm level plays an important role in power saving in a system. The integration of architecture and algorithm level can save immense power. Here, switched capacitance and switching activities lowering concepts are implemented to save power.

11.8.1 Lowering in Switched Capacitance

The primitive operations like ALU, memory, etc., should be optimized to diminish the power in an algorithm. The kind of operation should be considered with respect to optimizing the number of operations in an algorithm.

11.8.2 Lowering in Switching Activities

The power dissipation of processors is lowered by optimizing switching activities. A certain coding of the signals can be used in an exchange for binary code to optimize the switching activities. Gray code is used in microprocessor address lines to lower the net switching activities [30].

11.9 Power Estimation Technique

It is the technique to find the average power dissipation of the circuits. Here, the power analysis technique is explained at circuit, logic, architecture, and behavioral level.

11.9.1 Circuit Level Tool

Circuit level simulation of any design is performed and the average current is calculated for the average power calculation. This technique depends on the circuit input patterns. Statistical methods depend on the input pattern and monitor the power dissipation, which is based on the Monte Carlo concept [31]. The measurement around the existing simulation tool is the main merit of this technique.

11.9.2 Gate Level

The gate-level concept is implemented over the demerit of the circuit level. Here, probabilistic and event driven estimation approaches are used for power estimation. In the probabilistic approach, the pattern independent concept is used to estimate the power dissipation with respect to signal probabilities. It allows for overcoming the simulation technique demerits. This approach can be implemented as a fast logic synthesis power estimator with the merit of no supply simulation pattern. Event driven approach is used for gate-level power analysis for semi-custom design.

11.9.3 Architectural Level

The architectural level consists of functional blocks in which the complexity is low in comparison to gate and circuit levels. Here, many approaches are used for power analysis. The gate count concept is used for power dissipation at architectural level by depending on memory, logic count, busses, circuit, and layout style. The power factor approximation method is used for digital signal processing architecture to calculate power. The dual bit type method allows estimating the accurate power dissipation.

11.9.4 Behavioral Level

The behavioral level explains the input set and the function of a system. Here, the power estimation connects with the consumed energy to an algorithm execution. The final power dissipation is influenced by the behavioral level decision with different orders of magnitude [32].

11.10 Low-Power Flood Sensor Design

By considering the smart device design, low-power flood sensor can be used in our advanced home system [33]. It is used to check the availability of water on individual floors by giving an early signal. Here, anyone can access the remote to the sensor data by mobile networks or the internet. This design can be implemented in real life on a long term basis by lowering the complexity of the system [34]. This system can also be implemented in the industry by detecting the availability of water on the factory floor and lab, which can lead to water savings. The wireless flood sensor is designed according to low-power dissipation and efficiency [35].

11.11 Low-Power VCO Design

This work presents low-power VCO by using dynamic threshold MOSFETs (DTMOSs) [36] and active resistors. One pair of pMOS and nMOS transistors is linked with the cross-coupled pair to operate as an active resistor. DTMOSs are implemented here in the place of MOSFET by connecting the body to gate terminal. Therefore threshold voltage is reduced due to gate voltage variation with low-power operation [37]. The proposed VCO design is illustrated in Figure 11.4.

Here, the transistors $M_{N3,4}$, and $M_{P3,4}$ works as active resistors associated among the cross-coupled pairs $M_{P1,2}$ and $M_{N1,2}$. A capacitor bank with MOS switch topology is implemented in addition to MOS varactors C_{VAR1} and C_{VAR2} to increase the tuning range of VCO. The capacitor divider circuit of C_1-C_2 is implemented to manage the threshold voltage dynamically. DTMOS with low threshold voltage is implemented to optimize the power consumption [38, 39].

The proposed design is simulated in 180 nm CMOS process. This VCO consumes 0.15 mW from the 1.12 V power supply, which is much inferior to 0.42 mW and 0.72 mW of [37] and [40], respectively. This low-power consumption is achieved owing to the low value of V_{th}. Also, it results in the frequency range from 1.77 to 1.82 GHz.

11.12 Low-Power Gilbert Mixer Design

The mixer design focuses on different parameters, such as conversion gain, linearity, noise figure, and power consumption [2]. The proposed design results in low power by simulating in UMC 180 nm CMOS technology. Here, the bulk injection concept is used to lower the power consumption by the integration of the transconductance and switching stage into a single transistor with the elimination of parasitic effects.

The transistors implemented in the bulk injection concept act as a 4-terminal device. RF and LO signals are connected at the gate and body

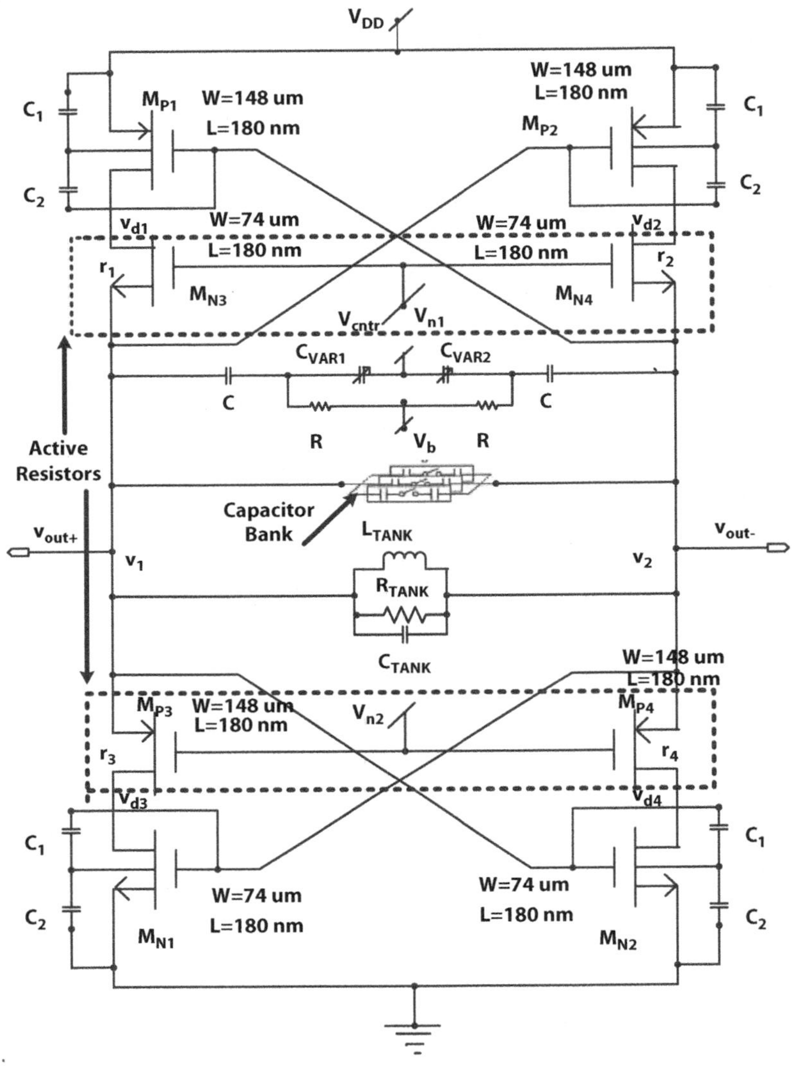

Figure 11.4 Schematic of the proposed VCO design [38].

parts of the device, respectively, to produce the IF signal at the drain side. The Gilbert cell single balanced mixer without and with the implementation of bulk injection technique is illustrated in Figures 11.5 (a) and (b) respectively [41]. It offers low power due to the subthreshold operation [42, 43]. So, the proposed low-power mixer can be implemented as a suitable block in the receiver front end design.

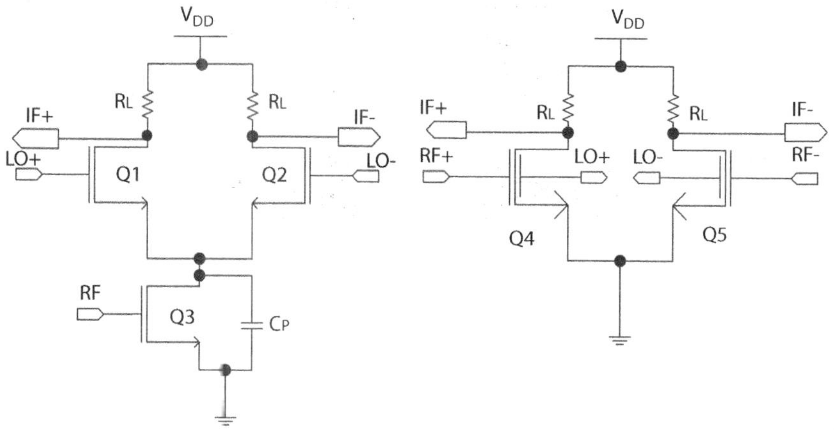

Figure 11.5 Schematic of (a) single balanced mixer (b) bulk injection mixer [42].

11.13 Conclusion

The mounting demand for computing with low cost has led to quicker, lesser, more reliable ICs, which need low power. The technology associated with these advances can be integrated into the advanced electronics systems to construct them extra functionality along with diminishing power requirements. This chapter focuses on the design of basic circuits and design aids for low power consumption, and it also provides different power optimization approaches at the system, algorithm, architecture, logic, and device levels. Here, few circuit designs like flood sensor, VCO, and mixer are discussed for low-power operations. For the demand of present wireless communications, like biomedical telemetries and mobile phones, these low-power design circuits are suitable owing to the better performance.

References

1. Piguet, C., *Technologies and Devices, Low-Power CMOS Circuits*, pp. 1–16, Taylor & Francis Group, Switzerland, 2006.
2. Leung, B., *VLSI for Wireless Communication. Prentice Hall Electronics and VLSI series*, 1st Ed, Pearson Education, Singapore, 2011.
3. Bellaouar, A. and Elmasry, M., II, Low-Power VLSI Design: An Overview, in: *Low-Power Digital VLSI Design*, pp. 1–12, Springer, Boston, MA, 1995.

4. Piguet, C., Are early computer architectures a source of ideas for low-power?, in: *IEEE Alessandro Volta Memorial Workshop on Low-Power Design*, IEEE Computer Society, pp. 4–4, March 1999.
5. Lipoff, S. and Little, A.D., Evaluation of New Battery Technology in Selected Applications. *IEEE Workshop on Low-Power Electronics*, Phoenix, 1993.
6. Rabaey, J.M. and Pedram, M., *Low Power Design Methodologies*, Kluwer Academic Publishers, En: Norwell, MA, 1996.
7. Stork, J.M.C., Technology Leverage for Ultra Low Power Information Systems. *IEEE Symposium on Low Power Electronics*, pp. 1–55, 1994.
8. []https://www.nap.edu/read/5905/chapter/6.
9. Yaeoub, G.Y. and Ku, W.H., An Enhanced Technique for Simulating Short-circuit Power Dissipation. *IEEE J. Solid-State Circuits*, 24, 3, 844–847, 1989.
10. Veendrick, H.J., Short-Circuit Dissipation of Static CMOS Circuitry and Its Impact on the Design al Buffer Circuits. *IEEE J. Solid-State Circuits*, 19, 4, 468–473, 1984.
11. Kakumu, M., Process and device technologies of CMOS devices for low-voltage operation. *IEICE T. Electron.*, 76, 5, 672–680, 1993.
12. Kawahara, T., Horiguchi, M., Kawajiri, Y., Kitsukawa, G., Kure, T., Aoki, M., Subthreshold current reduction for decoded-driver by self-reverse biasing (DRAMs). *IEEE J. Solid-State Circuits*, 28, 11, 1136–1144, 1993.
13. Mutoh, S., II, Douseki, T., Matsuya, Y., Aoki, T., Yamada, J., 1V high-speed digital circuit technology with 0.5/spl mu/m multi-threshold CMOS, in: *Sixth Annual IEEE International ASIC Conference and Exhibit*, IEEE, pp. 186–189, September 1993.
14. Bellaouar, A., Abu-Khater, I.S., Elmasry, M., II, Low-power CMOS/BiCMOS drivers and receivers for on-chip interconnects. *IEEE J. Solid-State Circuits*, 30, 6, 696–700, 1995.
15. Chandrakasan, A.P., Sheng, S., Brodersen, R.W., Low-power CMOS digital design. *IEICE T. Electron.*, 75, 4, 371–382, 1992.
16. Yuan, J.S., Lin, J., Alasad, Q., Taheri, S., Ultra-low-power design and hardware security using emerging technologies for Internet of Things. *Electronics*, 6, 3, 67, 2017.
17. Sasaki, K., Hanamura, S., Ueda, K., Oono, T., Minato, O., Sakai, Y., Toyoshima, H., A 15-ns 1-Mbit CMOS SRAM. *IEEE J. Solid-State Circuits*, 23, 5, 1067–1072, 1988.
18. Yabe, T., Matsuoka, F., Sato, K., Hayakawa, S., Matsui, M., Aono, A., Ochii, K., High-speed and low-standby-power circuit design of 1 to 5 V operating 1 Mb full CMOS SRAM, in: *Symposium 1993 on VLSI Circuits*, IEEE, pp. 107–108, May 1993.
19. Kitsukawa, G., Horiguchi, M., Kawajiri, Y., Kawahara, T., Akiba, T., Kawase, Y., Nishida, T., 256-Mb DRAM circuit technologies for file applications. *IEEE J. Solid-State Circuits*, 28, 11, 1105–1113, 1993.

20. Yamauchi, H., Suzuki, T., Sawada, A., Iwata, T., Tsuji, T., Agata, M., Inoue, M., A circuit technology for high-speed battery-operated 16-Mb CMOS DRAM's. *IEEE J. Solid-State Circuits*, 28, 11, 1084–1091, 1993.
21. Sakata, T., Itoh, K., Horiguchi, M., Aoki, M., Subthreshold-current reduction circuits for multi-gigabit DRAM's. *IEEE J. Solid-State Circuits*, 29, 7, 761–769, 1994.
22. Pedram, M. and Rabaey, J.M. (Eds.), *Power Aware Design Methodologies*, Springer Science & Business Media, New York, Boston, 2002.
23. Chao, K.Y., Low power considerations in floorplan design, in: *Proc. 1994 Intl. Workshop on Low Power Design*, Napa, California, pp. 45–50, 1994.
24. Tsui, C.Y., Pedram, M., Despain, A.M., Technology decomposition and mapping targeting low power dissipation, in: *Proceedings of the 30th International Design Automation Conference*, pp. 68–73, July 1993.
25. Tiwari, V., Ashar, P., Malik, S., Technology mapping for lower power, in: *Proceedings of the 30th International Design Automation Conference*, pp. 74–79, July 1993.
26. Keutzer, K. and Vanbekbergen, P., The impact of CAD on the design of low power digital circuits, in: *Proceedings of 1994 IEEE Symposium on Low Power Electronics*, IEEE, pp. 42–45, October 1994.
27. Chandrakasan, A. Sheng, S., Brodcrren, R.W., Low-Power CMOS Design. *IEEE J. Solid-State Circuits*, 27, 4, 472–484, 1992.
28. Ko, U., Balsara, P.T., Lee, W., A self-timed method to minimize spurious transitions in low power CMOS circuits, in: *Proceedings of 1994 IEEE Symposium on Low Power Electronics*, IEEE, pp. 62–63, October 1994.
29. Gerosa, G., Gary, S., Dietz, C., Pham, D., Hoover, K., Alvarez, J., Kahle, J., A 2.2 w, 80 mhz superscalar risc microprocessor. *IEEE J. Solid-State Circuits*, 29, 12, 1440–1454, 1994.
30. Su, C.L., Tsui, C.Y., Despain, A.M., Low power architecture design and compilation techniques for high-performance processors, in: *Proceedings of COMPCON'94*, IEEE, pp. 489–498, February 1994.
31. Huizer, C.M., Power dissipation analysis of CMOS VLSI circuits by means of switch-level simulation, in: *ESSCIRC'90: Sixteenth European Solid-State Circuits Conference*, vol. 1, IEEE, pp. 61–64, September 1990.
32. Mehra, R. and Rabaey, J., Behavioral level power estimation and exploration, in: *Proc. Int. Wkshp. Low Power Design*, pp. 197–202, April 1994.
33. Teixidó, P., Gómez-Galán, J.A., Gómez-Bravo, F., Sánchez-Rodríguez, T., Alcina, J., Aponte, J., Low-power low-cost wireless flood sensor for smart home systems. *Sensors*, 18, 11, 3817, 2018.
34. Hashim, N.M.Z., Hamdan, N.B., Zakaria, Z., Hamzah, R.A., Salleh, A., Flood Detector Emergency Warning System. *Int. J. Eng. Comput. Sci.*, 2, 8, 2332–2336, 2013.
35. Ghayvat, H., Liu, J., Mukhopadhyay, S.C., Gui, X., Wellness sensor networks: A proposal and implementation for smart home for assisted living. *IEEE Sens. J.*, 15, 12, 7341–7348, 2015.

36. Sun, J., Boon, C.C., Zhu, X., Yi, X., Devrishi, K., Meng, F., A low-power low-phase-noise VCO with self-adjusted active resistor. *IEEE Microw. Wirel. Compon. Lett.*, 26, 3, 201–203, 2016.
37. Assaderaghi, F., Parke, S., Sinitsky, D., Bokor, J., Ko, P.K., Hu, C., A dynamic threshold voltage MOSFET (DTMOS) for very low voltage operation. *IEEE Electron Device Lett.*, 15, 12, 510–512, 1994.
38. Rout, S.S., Acharya, S., Sethi, K., Design of a Low Power and Low Phase Noise VCO Using Active Resistor and DTMOS, in: *Advances in Electrical Control and Signal Systems*, pp. 671–679, Springer, Singapore, 2020.
39. Rout, S.S., Acharya, S., Sethi, K., A low phase noise gm-boosted DTMOS VCO design in 180 nm CMOS technology. *Karbala Int. J. Mod. Sci.*, 4, 2, 228–236, 2018.
40. Pepe, F., Bonfanti, A., Levantino, S., Samori, C., Lacaita, A.L., Suppression of flicker noise up-conversion in a 65-nm CMOS VCO in the 3.0-to-3.6 GHz band. *IEEE J. Solid-State Circuits*, 48, 10, 2375–2389, 2013.
41. Liang, K.H., Chang, H.Y., Chan, Y.J., A 0.5–7.5 GHz Ultra Low-Voltage Low-Power Mixer Using Bulk-Injection Method by 0.18-$\mu $ m CMOS Technology. *IEEE Microw. Wirel. Compon. Lett.*, 17, 7, 531–533, 2007.
42. Kim, M.G., An, H.W., Kang, Y.M., Lee, J.Y., Yun, T.Y., A low-voltage, low-power, and low-noise UWB mixer using bulk-injection and switched biasing techniques. *IEEE Trans. Microw. Theory Tech.*, 60, 8, 2486–2493, 2012.
43. Rout, S.S., Mohapatra, S.K., Sethi, K., Design of cascode mixer based on bulk injection and switched biasing techniques in 180 nm CMOS process for a high performance receiver front end. *J. Low Power Electron.*, 14, 1, 49–56, 2018.

12

GaN Technology Analysis as a Greater Mobile Semiconductor: An Overview

Biyyapu Sai Vamsi, Tarun Chaudhary*, Deepti Kakkar, Amit Tiwari and Manish Sharma

Dr B R Ambedkar National Institute of Technology Jalandhar

Abstract

The solution to the next critical step for an energy-efficient environment is in the use of modern technologies, such as wide-band semiconductors that make for greater energy efficiency, smaller sizes, lighter weight, lower total costs—or all of them together. Metal-Insulator-Semiconductor (MIS) High Electron Mobility Transistor (HEMT), GaN, a semiconductor heterostructures, is currently facing accelerated demand versus investment expense, providing tremendous potential to fuel economic growth of the silicon-based electronic industry. The Electron stability and high electrical field strength have shown tremendous promise for material-level multi-gigahertz communications and photonic applications in the past. The possibility of drawing on the inquiries presented here is very fascinating in ways focused on particular difficulties arising from completely mature markets and applications. This paper describes and analyzes the efficiency of the different epitaxial growth processes of insulator on semiconductor and that GaN devices as quasivertical structure in E-mode (Enhancement) are preferred due to greater convenience and advantages. The state-of-the-art technology uses thermally grown Al2O3 as a dielectric gate and passive layer for the surface. In addition to their popularity and credibility, these GaN machines are also being used in the area of power supply and motor control for industrial, commercial and automotive applications. The rising concern of environmental factors made the study of power electronic devices fabricate with lower leakage current and negligible power losses. Many high-voltage applications and dc/dc, ac/ac, ac/dc converter is provided. The converter is based on a cascaded connection of single-phase sub-multilevel converter units and full-bridge converters. As compared with the conventional converter, the number of voltage sources (dc), installation area,

**Corresponding author*: chaudharyt@nitj.ac.in

Suman Lata Tripathi, Souvik Ganguli, Tengiz Magradze, and Abhishek Kumar (eds.) Intelligent Green Technologies for Sustainable Smart Cities, (247–268) © 2022 Scrivener Publishing LLC

switches, and converter cost is reduced as the number of voltage steps increases. In order to calculate the magnitudes of the required dc voltage sources, some methods are provided. In ac/ac power conversion systems, a small number of asymmetrical power switching devices are used to convert input ac power to output ac power of constant/variable frequency. An ac to dc converter contains a bridge rectifier and boost circuit.

Keywords: GaN, MIS, HEMT, Al2O3

12.1 Introduction

In GaN device due to lack of high quality free-standing substrate, GaN hindering the fabrication of quasivertical devices. GaN-related alloys and heterostructures a 2-DEG[1](2D electron gas) devices, which can be used in direct current (DC /DC) converters, voltage traction inverters, renewable energies, and the most-sought electric vehicles (Figure 12.1).

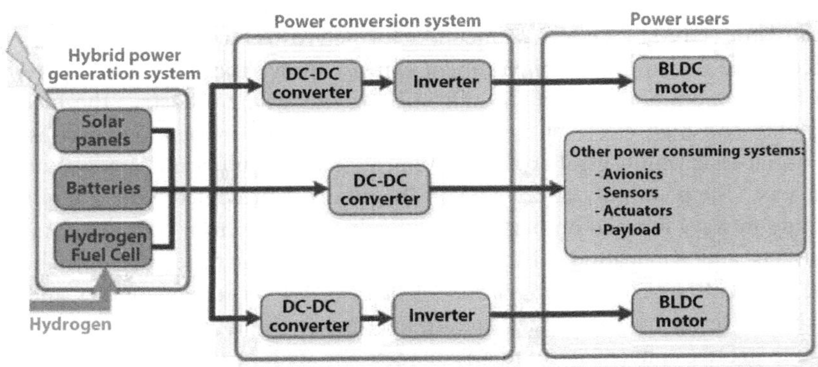

Figure 12.1 UAV scheme under development. A schematic view. (Ref. Development of GaN Technology-based DC/DC Converter for hybrid UAV).

[2], certain limitations of GaN include lateral area habitation and trapping of surface charges in high efficiency device fabrication is reduced by using quasi GaN growth strategies. This trench formation results in multidimensional trapping of charges resulting in capacitance of higher degrees of freedom. GaN devices can operate at very high voltages compared to Si devices which makes their usage feasible in high altitude and high temperature. These wide bandgap semi-conductor properties make GaN and likewise SiC to have low leakage currents and lower power losses and can

be used in DC/DC converters. Due to the thinness of layers in GaN compared to Si-based technology for same threshold voltage, they replace the technology of Si. GaN exhibits the critical value of field 1.5 times greater than of SiC and helps in achieving more power density. The switching transition from low to high termed as switching capacity is higher in GaN making it feasible for replacing other semiconductors. However, the limitation of use of GaN is that it cannot withstand higher temperatures and SiC will be the choice. Therefore, we should typically calculate the figure of merit of SiC and GaN based on application. The unique characteristics displayed by GaN, such as higher On-Resistance (Ron) and higher drain current and breakdown voltage of drain-source [3] makes it prime choice to Si and SiC materials. MULTILEVEL converters is a static, high power converters for medium to high-voltage applications, such as large Ve drives, FACTS devices, dynamic voltage restorers (DVR) and reactive power compensation. The multilevel converters give a desired stepped output voltage waveform by the proper arrangement of the power semiconductor devices from many lower dc voltage sources. The main advantage of multilevel converters is the use of good quality medium power semiconductor devices, which can also operate at reduced voltages. Consequently, the switching losses and voltage stress on power electronic devices are reduced significantly. Also, the output voltage has small voltage steps, which gives good power quality, low harmonic components and better electromagnetic compatibility. Multilevel converters are getting more and more famous in recent years.

The primary winding of transformer is connected between a pair of DC input terminals through an on/off switch and the secondary winding is connected between a pair of DC output terminals through a smoothing & rectifying circuit. The output voltage applied from the smoothing & rectifying circuit to the load is held constant by switching the input voltage through feedback control system. For intermittently operating the switch under light load there is a saw tooth generator circuit which provides a saw tooth voltage having a frequency less than that of the on/off operation of the switch.

An AC/AC power conversion system is presented here. In this, to convert input AC power to output AC power of constant/variable frequency, a small number of asymmetrical power switching devices are used.

An AC/DC converter consists of a bridge rectifier circuit. The boost circuit includes the diode, inductor and load capacitor connected in series and a shunting switch connected to shunt the load capacitor and diode. The control circuit for switching the shunting switch consists 1) a differential circuit, 2) a multiplier, and 3) a duty cycle generator in a feedback loop which is used to maintains a constant output voltage on the capacitor.

GaN-on-Si [4] technology is "normally off" devices require a non-negative gate voltage and results in generation of runaway currents in the form of leakage. Hence, metal-insulator-semiconductor (MIS) technology can be used, which give promising outputs. In radiofrequency application, the 5G wireless operators require higher-frequency efficiency and input power linearity.

12.2 Research and Collected Data

Its the GaN-on-SOI, these silicon on insulator design provides higher thickness of GaN device epitaxy, which results in higher voltage range and design of (quasi) vertical devices and for high current applications and the important use is that it provides low On-Resistance (Ron) architectures. The concerning issue is of current collapse, which is observed as a gradual increase in the Ron when the device is operating, which can deteriorate the performance of the device .so required calculations of isolation and insulation from microdust particles need to be taken care of before fabrication on silicon. But for high-power and low-cost transmissions GaN-on-SOI is compelling technology and for an energy efficient environment as well. To provide low leakage current and low turn-on voltage, the power diodes must be Schottky diodes with Gate Edge Transmission (GET) technology, which makes it feasible. GaN-on-IC technology is used to provide isolation and integration modules as embedded on chips as high side and low side switches.

Cascode [5] GaN model that are generally normally off device with higher threshold voltage (Vt) and higher Ron and internal capacitances (parasitic) compared to other GaN technologies. In contrast to d-mode GaN devices which lack anti-parallel diodes in its internal structure cascode model helps in reducing the reverse recovery dissipation during evanescent switching. This type of configuration is generally used for fail-safe operations in power electronics and to achieve this we cascade low voltage (LV) Si-based MOSFET with high-voltage (HV) GaN device. The higher Ron and parasitic capacitances play an important role in multiple switching networks, and it provides immunity over oscillations at gate terminal. Cascode devices show a smaller rise in temperatures compared to enhancement mode (E-mode) GaN device at higher voltages, which makes cascode devices more adaptable and in reverse I-V characteristics cascode devices provide more temperature linearity than E- mode devices. There is a need to limit the noise which can result in reverse breakdown or overshoot issues which requires caution in selection of signal and power needs.

MULTILEVEL converters is a static, high power converters for medium to high-voltage applications, such as large electric drives, FACTS devices, dynamic voltage restorers (DVR) and reactive power compensation. The multilevel converters give a desired stepped output voltage waveform by the proper arrangement of the power semiconductor devices from many lower dc voltage sources. The main advantage of multilevel converters is the use of good quality medium power semiconductor devices, which can also operate at reduced voltages. consequently, the switching losses and voltage stress on power electronic devices are reduced significantly. Also, the output voltage has small voltage steps, which gives good power quality, low harmonic components and better electromagnetic compatibility. Multilevel converters are getting more and more famous in recent years. The primary winding of transformer is connected between a pair of DC input terminals through an on/off switch and the secondary winding is connected between a pair of DC output terminals through a smoothing & rectifying circuit. The output voltage applied from the smoothing & rectifying circuit to the load is held constant by switching the input voltage through feedback control system. For intermittently operating the switch under light load there is a saw tooth generator circuit which provides a saw tooth voltage having a frequency less than that of the on/off operation of the switch. An AC/AC power conversion systems is presented here. In this, to convert input AC power to output AC power of constant/variable frequency, a small number of asymmetrical power switching devices are used. An AC/DC converter consists a bridge rectifier circuit. The boost circuit includes the diode, inductor and load capacitor connected in series and a shunting switch connected to shunt the load capacitor and diode. The control circuit for switching the shunting switch consists of the following: 1) a differential circuit, 2) a multiplier, and 3) a duty cycle generator in a feedback loop, which is used to maintains a constant output.

DC/DC CONVERTER

The dc/dc converter consists of a transformer having a primary winding connected across a dc/dc power supply through an on/off switch, a smoothing and rectifying circuit which is connected to the secondary winding of the transformer, an output voltage detector circuit, other detector circuit to detect the current flowing through the serial connection of the transformer primary and the switch, and a switch control circuit for on/off control of the switch according to the outputs from the switch current detector circuit and voltage detector circuit. One of the problems that is encountered in this type of dc/dc converter is the relationship between the switching frequency of the dc voltage and the load connected across the transformer secondary through the smoothing and rectifying circuit. The proportion of power loss at the switch to the power required by the load increased when the switching frequency at the time of a drop in power consumption of the load is left as high as under normal power consumption. The solution has proved not totally satisfactory. When the converter put to intermittent operation, the conducting period of the switch inordinately increased at the beginning of each operating period in order to compensate for a voltage drop in output voltage during the preceding nonoperating period. The prolonged closure of the switch invited a large current flow through the switch and the transformer. The result was the production of magnetostrictive noise by the transformer. The noted prior art converter relies for its intermittent operation upon on / off control of the voltage feedback by pulses, which are of rectangular in shape. These pulses are not exactly rectangular in shape but modified to suppress the magnetostrictive noise (ringing) of the transformer by resisting the increased current, which flows through the switch during intermittent operation. But no truly satisfactory suppression of transformer ringing has proved possible in this manner unless the feedback control circuit is designed with full consideration to such factors as fluctuations in the performance characteristics of its individual components. It is also a disadvantage that for intermittent operation of the converter, an additional pulse generator has had to be provided on the output side of the transformer.

AC/AC CONVERTER

This involves ac/ac energy conversion between two single or 3-phased systems using an intermediate dc circuit and finds utility in any number of power conversion applications like UPS, line conditioner, voltage restorer, motor drive and special power supply applications in which the output can provide variable voltage and or variable frequency ac electrical power. For this a nine device 3-phased ac/ac conversion system is provided, which have nine switches in a network forming 3 leg circuits with 3 switches in each leg circuit to convert ac input power at three input terminals to ac output power at three output terminals. The leg circuits are separately connected to the corresponding input terminal and an output terminal. First, switch coupled between the input terminal and a first dc terminal; second, switch between the input terminal and the output terminal; third, switch between the output terminal and a second dc terminal. A control system provides signals to the switches, which convert input ac electrical power from the input connection to output ac electrical power at the output connection.

Each leg circuit is provided with a set of signals to selectively place leg circuit in one of three states including, first state to couple the corresponding input and output terminals to the first dc terminal, second state to couple the input and output terminals to the second dc terminal, and third state to couple the input terminal to the first dc terminal and to couple the output terminal to the second dc terminal. In some cases, the control signals are pulse width modulated by the control system using sinusoidal pulse width modulation and the control system provides the switching control signals to convert input ac electrical power from the input to variable frequency output ac electrical power. The switches may be asymmetrical devices that can block voltage in only one direction and allow current flow in both the directions. Three-phase AC / AC conversion system, which consists a switching network including a total of nine switching devices able to block voltage only in one direction and allow current flow in both the directions. The converter has a control system, which is coupled with the switching network and controls the switching devices to convert input AC electrical power from a three-terminal input connection to output ac electrical power at a three terminal output connection. In some cases, the switching network comprises three leg circuits of three switches per leg, with each of the leg circuits being connected to a corresponding pair of input and output terminals. The control system in one example, which provides switching signals to the leg circuits in one of three states: 1) to selectively couple the corresponding input & output terminals to the first DC terminal in a first state, 2) to couple the input & output terminals to the second DC terminal in a second state, and 3) to couple the input terminal to the first DC terminal & to couple the output terminal to the second DC terminal in a third state.

Single-phase AC/AC converters with a single phase AC input having first & second input terminals, a single-phase AC output having first & second output terminals, and a DC circuit comprising first & second DC terminals and an energy storage component. Some implementation includes a total of six switching devices forming two leg circuits with a total of three switching devices in each leg circuit, where each leg is connected to a corresponding AC input terminal & a corresponding AC output terminal. In this case, 1) a first switching device of each leg is coupled between the corresponding input terminal and the first DC terminal, 2) a second switch is coupled between the input terminal and the output terminal, and 3) a third switch is coupled between the output terminal and the second DC terminal, where a control system provides switching control signals to convert the single-phase AC input power to output AC power at the output connection.

AC/DC CONVERTER

An AC to DC converter contains a boost circuit, which includes a diode and inductor connected in series with a shunting switch and a load capacitor connected to shunt the diode and load capacitor. The shunting switch is controlled by a control circuit having frequency higher than the frequency of the AC input. The control circuit also includes a differential circuit for comparing an output voltage signal to at least one reference voltage to generate a feedback signal. The feedback signal is just a multiplier, which is then applied to a multiplying circuit. The value of the multiplier is such that the voltage input signal and the product of the multiplier is proportional to the current passing through the inductor which is required to provide an output voltage of desired level. To generate a current command signal, the multiplier and the voltage input are applied to the multiplying circuit. To control the duty cycle of the shunting switch, a duty cycle generator responds to the current command signal. Because of the pulsating nature of the power input from the AC line, there is certainly some ripple voltage at the output at twice the line frequency. In the conventional converters, this ripple is fed back through the loop regulating the output and causes distortion in the input current waveform which causes a deviation from the desired unity power factor input. The differential circuit used to eliminate response of the feedback circuit to the ripple. The differential circuit responds very fast to levels of the output voltage which are outside of high & low limits of the amplitude. The differential circuit responds slowly to levels of the output voltage, which are in a dead band, which is in the high & low limits of amplitude. The difference between the high and low limits of the differential circuit is set to be about equal to and greater than peak to peak differences of an expected ripple in the output voltage. So, the normal output ripple lies entirely within the band and as a result no error signal is generated due to ripple. The differential circuit comprises an OP-AMP, which is driven to increase the multiplier if the output voltage signal is below the low limit or to decrease the multiplier if the output voltage is above the high limit and which holds the multiplier constant if the output voltage is between the low and high limits. The OP-AMP is driven by a pair of differential amplifiers through corresponding diodes. The differential amplifiers have a common reference as an input terminal, and other inputs to the differential amplifiers are then obtained from a voltage divider which attenuates the output voltage signals. The gain of the major feedback loop is controlled by a minor feedback loop consists the multiplying circuit and differential circuit. So, the incremental gain from the output voltage to the current command signal will be independent of the input line voltage when the DC output voltage goes outside the regulation band, Notch filter and a sample and hold circuit uses the dead band defined by high and low limits in the feedback circuit. The AC to DC converter provides nearly optimal regulation of the output voltage and power factor of over 99%. The circuit is more suitable for discrete or integrated implementation (Figure 12.2).

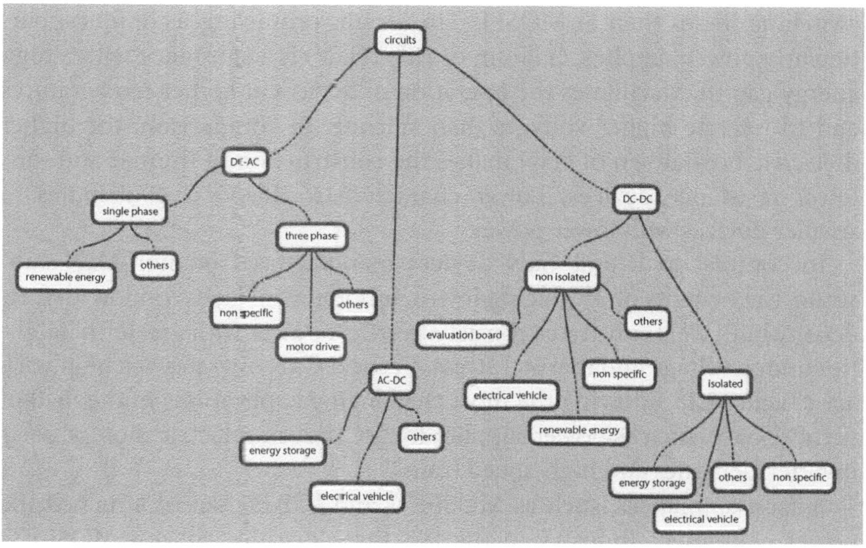

Figure 12.2 All the categories from previous reviewed papers. (Ref. Review of GaN HEMT Applications in Power Converters over 500 W).

12.3 Studies Reviewed and Findings

Here to overcome few of the abovementioned factors we use metal-insulator-semiconductor (MIS) HEMT [6] technology and certain observations in different fabrication methodologies are: High K and High Band gap (Eg) of major insulators (HfO2, CeO2, ZrO2) make perfect match with characteristics of GaN and higher values of gm (trans-conductance) can be achieved. GaN HEMT structures suffer from many disadvantages, few of which are to obtain high breakdown voltages. A large amount of space must be left between the gate and drain of the device, that can result in increased cost to density of current ratio. Additionally, HEMT structures are typically, depletion mode field effect transistors (FETs), which means that HEMT devices exhibit normally on operation and must be supplied a negative gate-bias in order to enter the blocking condition. New era Power technologies produce vertical power devices on GaN substrates using the homo-epitaxial GaN. Vertical GaN devices can switch and operate at higher speeds, leading to a new generation of powerful and more efficient electronic devices. 90% smaller than silicone are vertical GaN units. Capacity is associated directly with the device area. The smaller the unit, the lower the power and capacitance. The smaller the capacitance, the higher the level of switching. Vertical GaN delivers 67% on average lower switching losses than Si MOSFET in certain promising techniques, particularly power supplies. Gallium nitride (GaN) is a substance with a high energy gap that facilitates the operation of devices at higher temperatures and to tolerate higher voltages than silicone. In comparison, the higher dielectric breakdown of GaN makes the construction of thinner and thus lower resistance devices. Lower characteristic Rds(ON) contributes to smaller devices with lower power.

In contrast to lateral GaN devices manufactured on non-GaN substrate, the benefits of producing low-defect epitaxial layers on low-defect-density bulk GaN substrates consist of vertical power with greater reliabilities under-voltage and thermal stress. Vertical GaN operates at a high voltage (Table 12.1), allowing the most challenging applications in the field of vertical GaN, such as power supplies for DC servers, electric motors, solar inverters, engines and high-speed trains.

Silicon appliances, such as MOSFETs and IGBTs, have dominated the power equipment industry in the last three decades. Silicon MOSFETs recently saw only an incremental improvement in efficiency. Silicon is up to its bounds. There is not much that can be learned from silicon power systems depending on their material properties. At present, silicon carbide (SiC) is another replacement to silicon; however, but GaN usually has more desirable material characteristics.

GaN systems currently are manufactured with hybrid medium: thin silicon GaN films, or silicon carbide structures that build GaN-on-Si or

Table 12.1 GaN material properties differentiation between Si and SiC.

Parameter	**Si**	**SIC**	**GaN**
Bandgap(eV)	1.12	3.26	3.45
Critical electrical Field(Ec)	0.3	2.2	3.3–3.7
Baliga Figure of Merit	1	675	3000

Ref Vertical GaN Devices for a New Power Electronics. To fully exploit the properties of GaN, NexGen Power Systems Inc. is fabricating vertical power devices using homo-epitaxial GaN epitaxy on GaN substrates.

GaN-on-SiC HEMT structures (Figure 12.1). Lateral GaN-on-Si (or GaN-on-SiC) devices combine material and the CTE coefficient), which sacrifice both durability and efficiency, with a misaligned coefficient of thermally expanding. Furthermore, the channel is very close to the surface in a standard GaN HEMT system (in order of a few hundred nanometers), creating passiveness and cooling issues.

Due to the sensitivity to current conductions near the system surface, GaN-on-HEMTs are difficult to cool from the top. The buffer layers that distinguish Si from the GaN layer restrict the efficiency of the cooling on the bottom side. This ensures that sometimes custom packages must be produced to cool GaN-on-Si HEMTs and raise their costs even further.

Vertical GaN technology releases the full potential of superior material properties as GaN relies on GaN substrates produced homoepitaxially. Vertical GaN devices also use all three spatial dimensions: increased voltage breakdown by increasing the drift thickness, and low Rds(ON) efficiency by increasing the space of the device and essentially producing a 3-D device that is capable of decoupling the power and the breakdown of the drifting layer (Rds(ON)). High-performance power factor correction circuits are required for AC systems that significantly reduce the harmonic distortion. The high frequency of vertical GaN switching allows new control algorithms and offers smaller implementations and higher efficiency.

Vertical GaN devices are designed to transmit the current via the drift layer inside the bulk of the transistor. Thus, no dynamic Rds(ON) mechanism is available that is created by loads trapped by impurities in the surface interface. The extension into the channel of the depletion area of the gate source diode monitors current flow from the drain to the source. If avalanche voltages are exceeded, the reverse polarized gate source diode is the first to occur, which leads to an increase in the avalanche current and the channel opening and running.

Recently, there have been numerous efforts to alter the structure of a typical HEMT to produce enhancement mode FET devices (i.e. normally off operation). One such modification is the p-type GaN cap layer device. White lines indicate the path of electron conduction.

This structure facilitates normally off operation via trapping negative charges at the p type cap layer's interface with the i-AlGaN, this causes a discontinuity in the 2-D electron gas layer. Another device which modifies typical HEMT device operation is a cascode packaged HEMT device, manufactured by Transphorm. This device is rather amalgamation of typical GaN HEMT and low voltage Si-FET than just a structural pairing like p-cap GaN HEMT that uses structural modifications in GaN HEMT.

The benefit of this configuration is two-fold: gate control of the device is identical to the gate control of a low-voltage Si MOSFET, and the device, exhibits normally off operation. Important to note is that this device structure is not without setback, namely high-temperature operation is expected to be severely limited due to the inclusion of a Si device. Figure 12.3 HREM cross sectional image of device under no stress. Left to right snaps represent the source, gate and drain side. The horizontal line represents the junction between substrate and terminals. Standard III–V processing technology was used to fabricate the devices. Gates of 0.25 μm length were formed by etching a gate trunk in the SiN_x passivation layer followed by the deposition of evaporated Pt/Au metal. The trunk and the gate cap were defined by e-beam lithography. We generally prefer Al compounds such as Al2O3 with AlGaN but it results in higher forward bias leakage currents so we have to improve the energy gap between insulator and AlGaN to reduce the leakage currents. In field of power conversion, rapid trench MOSFET technologies involving semi-quasi (vertical) GaN technology are growing at rapid pace. [7] (Figure 12.4) in Al2O3/GaN MOS

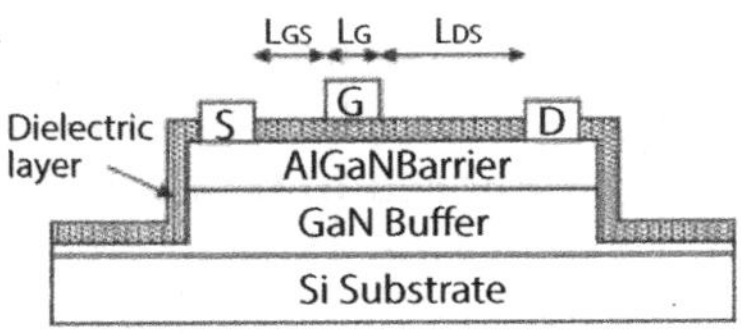

Figure 12.3 Cross sectional view of MIS-HEMT (Metal-insulator-semiconductor HEMT). (Ref. AlGaN/GaN MIS-HEMT with PECVD SiNx, SiON, SiO2 as Gate Dielectric and Passivation Layer).

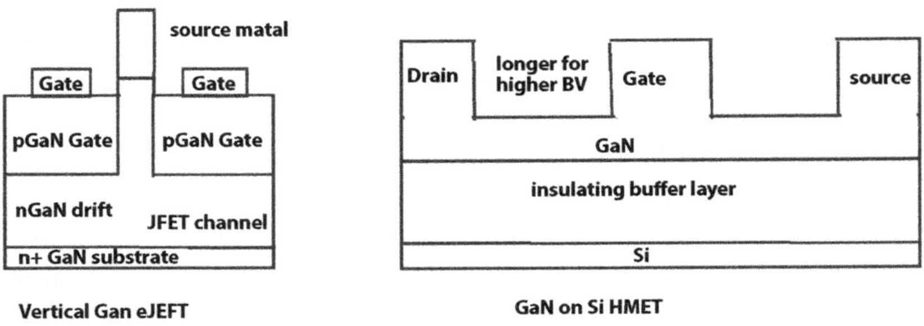

Figure 12.4 Vertical GaN eJFET and GaN-on-Si HEMT diagram. (Ref. Vertical GaN Devices for a New Power Electronics. To fully exploit the properties of GaN, NexGen Power Systems Inc. is fabricating vertical power devices using homo-epitaxial GaN epitaxy on GaN substrates.)

structures, trenches and trappings at different locations, which are responsible for change in threshold voltage is due to degeneration of substrate state dielectric and states juxtapose at Al2O3/GaN interface and frontier border sites. The other fabrication technology is that of SiNx in AlGaN/GaN [8] MIS HEMTs, which also causes leakage current in forward bias due to contamination. To reduce this, we need to improve the complexity of epitaxial method and cost of cleaner environment increases much. So metal organic chemical vapor deposition (MOCVD) can be used and can create less dangling bonds. In-situ methods are used to achieve single crystalline structure and can be observed under transmission electron microscope (TEM) of very high resolution [9]. Monolithic power amalgamation exploits the high-frequency efficiency operations of GaN which shows superior performance than GaAs. In HEMTs. The voltage applied at gate terminal is negative and which is not fault-proof thereby requiring greater safety during operation making GaN hold back better performance at very high-voltage and high-power DC/DC conversion devices. Another method using atomic layer deposition (ALD) to deposit SiO2 on GaN and AlGaN using tris (dimethylamino silane) as a Si-source but higher leakage currents are observed and hence we need to be very careful in choosing the starting materials. ALD GAN The sample used in this letter is a commercial Al0.26Ga0.74N/GaN HEMT wafer grown by MOCVD on (111) silicon substrate. It includes a 1.8-μm GaN buffer, a thin AlN (~1-nm) interface enhancement layer, an 18-nm un-doped AlGaN barrier, and a 2-nm undoped GaN cap. The fabrication processes started with mesa etching. Ohmic contacts were formed by an alloyed Ti/Al/Ni/Au metal stack. The specific contact resistance is 0.7 $\Omega \cdot$ mm. Ni/Au was used for the gate HUANG *et al.*: effective passivation of AlGaN/GaN HEMTs by AlD-grown AlN thin film electrode Atomic force microscopy confirms that the rapid photothermal processing (RPP) process is reliable in removing the surface oxide layer. For the ALD-AlN passivation of S3, a 4-nm AlN layer was deposited immediately after the RPP at 300 °C, using N2–H2 and trimethylaluminum (TMA) plasma precursors as the N and Al sources. The growth rate is ~0.6 Å/cycle. For all the devices characterized in this letter, the gate length *LG* and gate–source separation *LGS* are 1.5 and 1 μm, respectively. The gate–drain separation *LGD* is designed to be 2, 5, and 10 μm. All the measurements were performed with the substrate grounded unless specified otherwise. An atomically sharp interface between the ALD-AlN and the AlGaN/GaN HEMT is observed, as shown in the high-resolution transmission electron-microscopy (HRTEM) Regional polycrystalline domains appear at the top portion of the AlN layer as discussed in Figure 12.5 and Figure 12.6. Atomic concentrations of Al, N, O, Ga, and C were measured

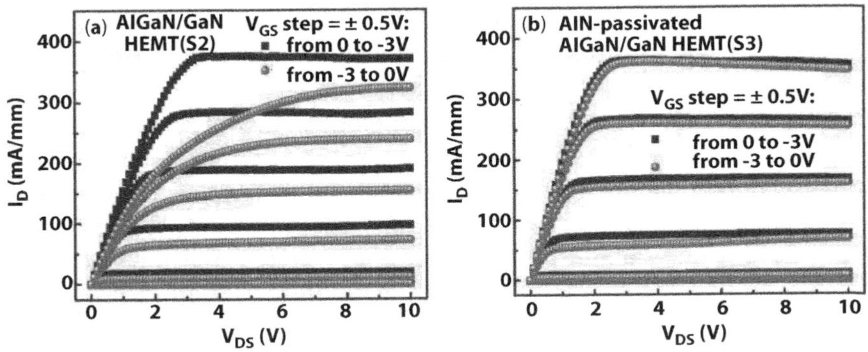

Figure 12.5 Output characteristics of AlGaN/GaN HEMTs with $LGD = 5\ \mu m$. (a) S1. (b) S3. (Ref. AlN passivation by plasma-enhanced atomic layer deposition for GaN-based power switches and power amplifiers).

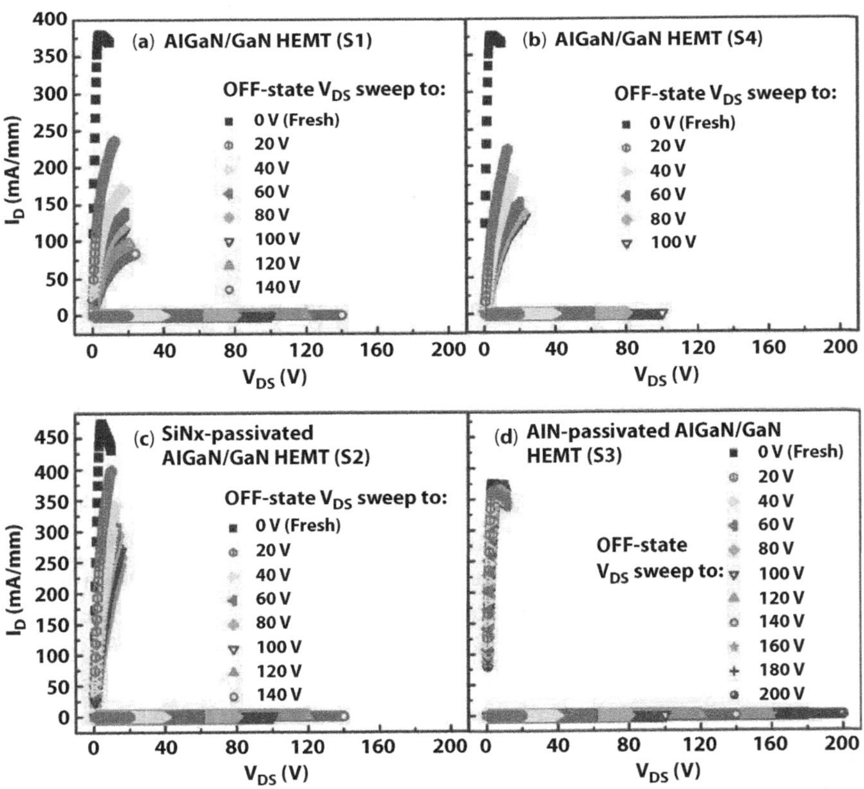

Figure 12.6 Transient output characteristics of AlGaN/GaN HEMTs with $LGD = 5\ \mu m$. (a) S1. (b) S4. (c) S2. (d) S3.

by X-ray photoelectron spectroscopy (XPS). Near the AlN–GaN interface, low oxygen concentration (~8%) was observed compared to 41% nitrogen. The higher concentration of surface oxygen denotes the oxidation of AlN surface due to atmospheric exposure. In the XPS study, the Ga 2*p*3/2 core-level spectrum was used to monitor the Ga-related bonds. It is confirmed that Ga-O bonds are effectively removed by the RPP process. The detailed analysis of the XPS experiment will be presented elsewhere. Figure 12.7 illustrates transient characteristics of output.

The source–drain bias is always swept from low to high at the same sweeping rate, while the gate is stepped up and down. The minimum gate bias (−3 V) is lesser the threshold voltage (−2.1 V) of S1 and S3.

The switching from the OFF to the ON state ($VGS = 0$ V) is a single event with a switching time of 70 ms, and the accumulative OFF-state time to sweep $VDS = 200$ V is 46 s

In InAlN/GaN, the InAlN acts as electron supplier and similar improved Reverse Bias(RB) characteristics and low leakage currents at low gate biases are observed but there is a remarkable leakage current at high RB mainly originating from the intrinsic layer due to large electric field in the InAlN barrier layer. [10] using PECVD we deposit insulation layers onto HEMT which shows certain properties unique to the deposited layer such as maximum breakdown voltage is shown by SiON MIS HEMT and on-off current ratio. SiNx layer on MIS HEMT displayed greater leakage current density but performance is well in current collapse. It also displays lowest deep trap and highest shallow trap densities. The MIS diode early collapse is observed in SiO2-based MIS HEMTs which possess minimal leakage current densities and displays largest density of deep traps. The spectrum of photoluminescence confirms the factors causing current collapse. Due to lower deposition rates of ALD, Al2O3 and HfO2 are not suitable for providing passivation even though they show greater decrease in current collapse and gate leakage currents. GaN HEMT [11] sees applications in power

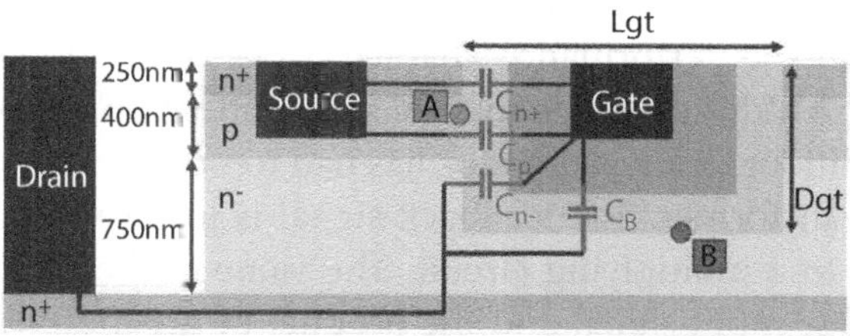

Figure 12.7 Cross-sectional schematic of trench-based vertical MOS device. (Ref. Modeling of gate capacitance of GaN-based trench-gate vertical metal-oxide-semiconductor devices).

converter circuits which include DC/DC, AC/DC and DC/AC converters. Isolated DC/DC converter comprises of phase shifted full bridge converter, Dual active bridge and LLC resonant converter. The transformer in DC/DC converter provides galvanic isolation which is not so efficient but can provide promising output characteristics with Zero Voltage Switching or Zero Current Switching. But in non- isolated DC/DC converters we omit the disadvantages of transformer by eliminating the galvanic isolation and the mandatory concern is of DC current injection.

In renewable energy–based power generation use of GaN HEMT in unique converter topologies which include multi input topologies like interleaved converters because it is needed to assimilate distributed generation systems. One of the problems that is encountered in this type of dc / dc converter is the relationship between the switching frequency of the dc voltage and the load connected across the transformer secondary through the smoothing and rectifying circuit. The proportion of power loss at the switch to the power required by the load increased when the switching frequency at the time of a drop in power consumption of the load is left as high as under normal power consumption. The solution has proved not totally satisfactory. When the converter put to intermittent operation, the conducting period of the switch inordinately increased at the beginning of each operating period in order to compensate for a voltage drop in output voltage during the preceding non-operating period. The prolonged closure of the switch invited a large current flow through the switch and the transformer. The result was the production of magnetostrictive noise by the transformer. The noted prior art converter relies for its intermittent operation upon on / off control of the voltage feedback by pulses which are of rectangular in shape. These pulses are not exactly rectangular in shape but modified to suppress the magnetostrictive noise (ringing) of the transformer by resisting the increased current which flows through the switch during intermittent operation. But no truly satisfactory suppression of transformer ringing has proved possible in this manner unless the feedback control circuit is designed with full consideration to such factors as fluctuations in the performance characteristics of its individual components. It is also a disadvantage that for intermittent operation of the converter an additional pulse generator has had to be provided on the output side of the transformer. The feedback signal is just a multiplier which is then applied to a multiplying circuit. The value of the multiplier is such that the voltage input signal and the product of the multiplier is proportional to the current passing through the inductor which is required to provide an output voltage of desired level. To generate a current command signal, the multiplier and the voltage input are applied to the multiplying circuit. To control the duty cycle of the shunting switch a duty cycle generator responds to the current command signal. Because of the pulsating nature of the power input from the AC line, there is certainly some ripple voltage at the output at twice the line frequency. In the conventional converters, this ripple is fed back through the loop regulating the output and causes distortion in the input current waveform which causes a deviation from the desired unity power factor input. The differential circuit used to eliminate response of the feedback circuit to the ripple. The differential circuit responds very fast to levels of the output voltage which are outside of high & low limits of the amplitude. The differential circuit responds slowly to levels of the output voltage which are in a dead band which is in the high & low limits of amplitude. The difference between the high and low limits of the differential circuit is set to be about equal to and greater than peak to peak differences of an expected ripple in the output voltage. So, the normal output ripple lies entirely within the band and as a result no error signal is generated due to ripple. The differential circuit comprises an OP-AMP which is driven to increase the multiplier if the output voltage signal is below the low limit or to decrease the multiplier if the output voltage is above the high limit and which holds the multiplier constant if the output voltage is between the low and high limits. The OP-AMP is driven by a pair of differential amplifiers through corresponding diodes. The differential amplifiers have a common reference as an input terminal, and other inputs to the differential amplifiers are then obtained from a voltage divider which attenuates the output voltage signals. The gain of the major feedback loop is controlled by a minor feedback loop consists the multiplying circuit and differential circuit. So, the incremental gain from the output voltage to the current command signal will be independent of the input line voltage when the DC output voltage goes outside the regulation band, Notch filter and a sample and hold circuit uses the dead band defined by high and low limits in the feedback circuit. The AC to DC converter provides nearly optimal regulation of the output voltage and power factor of over 99%. The circuit is more suitable for discrete or integrated implementation. The average power obtained from an AC supply through an AC to DC converter is always less than the product of the RMS current and RMS voltage. power factor is the ratio of the average power to the product of RMS voltage and current. Commonly the switching power supply has an input power factor of 65%; this means that the power drawn from the utility line is 65% of the product of the voltage and current in the utility line and only 65% can be obtained with unity power factor. The voltage supplied from an outlet is a sinusoidal, but the system commonly draws non sinusoidal current. Unity power factor is obtained by synchronizing sinusoidal current to the sinusoidal voltage. Assuming that the current is sinusoidal also it limits the harmonic current. One case of AC to DC converter which provides the power factors approaching unity is a full wave rectifier which is followed by a boost converter. In the boost converter, the output of the full wave rectifier is applied to a series connection of a load capacitor, diode and an inductor. The charge carried by the load capacitor is the supply for the computer or other device which utilizing the power. A shunting switch is coupled from a node between the diode and inductor to ground. The shunting switch is alternately on and off at a high frequency, such as 40 kHz. When the switch is closed, the inductor is charged by current from the bridge rectifier to ground. When opened, the inductor discharged and drives a current spike through the diode to charge the capacitor. In this type of system, the duty cycle of the switch can be controlled in such way that the current strictly follows the voltage applied from the bridge rectifier in waveform. To control the duty cycle of the shunting switch there is a feedback circuit which assures that the output voltage stored on the load capacitor is retained at a required level. Rectifiers or AC/DC converters are profoundly used in adjustable speed drives and in UPS (Uninterrupted power supply) systems. The major disadvantages in this type of line frequency-based converters is that they are generally large sized, low quality of output power due to harmonics in current, slow voltage regulation at output and low efficiency. These factors make manufacturers tend to use pulse width modulation-based AC/DC converters with correction of power factor capabilities which work on switch-mode techniques.[12] The significant rise in the use of flexible LEDs, a rat race between the semi-conductors can be seen. It depends on multiple factors, such as consumption of power, lifetime, stability to changes in temperature. any researchers studied the flexible type conventional micro LEDs before the intervention of Internet of things (IoT) which revolutionized the living style and more comprehensive techniques in fabricating flexible micro LEDs based on GaN are explored. Figure 12.8 elucidates the fabrication procedure. The lightness and ultra-sensitivity of these GaN-based micro LEDs make them capture smallest mechanical distress due to bending. An experiment is carried out on mice where flexible micro LEDs are inserted into their cortex without damaging the neuro and motor activities.

They use monolithic fabrication technique ,and Ag (silver) nanowire is used. The exceptional characteristics shown by these flexible GaN-based micro LEDs are great optical power, longer lifetime, and greater stability to mechanical stress and temperature changes.

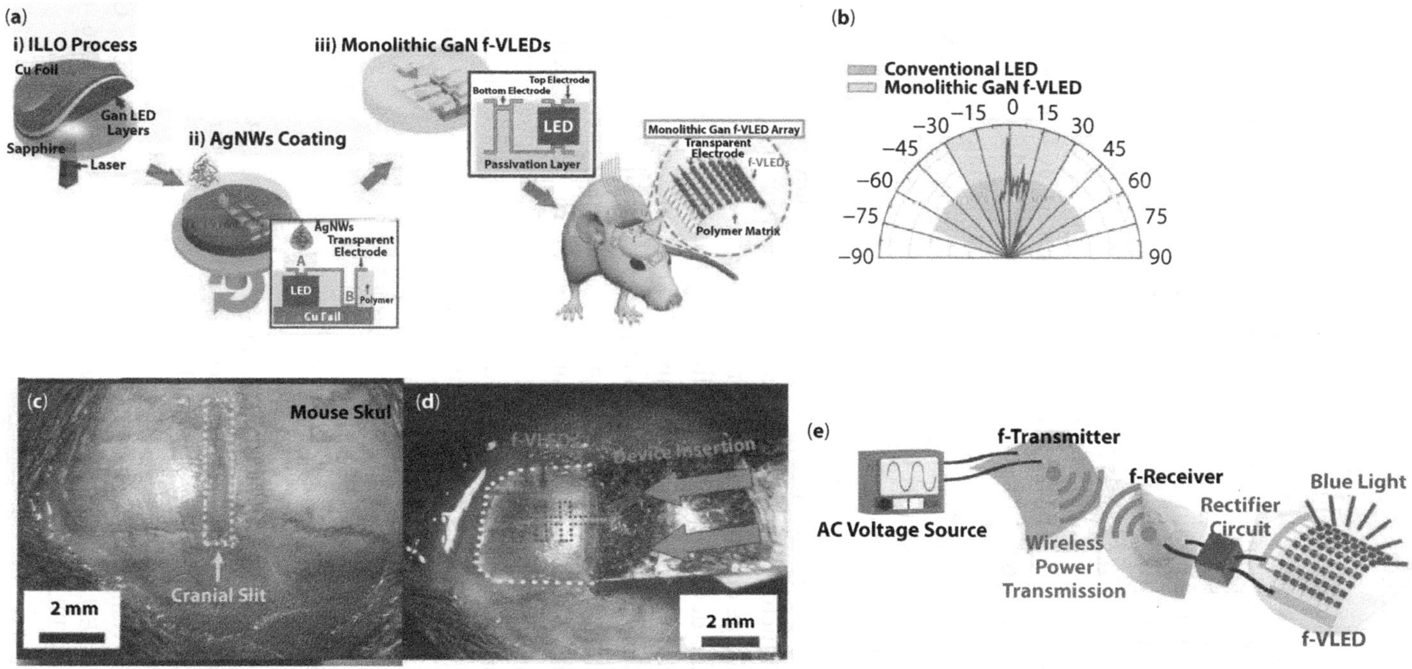

Figure 12.8 (a) The fabrication procedure of monolithic GaN with a biomedical application. b) The angular distribution of light in monolithic blue f-VLED. Blue region represents the angle of light emission of GaN f- VLED (monolithic). c) A slit is made into the cranial part of skull of rodent. d) A GaN f-VLED is inserted into the sub-cranial incision. e) A resonance coupling of inductance of system of GaN-based f-VLED schematic. (Ref. Monolithic Flexible Vertical GaN light-emitting diodes for a transparent wireless brain optical Stimulator).

12.4 Conclusion

By exploiting the vast uses and applications of the GaN-based compound devices, which takes the future by replacing many Si-based devices but need to be processed and manufactured with utmost precision that can make them a bit expensive.There is a need to find the supporting networks for circuit spares and minimization of area of board to enhance switching properties. A wide study can be done in understanding the capacitance distributions in vertical GaN-based FETs. With vast applications in power electronics and semiconductor physics of iii-iv compounds. The never-ending applications of MIS HEMTs and 2-DEG devices in DC/DC converters in EVs. Power linearity of the GaN devices make it useful in communication devices of 5G wireless network. In power switching and RF applications we saw good results using ALD-Al2O3 layered AlGaN/GaN MIS HEMT. In AC/DC converters we can use for energy storage and in EVs and others. In DC/AC converters, RE (Renewable energy) source devices use single phase converters and Motor drivers us three phase converter. This paper reviewed the applications from communication part and physics of semiconductor part and the areas of power electronics where efficiency and leakage currents play major role. The future GaN applications for high-voltage devices depend heavily on improving the quality of the material and developing vertical systems based on bulk GaN. In this broad scenario, the share of the SiC and GaN in power electronics is expected to expand as each material will enter the market at different speeds in different applications. But GaN is still a distance from SiC in power electronics. The material is actually less mature and therefore less adopted at present. For this reason, it must still be justified that GaN devices are included in power electronic products. Reliability is a big challenge for GaNs application in power systems in this context.

References

1. Tripathy, S., Lin, V.K., Dolmanan, S.B., Tan, J.P., Kajen, R.S., Bera, L.K., Han, S., AlGaN/GaN two-dimensional-electron gas heterostructures on 200 mm diameter Si (111). *Appl. Phys. Lett.*, 101, 8, 082110, 2012.
2. Li, H., Zhang, X., Wen, L., Brothers, J.A., Yao, C., Han, C., Liu, L., Xu, J., Puukko, J., Wang, J., Paralleled Operation of High-Voltage Cascode GaN HEMTs. *IEEE J. Emerg. Sel. Top. Power Electron.*, 4, 815–823, 2016, doi: 10.1109/JESTPE.2016.2557316.

3. Chen, K.J., Häberlen, O., Lidow, A., lin Tsai, C., Ueda, T., Uemoto, Y., Wu, Y., GaN-on-Si power technology: Devices and applications. *IEEE Trans. Electron Devices*, 64, 3, 779–795, 2017.
4. Roccaforte, F., Fiorenza, P., Greco, G., Nigro, R.L., Giannazzo, F., Iucolano, F., Saggio, M., Emerging trends in wide band gap semiconductors (SiC and GaN) technology for power devices. *Microelectron. Eng.*, 187, 66–77, 2018.
5. Hashizume, T., Nishiguchi, K., Kaneki, S., Kuzmik, J., Yatabe, Z., State of the art on gate insulation and surface passivation for GaN-based power HEMTs. *Mater. Sci. Semicond. Process.*, 78, 85–95, 2018.
6. Geng, K., Chen, D., Zhou, Q., Wang, H., AlGaN/GaN MIS-HEMT with PECVD SiNx, SiON, SiO2 as gate dielectric and passivation layer. *Electronics*, 7, 12, 416, 2018.
7. Lee, H.E., Choi, J., Lee, S.H., Jeong, M., Shin, J.H., Joe, D.J., Lee, K.J., Monolithic flexible vertical GaN light-emitting diodes for a transparent wireless brain optical stimulator. *Adv. Mater.*, 30, 28, 1800649, 2018.
8. Ma, C.T. and Gu, Z.H., Review of GaN HEMT applications in power converters over 500 W. *Electronics*, 8, 12, 1401, 2019.
9. Borga, M., Mukherjee, K., De Santi, C., Stoffels, S., Geens, K., You, S., Meneghini, M., Modeling of gate capacitance of GaN-based trench-gate vertical metal-oxide-semiconductor devices. *Appl. Phys. Express*, 13, 2, 024006, 2020.
10. Martinez-Heredia, J.M., Colodro, F., Mora-Jiménez, J.L., Remujo, A., Soriano, J., Esteban, S., Development of GaN Technology-Based DC/DC Converter for Hybrid UAV. *IEEE Access*, 8, 88014–88025, 2020.
11. Mukherjee, K., Borga, M., Ruzzarin, M., De Santi, C., Stoffels, S., You, S., Meneghini, M., Analysis of threshold voltage instabilities in semi-vertical GaN-on-Si FETs. *Appl. Phys. Express*, 13, 2, 024004, 2020.
12. Sun, R., Lai, J., Chen, W., Zhang, B., GaN power integration for high frequency and high efficiency power applications: a review. *IEEE Access*, 8, 15529–15542, 2020.

13

Multilevel Distributed Energy Efficient Clustering Protocol for Relay Node Selection in Three-Tiered Architecture

Deepti Kakkar*, Gurjot Kaur and Aradhana Tirkey

Dept. of Electronics and Communication Engineering, NIT, Jalandhar, India

Abstract

Functioning of WSNs is constrained by the limited battery power of sensor nodes as in most of the applications, it is impractical to replace or recharge the battery. Preserving the energy of the network, in order to maximally make the network work well with a long life, is a crucial design concern. Researchers have propounded various methods for relaying data via relay nodes in three-tiered architecture for clustered heterogeneous WSNs, relay node with most energy level and relay with minimum ratio of distance over energy also called rate forward. These methods compete with each other in termsof number of alive nodes and conserving the energy of relay nodes during subsequent rounds. The missing factor in these methods is the initial energy of node, which must be focused for balancing the load of data transmission by relay nodes. To overcome these limitations, genetic algorithm (GA) approach in multilevel DEEC protocol for relay node selection in three-tiered architecture is proposed. In this work, the selection for relay node is optimized in terms of initial energy of the node, residualenergy of node, distance of node from BS, and its probability for getting selected as CH. With optimized relay, the lifetime of the network is observed to be enhanced considerably. It also contributes a lot in decreasing the energy consumption of the network. This technique may be practically implemented with great ease.

***Keywords*:** WSN, clustering, heterogenous networks, genetic algorithm, energy efficient

**Corresponding author*: kakkard@nitj.ac.in

Suman Lata Tripathi, Souvik Ganguli, Tengiz Magradze, and Abhishek Kumar (eds.) Intelligent Green Technologies for Sustainable Smart Cities, (269–290) © 2022 Scrivener Publishing LLC

13.1 Introduction

13.1.1 Overview

Various advancements in micro-electro-mechanical systems (MEMS) have facilitated the development of tiny, less power, small cost, multitasking sensors in both technical and economic ways. These sensors measure ambient conditions in the environment surrounding them as motion, light, sound, temperature etc. Into a solitary small sensor gadget, communication, computation, and sensing capabilities are combined through the emerging field of wireless sensor networks (WSN). In order to expand the scope of internet into the real world, these devices can construct a sea of connectivity through advanced web networking protocols. Deployment of large numbers of tiny nodes, which can amass and arrange themselves, aid to the power of WSNs. The organization of several devices, although the capabilities of a single device are ostensible, offers essential innovative technological potentials. For these sensor devices, utilization scenarios range from monitoring of ecological conditions steadily to universal processing situations and up to well being of structures or hardware in situ checking. One of the clearest utilizations of wireless sensor system innovation is to monitor remote situations for low recurrence information patterns [1]. In case of a chemical plant, for instance, hundreds of sensor devices could involuntarily analyze the spillage of undesired chemical [2]. Deployment costs would be minimal for such kind of wireless monitoring systems, unlike conventional wired systems. A quarter-sized sensor could be essentially placed by installers at each recognizing point, rather than deploying a large number of feet of wire guided through cautious channel. Figure 13.1 shows one wireless sensor device [3]. With no redo or obscured design, the system could be additionally be extended by simply including more gadgets. With the closest high-power control tower or Base Station (BS), wireless sensor nodes need no direct communication but just with their local neighboring nodes, unlike customary wireless devices. Each individual sensor or actuator contributes being a part of the overall infrastructure instead of depending on a pre-deployed infrastructure. To transport data among several numbers of tiny devices implanted in a multi-hop fashion, networking protocols work in peer to peer manner providing a mesh-like interconnection. To introduce new nodes or enlarge to swathe a larger geographic area, the flexible mesh architectures envision adapting dynamically. To compensate for node failures in addition, the system can autonomously adapt.

Figure 13.1 A wireless sensor device.

13.1.2 Routing Challenges and Design Issues

WSN design is affected by various factors, inclusive of scalability, network topology, transmission media, operating conditions, data delivery, production expenses, fault tolerance, quality of service and power consumption [1]. These factors must be considered for satisfactory communication in WSNs [4]. The major challenges are:

1. Scalability: WSNs consist of hundreds or even greater number of nodes. Routing protocol should be able to handle the growing amount of load in network.
2. Transmission Media: In multi-hop WSN, connected nodes communicate through a wireless channel. Usually the required bandwidth for communication of sensor data is low, in order of about 1 to 100 kb/s. The key design challenge is in media access control (MAC), related to transmission media.
3. Fault Tolerance: WSNs are susceptible to failures, some of the nodes may stop working by some physical obstacles or damage and lack of energy supply. The routing protocols should be robust, failures of some nodes should not hamper the network operation.

4. Power Consumption: As lifetime of WSNs depend on the energy resources of sensor nodes, the energy usage concerns have great impact on network functioning. The power expended during transmission has high consumption of energy. Direct communication has higher energy consumption than the multi-hop communication. On the other hand, multi-hop communication needs extra topology management and MAC protocols.
5. Quality of Service (QoS): Some specific time critical applications require QoS. The significance of data gets expired after some time period. For such applications, routing protocols should be designed with low latency.

13.1.3 Heterogeneous Wireless Sensor Networks (HWSNs)

Following can be the resource heterogeneities among sensor nodes viz. energy heterogeneity, link heterogeneity and computational heterogeneity [5]:

1. Computational Heterogeneity: It implies that advanced microprocessor modules and extensive memory storage are available in heterogeneous nodes than the normal nodes [6]. The heterogeneous nodes can give intricate information processing and long tenure of storage for accomplished data, with the intense computational assets.
2. Link Heterogeneity: It means that higher bandwidth capabilities and powerful transceiver are there in heterogeneous node and are capable of working with expensed network unlike the normal nodes. More reliable transmission of data packets can be imparted with the aid of link heterogeneity [6].
3. Energy Heterogeneity: The sensor nodes have different percentage of battery powers equipped internally. It implies that the heterogeneous node is sequentially powered or is equipped with replaceable power supply. The most crucial one is energy heterogeneity among above three sorts of resource heterogeneity, because both computational heterogeneity and link heterogeneity will devour more resource of network energy [5].

As shown in Figure 13.2, there are two types of sensor nodes used in the WSN, the normal sensor nodes and the super-nodes having some extra

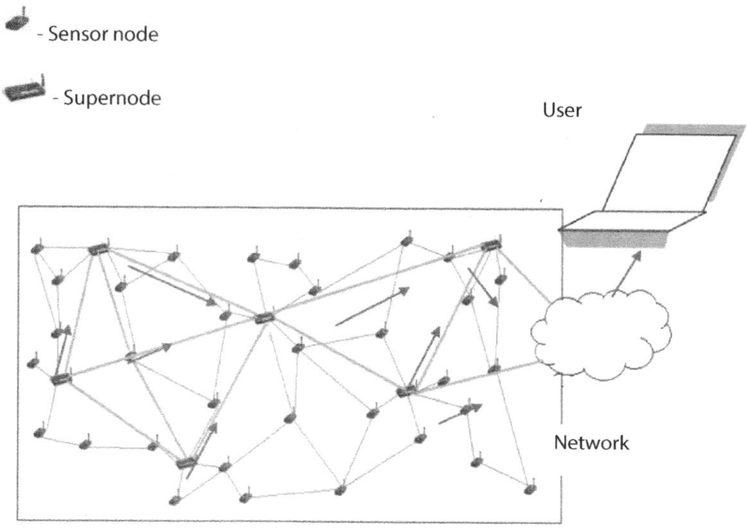

Figure 13.2 Illustration concept of HWSN.

percentage of battery power than normal nodes [7]. In this way, this is a two-level heterogeneous WSN. Further, this can be extended to three-level or multilevel where each node in the network has different percentage of battery power. This enhances lifetime of the network and hence better performance. Also heterogeneity of the network can be best exploited in terms of clustering the network [5].

13.1.3.1 Clustering in WSN

Clustering involves different groups of the sensor nodes being apportioned called as clusters. Cluster Head (CH) administers the entire cluster and other nodes are referred to as cluster member nodes. Specifically, cluster nodes do not communicate directly with the BS. The processed and collected data by the sensing nodes is surpassed to the CH. After having received the data from cluster nodes, CHs aggregate it and BS receives the transmitted data. Prolonged network lifetime is the ultimate result of clustering. Consequently the energy consumption gets minimized and to BS, increased number of useful message is imparted [8]. Clustering has numerous advantages [9]. Some of these are:

1. A CH is able to perform data aggregation in its cluster and decreases the number of redundant packets in the network.

2. By localizing the route set up within the cluster, the size of the routing table stored and managed at the individual nodes is reduced with clustering.
3. Sensors care only for connecting with their respective CHs and thus clustering reduces overhead for topology maintenance.
4. It confines the scope of inter cluster interactions to CHs by conserving the communication bandwidth and evades repetitive trade of messages among sensor nodes.
5. The cost of energy consumption can be reduced by CHs by scheduling various activities in the cluster.
6. By executing optimized management strategies with CHs, the battery life of the individual sensors can be extended and ultimately the network lifetime is improved.

13.1.4 Relay Node Selection Scheme

Relay nodes have been often used together with the clustering protocols to enhance the network lifetime. Tuah *et al.* [6] have highlighted the use of relay node for intra- cluster communication between cluster and BS. This has contributed to a great extent in efficiently utilizing the network energy as well has improved the performance of the network. Two-layer communication architecture is followed by a single hop clustering protocol that consists of sensor nodes and CH. Sensor nodes send their data to the elected CHs of respective clusters in the first layer. CHs perform data collection and aggregation, and then finally transmit the data to the BS in the second layer. Two- layer communication is followed by most protocols that use direct data communication for CH nodes to the sink. For some scenarios, where BS is located at a far distance from CHs, the energy consumption is higher for such a long distance data transmission for CHs as compared to the closely located CHs for similar kind of data traffic patterns [10]. Adding to this, the work of CH nodes needs to receive, process, and transmit data to the base station that will result in high power consumption in network. As a solution to it, the use of relay nodes for forwarding data packets to the sink from CHs will help in reducing the transmission cost suffered by distant located CH nodes. In Figure 13.3 three-layer communication architecture has been illustrated [11]. The motivation is to minimize the energy consumption while transmitting data between CHs and base station for an expensed large scale network. The relay node has been chosen among the CH nodes in the second layer and is located at the third layer of communication. Between CH node and the BS, it explicitly acts as forwarding

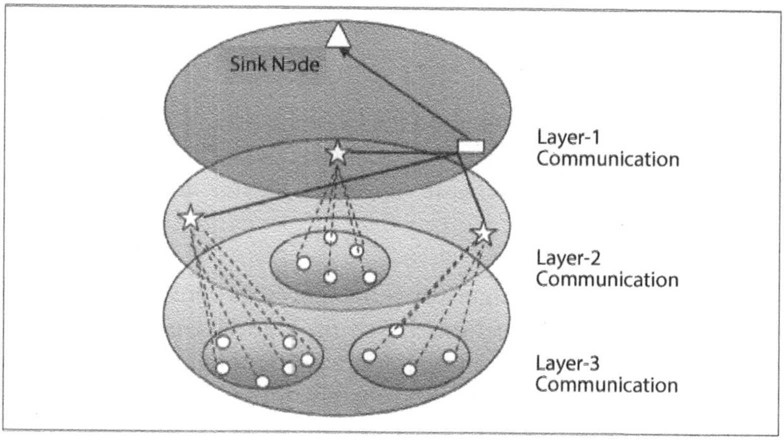

Figure 13.3 Three-layer Communication Architecture.

manager. The architecture also benefits in trivial data forwarding, that greatly decreases the related overhead of routing protocols in multi-hop communication.

13.1.5 Genetic Algorithm

Genetic Algorithm (GA) is one of the nonlinear optimization techniques being utilized for CH selection in routing protocols now days. We have used this efficient algorithm for appropriate relay node selection as it estimates an optimal solution by generating different individuals [12–14]. In GA, we have a pool of potential solutions which further go through the recombination (mutations) and reproduce the offspring. Each solution is assigned a fitness value and the fitter individuals are given greater chances to mate and create fitter offspring.

Repeating the above steps for specified number of generations and until the stopping criterion is satisfied, the algorithm gives the optimum solution to the problem.

13.1.6 Problem Definition and Motivation

Clustered hierarchical WSNs have their own advantages like scalability and reliabilityof the system, conserving the communication bandwidth, reducing topology overhead, etc. WSNs have wide area of applications including medical, environmental, military, smart monitoring areas. Wireless sensor is solely dependent upon the battery source to run through its lifetime. The key factor is optimization of energy usage in the network and the various ways are listed below [6]:

1. Formulating new equation for CH selection.
2. Use only residual energy for CH selection.
3. Use relay node scheme for inter-cluster long distance communication.

Various relay selection algorithms have been proposed for three-layer network architecture. Algorithms based on highest energy level method and minimum rate forward parameter are few among them. Since both of these algorithms are competing each other for subsequent rounds, it can't be said that which one is better. In order to improve the existing performance of network using these previous techniques, GA based relay selection is proposed that outperforms both the earlier methods. The parameters that are used for decision of relay node are node's initial energy, residual energy, distance to base station and the probability of getting selected as CH. It is designed to assure minimum energy cost as well as saving the nodes with comparatively low residual energy.

13.1.7 Proposed Work

The prime objective of proposed work is to design a relay node selection algorithm for improved performance of heterogeneous WSNs based on GA. The key objectives are:

1. Reduce the energy consumption of the network using the advantage of multi hop inter cluster communication.
2. Boosting the performance of the network by enhancing lifetime.
3. Raise the throughput of the network by extending the working period of the network.

13.2 Implementation of Proposed Relay Node Selection Based on GA

13.2.1 Network Model

The sensor network consists of 100 nodes and is randomly distributed in a two- dimensional area of 100 x 100 sq meters. It has BS at a position (50,100) m away from the network. All sensor nodes are immobile equipped with restricted power supply and recharging or replacing of batteries is not possible at any point after the network has been activated. The distances among the sensor nodes and from BS, are estimated through received signal strength indicator (RSSI). BS schedules the transmission and reception of data packets based on TDMA scheduling and Code Division Multiple Access (CDMA) scheme is employed to avoid collision inside and outside the clusters. The network model being used is hierarchical clustered heterogeneous wireless sensor networks which is similar to that of [15, 16]. All these models consider the concept of rounds (iterations).

13.2.2 Heterogenous Network Model

For a network used in DEEC protocol [15], the average probability to be a CH during n_i rounds is given in equation (13.1) that is actually based on residual energy of node s_i at round r.

$$P_i = \frac{1}{n_i} \tag{13.1}$$

where n_i denotes the number of rounds. A node s_i becomes CH, also called rotating epoch. When nodes posses same amount of energy at each round, probability P_i is chosen to be P_{opt}, which is the optimum probability for heterogeneous nodes to become CHs, which makes sure that there are P_{opt} N cluster heads every round and all the sensor nodes die at approximately same time as described in [17]. But when there are nodes with different amounts of battery energy, P of nodes with extra energy should be greater than the priory value of P_{opt}. The basic threshold probability is given in equation (13.2), to decide whether a node is eligible to become a cluster head.

$$T(S_i) = \begin{cases} \dfrac{P_i}{1 - P_i \times \left(r \times mod \dfrac{1}{P_i} \right)} & if\ s_i \in G \\ 0 & Otherwise \end{cases} \tag{13.2}$$

where s_i denotes a sensor node, r is current round number; G is the set of eligible nodes which can become cluster heads in r rounds. In every round, node s_i engenders a random number between 0 and 1. If this produced number is less than T (s_i), it is elected as cluster head during present round. As stated earlier that n_i is actually based on residual energy of node s_i at round r. For DEEC protocol n_i is given by equation (13.3).

$$n_i = \frac{E(r)}{P_i \times E_i(r)} = n_{opt} \frac{E(r)}{E_i(r)} \tag{13.3}$$

Probability P_i gets changed for the heterogeneous models where there are normal nodes, advanced nodes having some extra energy than the normal nodes, thus it is a two level heterogeneity. The weighted probabilities are given in equations (13.4) and (13.5).

$$P_{adv} = \frac{P_{opt}}{1+ak} \tag{13.4}$$

$$P_{nrm} = \frac{P_{opt}(1+a)}{1+ak} \tag{13.5}$$

Corresponding probabilities P_i for normal and advanced nodes are given in equation (13.6).

$$P_i = \begin{cases} \frac{P_{opt}(1+a)}{(1+ak)} \times \frac{E_i}{E_r} & \textit{if } s_i \textit{ is normal mode} \\ \frac{P_{opt}}{(1+ak)} \times \frac{E_i}{E_r} & \textit{if } s_i \textit{ is advanced mode} \end{cases} \tag{13.6}$$

where P_{opt} is optimum probability for CH selection, k is fraction of total number nodes that are advanced, a is extra energy factor for the advanced nodes, is current energy level of node, is average energy of network in round r. The above model can be extended to multi-level heterogeneous networks as described in equation (13.7) [13].

$$P(s_i) = \frac{P_{opt} N(1+a_i)}{N + \sum_{i=1}^{N} a_i} \tag{13.7}$$

where $P_{(si)}$ is the weighted probability and thus P_i for multi-level heterogeneous network is given by equation (13.8) as described in [1].

$$P_i = \frac{P_{opt} N(1+a_i)E_i(r)}{N+\sum_{i=1}^{N} a_i E(r)} \tag{13.8}$$

13.2.3 Radio Energy Dissipation Model

The energy model for the proposed work has been adopted from [9, 13], [18], [19]. The transmitter dissipates energy to run the radio transmit electronics and power amplifier, and the receiver dissipates to run radio receiver electronics as illustrated in Figure 13.4.

13.2.4 GA-Based Relay Node Selection

After selection of CHs and formation of corresponding clusters, now comes the selection of relay node. We find the appropriate relay node based on the GA approach. The flowchart of the proposed approach is given in Figure 13.5. The relay node selection process starts with first measuring initial and final energy levels, distance from base station and measuring threshold probability ($T(s_i)$) for becoming CH, for each selected CH node in the current round. Steps are discussed below:

1. Population Initialization: The initial population for the algorithm is the selected number of CHs in a given round.
2. Evaluation of Fitness Function: The proposed fitness function is given in equation (13.9)

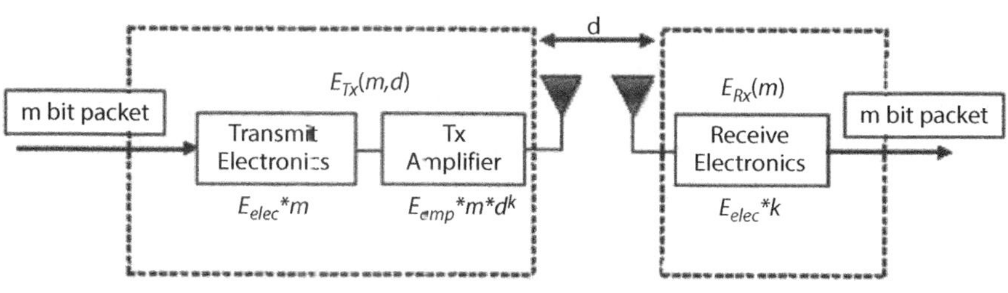

Figure 13.4 Radio Energy Dissipation Model.

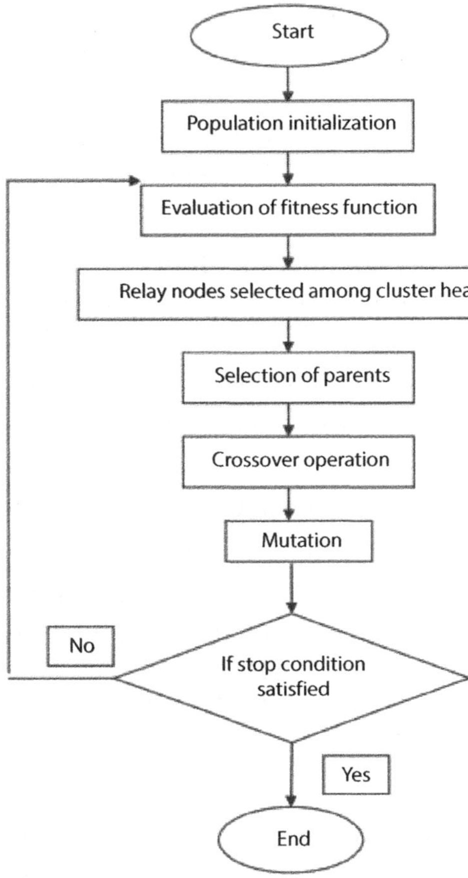

Figure 13.5 Flowchart of proposed approach.

$$F(i) = \frac{E_{(initial)}(i)}{E_{(CH)}(i)} \times \frac{E_{ch\_bs}(i)}{T(s(i))} \tag{13.9}$$

where $E_{initial}$ (i) is initial energy, E_{CH} (i) is residual energy level of ith cluster head node, D_{ch_bs} (i) is distance of the CH from BS, $T(s(i))$ is threshold as given by equation (13.2). These four parameters are the decision variables used. The fitness function is evaluated for each individual CH. We need to minimize the value of fitness function as $E_{initial}$ (i) is a constant value for CHs and E_{CH} (i), D_{ch_bs}(i), T(s(i)) change for each of the CHs in each iteration; E_{CH}(i) and D_{ch_bs}(i) have negative relation as that relay node is desired with maximum residual energy and at minimum distance from BS. The

$T(s(i))$ value is desired to be higher than that of random number generated by the nodes for satisfactory selection of CHs. All these parameters tend to minimize the value of fitness function [13]. Thus in order to obtain the appropriate relay node to maximize the network lifetime and also reduce energy consumption of the network, the mechanism aims at minimizing the value of fitness function.

3. Relay Nodes Selected among CHs: The node with minimum value of fitness i.e. the best fitness according to the given fitness function will be selected for new population.
4. Selection of Parents: Rank selection method is used for selecting parents to produce offspring for next generation. According to the fitness scores, the node with minimum value will be ranked as one, the second minimum as two and so on.
5. Crossover Operation: For individuals with continuous variables, Haupt's method is adopted for crossover operation. The parent chromosomes are p_1 and p_2 with genes or decision variables as x and y and are given in [20]:

$$p_1 = [x_1, y_1] \; ; p_2 = [x_2, y_2] \tag{13.10}$$

Randomly one parameter is selected for point of crossover. With given probability of crossover say β, crossover occurs as shown below:

$$x_{new1} = (1 - \beta)\, x_1 + \beta\, x_2 \tag{13.11}$$

$$x_{new2} = (1 - \beta)\, x_2 + \beta\, x_1 \tag{13.12}$$

Remaining parameters y_1 and y_2 are same. So the new offspring are:

$$offspring_1 = [x_{new1}, y_1] \tag{13.13}$$

$$offspring_2 = [x_{new2}, y_2] \tag{13.14}$$

6. Mutation: Any random parameter value of the offspring is replaced by some value specified in the available range as per the given mutation probability.

If the stopping criterion is reached i.e. getting same individual with minimum fitness value which implies the best fitness, more than once for a given generation, the process is stopped. The final individual is selected as relay node. If it is not satisfied, then the loop returns to step 2, evaluating fitness function and the whole process goes on.

13.2.5 Steady State Phase or Data Communication Phase

After performing the above operations for each round of the network, we get a unique relay node which is the best to forward the packets from the cluster heads to the base station. In the data communication phase, the cluster members send the sensed information in form of data packets to the corresponding cluster heads based on TDMA scheduling and CDMA scheme is employed to avoid collision inside and outside the clusters. Further, the cluster heads receive the data then aggregate them and further transmit all data to the selected relay node which finally forwards the packets to the base station. One round ends until base station receives packets from the relay node.

13.3 Results of Simulation For Energy Consumption, Lifetime and Throughput of Network

13.3.1 Simulation Setup

For actualizing ongoing application, a test system is required to analyze the network system execution. The range of network system and location of base station assume an imperative part in choosing the network lifetime. Table 13.1 depicts the simulation parameters used. The network size is assumed as 100m × 100m in area and there are 100 sensor nodes that are randomly distributed in the two-dimensional area. The location of base station is at (50,100) m. The simulation parameters are adopted from [13, 14, 16]. The parameters used for GA are shown in Table 13.2.

Starting with the creation of network model, network parameters like size of sensing area, positions of nodes that are randomly distributed, position of base station, total number of rounds etc. are made. Then, framing of radio energy dissipation model is carried out such as initializing energy parameters of the sensing nodes, initial energy, transmission energy and receiver energy etc.

Table 13.1 Parameters used in the HWSN simulation.

Parameters	Value
Sensing area of network	100m × 100m
Location of BS	(50,150) m
Number of nodes	100
Packet size	4000
P_i	0.1
Einitial	0.5 J
Eelec	50 nJ/bit
Efs	10 pJ/bit/m^2
Emp	0.0013 pJ/bit/m^4
EDA	5 nJ/bit/signal
a(energy factor)	4

Table 13.2 Parameters used for genetic algorithm.

Parameters	Value
Crossover probability	0.7
Mutation probability	0.01
Maximum iteration	100
Selection type	Rank selection

After the parameters are decided, for given number of rounds, the operation of cluster head selection begins, followed by cluster formation as per the given criteria of threshold. Then the selection of relay nodes is done based on the highest energy level of cluster heads, minimum energy over distance rate and by using genetic algorithm based relay node selection. Data communication phase starts with the transmission of data from sensing nodes to the respective cluster heads and then from cluster heads to the selected relay node and finally to the BS. The results are obtained for network lifetime, residual energy consumption and throughput of the network for above mentioned methods and are compared with each other.

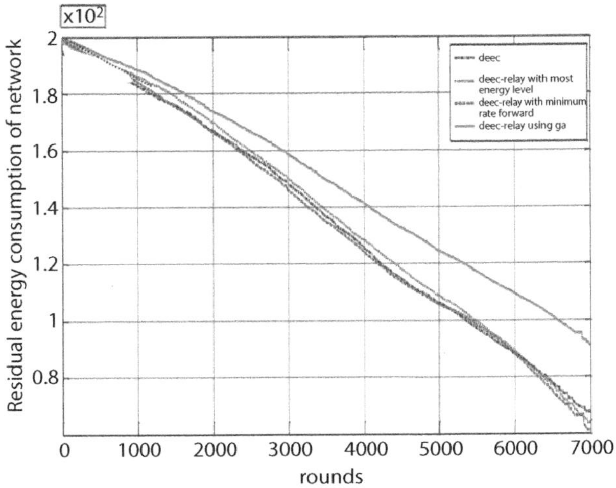

Figure 13.6 Comparison of residual energy consumption of the network for DEEC, DEEC-relay with most energy level, DEEC-relay with minimum energyover distance and DEEC-relay using GA.

13.3.2 Comparison of Residual Energy Consumption

The percentage of remaining energy of the network, when nodes die gradually is defined as residual energy consumption. Figure 13.6 shows the total remaining energy of the network when nodes die gradually during procession of rounds. The energy of sensor nodes keeps on decreasing with more number of iterations. Every time a round is completed, then further communication is being established by remaining energy of nodes. The comparison is shown among DEEC protocol, DEEC-relay with most energy level, DEEC-relay with minimum energy over distance parameter and DEEC-relay using GA. Although the two methods for relay selection has advantage over DEEC because of use of relay node that helps in balancing load for far off located CHs, but with the implementation of GA based relay, higher improvement is attained because the initial energy of nodes and threshold probability of CHs have been considered for efficiently balancing load in the network. As initial energy of the network assumed in the simulation is about 200 joule, it keeps on decreasing with the progression of network [11].

13.3.3 Comparison of Lifetime of Network

The instantaneous measure of lifetime of network reflects of number of alive nodes and that of each heterogeneous node that has not expended all of its energy. Lifetime of a sensor network is basically decided by the fact how long it has taken in terms of rounds to have First Node Dead (FND) and Half Node Dead (HND) [14]. Further, when the rest of nodes die they tend to lead the network towards instability. Figure 13.7 shows the comparison between the proposed DEEC-relay using GA and DEEC, DEEC-relay with most energy level, DEEC-relay with minimum energy over distance. It can be seen that DEEC-relay using GA has shown much more enhancement in lifetime of network as compared with DEEC, DEEC-relay with most energy level and DEEC-relay with minimum energy over distance. There is an obvious increase in network lifetime and number of alive nodes prevailing for more number of rounds in their lifetime.

The communication rounds of DEEC- relay using GA are much higher than that of DEEC, DEEC-relay with most energy level, DEEC-relay with minimum energy over distance. The lifetime of the network has been enhanced by about 28% using the DEEC-relay using GA method than the original DEEC protocol; further, the proposed method has enhanced about 7% and 9% lifetime than DEEC-relay using the most energy level method and DEEC- relay using the minimum energy over distance method respectively measured at the FND. Thus, it provides a better stable region than the previous algorithms. While for measuring lifetime at HND, the proposed method outperforms the DEEC protocol by 32%, DEEC-relay using the most energy level method by 22% and DEEC-relay using the minimum energy over distance method by 31 %. Thus, from the

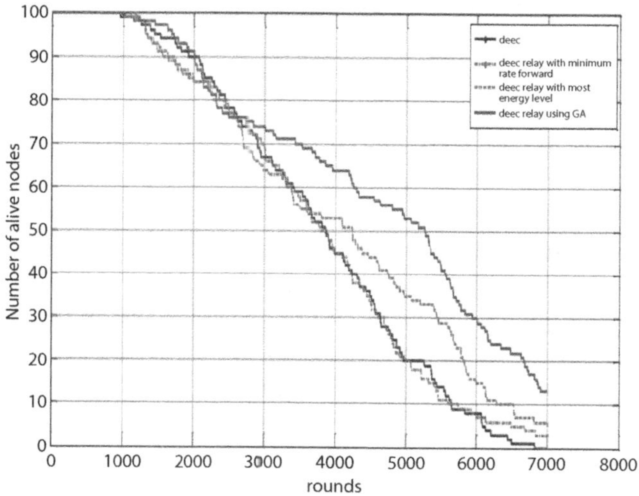

Figure 13.7 Comparison of proposed work with previous works in terms ofnumber of alive nodes per round.

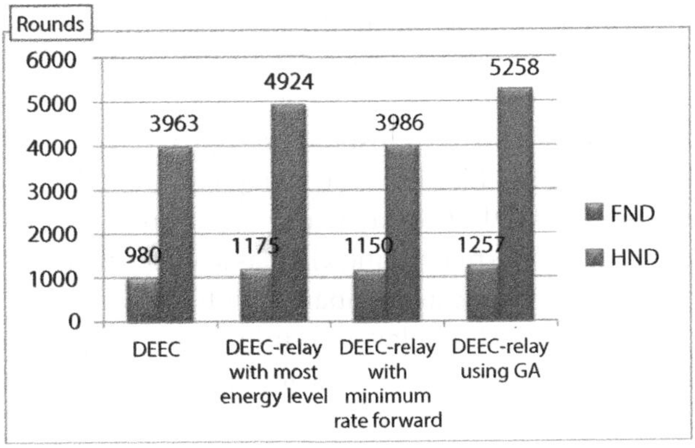

Figure 13.8 Comparison of lifetime of the network in terms of FND and HND.

experimental results, DEEC-relay using GA has effectively prolonged the lifetime of network. The above results have also been shown in terms of FND and HND in subsequent rounds in Figure 13.8.

13.3.4 Comparison of Throughput at BS

The most important metric for evaluating performance of the network is throughput that is defined as the total number of packets that are received at BS either from CHs or relay node of the network. Each packet of message is assumed to be of 4000 bits. Throughput is measured at BS on completion of each round. Consequently with increasing number of rounds, throughput also increases [16]. As throughput is one of the important performance metrics to evaluate performance of a network, Figure 13.9 gives a comparison of the proposed method DEEC-relay using GA with DEEC, DEEC-relay with most energy level and DEEC-relay with minimum energy over distance algorithms. For starting rounds, the throughput values are almost similar for all the algorithms, but as the rounds progress, throughput value for DEEC are smaller as compared to improved protocols because of lack data transmission using relay node as a result of shortened lifetime. DEEC-relay with most energy level and DEEC-relay with minimum energy over distance algorithms have improved data rates than DEEC by utilization of relay nodes for data transfer to BS. Moreover, DEEC-with most energy level has better throughput than DEEC-relay with minimum energy over distance algorithm because energy levels of the CHs are dominating over their distances from the BS. And finally, DEEC-relay using GA outperforms all the previous algorithms by selecting appropriate relay node among selected CHs considering parameters which are initial energy of

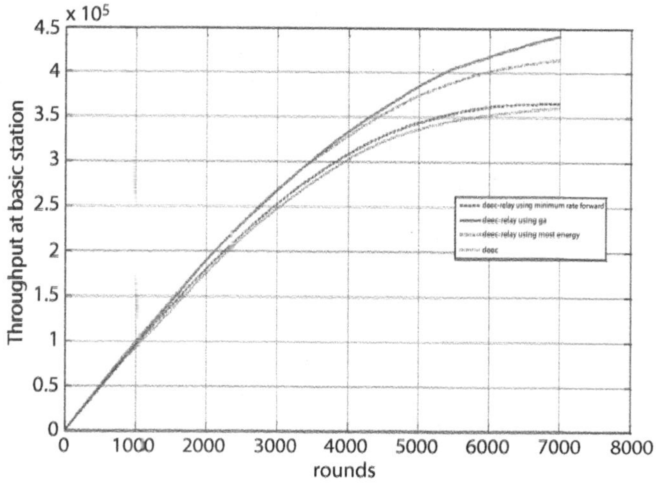

Figure 13.9 Comparison of throughput of the network measured at BS.

CH node, residual energy of CH node, threshold probability of CH selection and the distance to the BS from CH node. This criterion for selection of relay node based on GA attains most appropriate relay node that efficiently balances the load in the network thereby conserving the residual energy of network. This results in extended working period i.e. increased number of rounds of network and consequently number of data packets received at BSis increased considerably.

13.4 Conclusion and Future Scope

This chapter focuses on adaptive energy conservation and optimization in three- communication architecture of HWSNs. The GA based selected relay node is determined to be highly energy efficient and the network has shown improvement in its lifetime. The residual energy has also been conserved maximally during each round for driving the network for further rounds of iteration. When the lifetime of the network is measured at FND and HND, the communication rounds of DEEC-relay using GA are higher than that for DEEC, DEEC-relay with most energy level, DEEC-relay with minimum energy over distance. The lifetime of the network has been enhanced by about 28% using the DEEC-relay using GA method than the original DEEC protocol; furthermore, the proposed method has enhanced about 7% and 9% lifetime than DEEC-relay using the most energy level method and DEEC-relay using the minimum energy over distance method respectively measured at the FND. Thus , it provides a better stable region than the previous algorithms. While for measuring lifetime at HND, the proposed method outperforms the DEEC protocol by 32%, DEEC- relay using the most energy level method by 22% and DEEC-relay using the minimum energy over distance method by 31 %. Along with the increased lifetime of the network, load balancing has been achieved as well, that would make it suitable for various real time applications such as far field monitoring, wide area monitoring, remote monitoring etc.

The throughput of network has also been improved considerably by the DEEC-relay using GA algorithm. The results of proposed scheme are simulated and evaluated in MATLAB. The simulation results show that the proposed scheme that is DEEC-relay using GA has improved results as compared to previous methods.

In the future, the work may be extended for appropriate relay node selection, by further optimizing the energy parameters along with distance parameter. The fitness function can be improved by considering the intercluster and intra cluster distances along with the energy parameters. Further research could be carried out for investigating the problem of placement of relay node for fault tolerance and mobility of the nodes. The network scenario could be further extended for three-dimensional deployment of nodes for some space monitoring requirements. Different optimization techniques may also be implemented for appropriate relay selection for dynamically changing environment of network.

References

1. Akyildiz, I.F., Su, W., Sankarasubramaniam, Y., Cayirci, E., Wireless sensor networks: A survey. *Comput. Netw.*, 38, 393–422, 2002.
2. Hill, J.L., *System Architecture for Wireless Sensor Networks*, Ph.D. Dissertation, Univerisy of California, Berkeley, 2003.
3. url- http://www-micrel.deis.unibo.it/sitonew/
4. Al-Karaki, J.N. and Kamal, A.E., Routing techniques in wireless sensor networks: A survey. *IEEE Wirel. Commun.*, 11, 6, 6–28, 2004.
5. Yarvis, M., Kushalnagar, N., Singh, H., Rangarajan, A., Liu, Y., Singh, S., Exploiting heterogeneity in sensor networks, in: *Proceedings IEEE 24th Annual Joint Conference of the IEEE Computer and Communications Societies*, vol. 2, IEEE, pp. 878–890, March 2005.
6. Tuah, N., Ismail, M., Jumari, K., An energy-efficient node-clustering algorithm in heterogeneous wireless sensor networks: A survey. *J. Appl. Sci. (Faisalabad)*, 12, 13, 1332–1344, 2012.

7. Awada, W. and Cardei, M., Energy-efficient data gathering in heterogeneous wireless sensor networks, in: *2006 IEEE International Conference on Wireless and Mobile Computing, Networking and Communications*, IEEE, pp. 53–60, June 2006.
8. Katiyar, V., Chand, N., Soni, S., Clustering algorithms for heterogeneous wireless sensor network: A survey. *Int. J. Appl. Eng. Res.*, 1, 2, 273, 2010.
9. Akkaya, K. and Younis, M., A survey on routing protocols for wireless sensor networks. *Ad Hoc Netw.*, 3, 3, 325–349, 2005.
10. Tuah, N., Ismail, M., Jumari, K., Energy efficient algorithm for heterogeneous wireless sensor network, in: *2011 IEEE International Conference on Control System, Computing and Engineering*, IEEE, pp. 92–96, November 2011.
11. Tuah, N. and Ismail, M., Extending lifetime of heterogenous wireless sensor network using relay node selection, in: *2013 International Conference of Information and Communication Technology (ICoICT)*, IEEE, pp. 17–21, March 2013.
12. Goldberg, D.E., *Genetic Algorithm in a Search, Optimization and Machine Learning*, Addison-Wesley Longman Publishing Co. Inc., USA, 1989.
13. Norouzi, A. and Zaim, A.H., Genetic algorithm application in optimization of wireless sensor networks. *Sci. World J.*, vol., Article ID 286575, 15, 2014.
14. Pal, V., Singh, G., Yadav, R.P., Cluster head selection optimization based on genetic algorithm to prolong lifetime of wireless sensor networks. *Procedia Comput. Sci.*, 57, 1417–1423, 2015.
15. Smaragdakis, G., Matta, I., Bestavros, A., SEP: A stable election protocol for clustered heterogeneous wireless sensor networks, in: *Second international workshop on sensor and actor network protocols and applications (SANPA 2004)*, vol. 3, August 2004.
16. Qing, L., Zhu, Q., Wang, M., Design of a distributed energy-efficient clustering algorithm for heterogeneous wireless sensor networks. *Comput. Commun.*, 29, 12, 2230–2237, 2006.
17. Heinzelman, W.B., Chandrakasan, A.P., Balakrishnan, H., An application-specific protocol architecture for wireless microsensor networks. *IEEE T. Wirel. Commun.*, 1, 4, 660–670, 2002.
18. Kumar, D., Aseri, T.C., Patel, R., EEHC: Energy efficient heterogeneous clustered scheme for wireless sensor networks. *Comput. Commun.*, 32, 4, 662–667, 2009.
19. Alla, S.B., Ezzati, A., Hssane, A.B., Hasnaoui, M.L., Hierarchical adaptive balanced energy efficient routing protocol (HABRP) for heterogeneous wireless sensor networks, in: *2011 International Conference on Multimedia Computing and Systems*, IEEE, pp. 1–6, April 2011.
20. Haupt, R.L. and Haupt, S.E., Practical genetic algorithms. John Wiley & Sons, 2004.

14

Privacy and Security of Smart Systems

K. Suresh Kumar, D. Prabakaran, R. Senthil Kumaran* and I. Yamuna

IFET College of Engineering, Villupuram, Tamil Nadu, India

Abstract

A smart city is a digitized urban area with a collection of electronic sensor nodes for data collection to maintain the resources and various features efficiently. The smart systems support the automated functions of smart cities or smart homes and their facilities to monitor the functions of any closed infrastructure. The data collected from various sensor nodes and measuring devices are employed for monitoring and maintaining the resources across the city. Security is the major aspect of the smart city as the collected data must be treated securely to pretend unauthorized access by the attackers. The security concerns motivated me to present this chapter for discussing the security and privacy concerns and solutions of smart systems. This chapter discusses the various security issues of smart systems and the existing privacy policies. This chapter analyzes the various vulnerabilities of the smart systems and real-time case studies have been discussed to identify the root cause of security breaches. In addition, this chapter provides the mathematical approach to mitigate the security breached and suggestions to enhance the privacy policies of the smart systems.

Keywords: Internet of things (IoT), privacy, security, smart city, smart systems, vulnerability

14.1 Smart Systems—An Overview

Smart Systems are the system that can operate autonomously in nature. It can be able to interact with the environment through various sensors or some miniaturized devices. The recent evolution of electronic devices results in developing a variety of flexible and user interactive systems for

**Corresponding author*: sen19841@gmail.com

Suman Lata Tripathi, Souvik Ganguli, Tengiz Magradze, and Abhishek Kumar (eds.) Intelligent Green Technologies for Sustainable Smart Cities, (291–316) © 2022 Scrivener Publishing LLC

smart living technology. Nowadays, almost all electronic appliances are controlled remotely through electronic devices, such as smartphones, personal device assistants (PDA) devices, and laptops, which pass our day-to-day life easier.

Smart systems are set to empower the various domain of communication with a connected world. It can accept various types of inputs and responses and act accordingly based on the specific application [4]. This also helps to eliminate the man-made errors for control and monitoring of certain critical mechanisms. Smart systems can also offer an increased level of safety and security, vehicle control, healthcare field for monitoring and treatment, agriculture, manufacturing process, and also in day-to-day life working environments. Smart systems are the new generation systems that combine the various latest technologies like Artificial Intelligence, Machine Learning, and the Internet of Things (IoT). These combined techniques collect the information from various sensors and also from certain machines leads to becoming the integral part that enables the digital world to be integrated into definite networks. As per the Harbor research survey, Smart Systems, Services, and Internet of Things platform market potential can be increased from $75 billion in the year 2017 to as much as a $300 billion potential market by the year 2023.

Smart systems and IoT platforms can be divided into four classifications. They are the basic device management functions and services, devices connectivity issues, capability in data management and transformation, a software platform to integrate with the hardware component.

Due to the newly emerging techniques of big data, sensor networks, and embedded systems, Artificial intelligence and the Internet of things combination leads to changes in the regular process of life across various domains.

14.2 Security and Privacy Challenges in Smart Systems

The technology that was accepted by most of the users is due to the most exciting and the smart method to run daily life is the Internet of Things (IoT). In this technique, user information can be shared over the network. Hence, it is easily possible to misuse the data, which is shared in the network. The possibility of describing user activities and tracing user data is termed privacy loss. Internet of things changes the world into full automation [5]. As the technology booms, it is essential to provide security to the corresponding system. As our life moves into automation, at the same time, the vulnerabilities present in the network/ system can also be assessed critically.

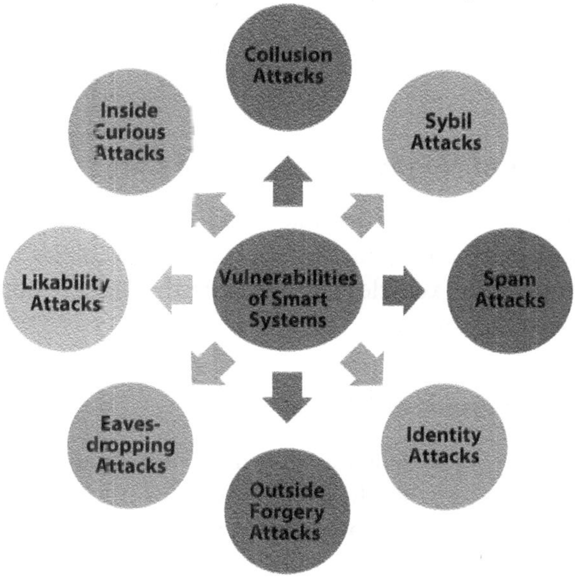

Figure 14.1 Vulnerabilities of smart systems.

In general, the threat may be defined as the several actions that can influence a network or a system using a natural or in an artificial manner, refer Figure 14.1. When necessary actions are not taken, these threats may fail. Hence it is essential to focus on the threats and the challenges involved in a smart system. The smartness concept is not only active for a specific domain, it was spread to various applications like smart cities, smart homes, smart industry, and smart service systems. The most common characteristic in these various applications of technology is to support smart capabilities to some systems. However, the related works reveal few definitions that may also not be directly linked with the technique to smart systems. For example, in business and management domain defines smart as the flow of information to a financier through a company.

With the rapid growth in today's technology, it is important to keep smart devices operating safely. It is also essential to note the smart devices when they are interconnected. Some of the major factors for the smart hacking data breach, Smart devices control, and financial loss. Even though developments in smart systems have contributed for the growth of society, almost all the smart application is susceptible for hacking. The following schemes show the latest issues created from the quickly growing smart applications.

14.2.1 Botnet Activities in Smart Systems

An IoT botnet is a group of cooperated heterogeneous IoT devices like cameras, routers, DVRs, wearables, and other embedded technologies, infected with malware and causes distributed-denial-of-service (DDoS) attack [1, 3].

14.2.2 Threats of Nonhuman-Operated Cars

The major aim of operating the autonomous vehicle is to reduce traffic accidents on the roads. But, hackers can hack the vehicles to operate the wrong functions of the vehicles like applying the brakes, stop the engine, and also the steering control.

14.2.3 Privacy Issues of Virtual Reality

The Virtual Reality (VR) scheme is used by individual persons and also by various organizations like healthcare services, the engineering industry sector, etc., The communication between the VR devices and the data stored by the sensors all pose threats of privacy leakage [2].

14.3 Case Studies—Security Breaches in Smart Systems

The smart systems are composed of various smart devices interfaced with sensors, with an objective function of operating over the internet without having any direct contact of operation. The Internet is the backbone of the smart systems, which possess an insecure channel leads to more probability of security attacks [6]. The adversaries tend to track for the vulnerable ports or devices in the smart systems so that to succeed in a security breach to gain unauthorized access to the smart systems. The hackers, on gaining unauthorized access, execute the plan of invading the entire smart systems, which may lead to any criminal or illegal activities. In most scenarios, the security breach [7] in the smart system remains unexplored and most cases were not reported. The statistics of security attacks year-wise in terms of malware (in millions) are depicted in Figure 14.2.

Some of the known case studies for the security breaches in smart systems are breaching Smart surveillance cameras, hacking smart televisions, hacked smart bulbs, vulnerable security homes, and identity stealing using smart coffee machines.

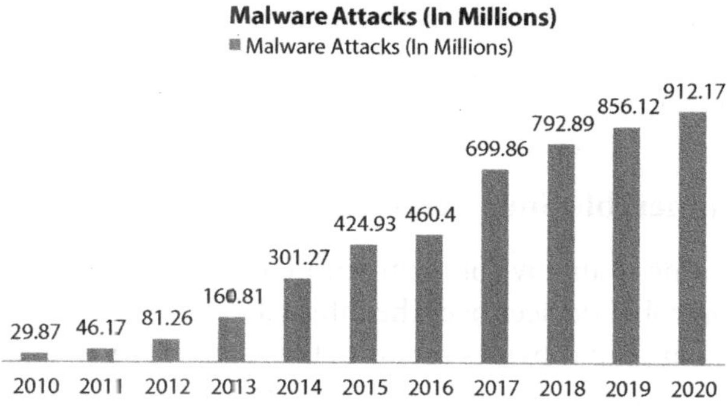

Figure 14.2 Malware attacks on smart systems (in millions).

14.3.1 Breaching Smart Surveillance Cameras

The Surveillance cameras will be the ultimate target for the adversaries to invade, such that hoist the illegal activities. The smart surveillance cameras capture and extend their operation to alerting the authorized persons about the suspicious movements captured in the surveillance camera. The thieves or any criminals involving in illegal activities tend to take control over the surveillance cameras. To perform this action, the adversary tends to search for the vulnerable ports to execute the malicious algorithms to take control over the entire smart system and smart surveillance cameras.

14.3.2 Hacking Smart Televisions

The smart television, which can be connected to the internet and other smart devices, will remain unnoticed by the administrator of the smart systems with the thought that it has no sensitive information. The attackers targets such devices as weaker ports as the Smart Television is vulnerable to varieties of threats. Taking the control of smart television, the attacker can able to view the activities of the authorized user, which is termed to be eavesdropping attack.

14.3.3 Hacked Smart Bulbs

The surprising aspect of the security breach of a smart system is the hacking of smart bulbs, which can be controlled using WiFi and are operated through a pre-programming concept. The administrators often ignore providing security to such devices with the thought of no sensitive information stored in the smart bulbs. The adversary concentrates on such weaker ports of smart systems through which the entire smart systems can be invaded to get illegal access to data.

14.3.4 Vulnerable Smart Homes

The smart home is an environment with connected devices that provides complete control of devices over the internet. The internet is the backbone of smart systems and provides access to the users via various access points. The security breach in the smart home exposes the personal data to the adversary and the risk factor is considered to be a maximum level. The administrators of the smart home must re-examine the security policy and privacy preservation policies in the smart home to mitigate the risk of vulnerabilities.

14.3.5 Identity Stealing using Smart Coffee Machines

The smart coffee machines are controlled by the smartphone of the administrators when exposed to a security breach will ease the stealing of the identity of the user. The hacker invading the vulnerable port of the coffee machine will create an attacking channel to reach the smartphone of the administrator. The sensitive information in the smartphone, mentioning the user account, login credentials, and secure messages will be prone to attack and can be accessed illegally by the hackers.

14.4 Existing Security and Privacy Protection Technologies

This section highlights the existing security and privacy protection methodologies and algorithms that secure the smart systems from the vulnerabilities and security threats. It is essential to have a rigid algorithm and methodology for preserving the infrastructure of the smart systems.

Table 14.1 illustrates the various technologies like cryptographic algorithms, Biometrics, Blockchain technology, Machine Learning Technology, Deep Learning Technique, Artificial Intelligence, and Game Theory, which are the existing technology that steps ahead in providing a better level of security to the infrastructures of the smart systems. This section is preceded by explaining the notable algorithms and procedures involved in the aforementioned security providing technologies.

Table 14.1 Existing security and privacy protection technologies.

Disciplines	Technologies	Application scenario
Cryptography	Two Level Authentication Key Exchange Scheme RFID Homomorphic Encryption	Smart Transportation Smart Shopping Smart Grid
Biometrics	Noise Cancellation process Key Negotiation Protocol	Mobile sensors Storage Devices
Blockchain	Smart Home Architecture Decentralized Block chain based framework Block chain technique with distributed architecture	Smart Home Smart Transportation IoT Architecture
ML, DL and AI	K Means clustering for Privacy Preserving Bayesian Linear Regression model SVM based authentication system	Social Networks Mobile devices Smart Phones
Game Theory	Zero Sum game Bayesian game indeterminate information	Wireless Networks Honeypot-enabled network

14.4.1 Cryptography

Cryptography is a science to execute secured communication over the insecure channel by converting the understandable plain text to non-understandable cipher text using secret keys. Cryptography thus provides confidentiality, integrity, and availability of service, which is shortly termed as CIA. The cryptographic algorithms possess a strong mathematical foundation to realize the security practices through implementing authentication and authorization process to verify the service requesting users. The cryptographic algorithms are classified into two sections based on the number of keys employed for the authentication process. The symmetric key and public key cryptography, along with its multiple algorithms, provide a strong backbone for the security of smart systems. This section describes the notable and secured cryptographic algorithms employed in securing the smart system resources. Some of the highlighted cryptographic algorithms, which outperform security principles are data encryption standard (DES), triple data encryption standard (R-DES), RSA algorithm, advanced encryption standard (AES), and Blowfish algorithms. The smart system users will be requested to self verify the authentication process by providing authentication credentials, which will be communicated to the authentication server of the smart system by performing a 132 to 448-bit encryption scheme along with any of the aforementioned algorithms. The one-time registration (OTR) process performed in establishing authentication mechanism using Feistel Structure (Advanced Encryption Standard) is portrayed in Figure 14.3.

The rigidity of the aforementioned proposed model depends on the block cipher principle and the number of bits used as the key for performing the encryption and decryption process. The key length defines the number of rounds to be performed in the Feistel structure thus provides the inclined level of security as the key length (in bits) is increased. The users of the smart systems have to perform dual registration namely the user registration and the device registration. The grant of access to the new user must be registered by adding the user credentials along with the definition of a level of privilege to the corresponding users. In addition, the inclusion of a new device must be registered and approved by the smart system administrator such that it be available to the users. These registration processes are performed by the One Time Registration process performed in secured Feistel Encryption policy. The probability of adversary accessing the smart system providing the credentials of the authenticated user is more and hence,

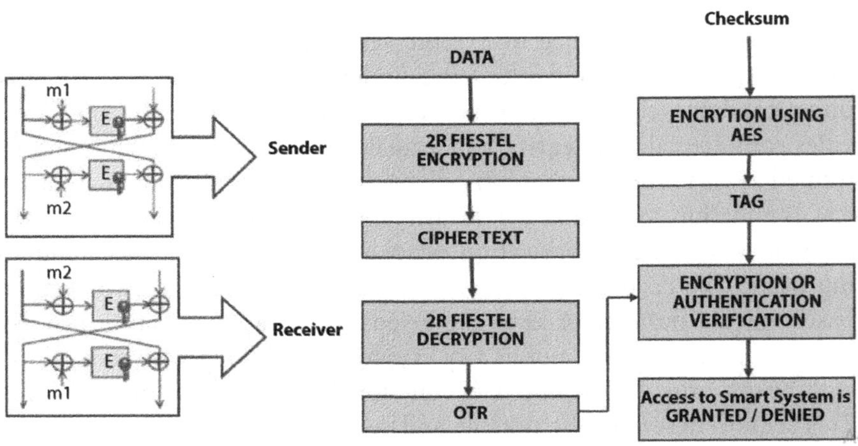

Figure 14.3 One time registration process using Fiestel structure.

it is advisable to incorporate a multi-level authentication scheme to reduce the probability of succeeding the unauthorized access by the adversary.

14.4.2 Biometric

The biometric is the unique identity of a person with minimal possibility of duplication during the authentication process. The biometric identities [8] that employed an authentication process are classified into Psychological biometric and behavioral biometric, which can be further classified as fingerprint, face recognition, Palm print recognition, Audio/Voice authentication, Individual Signature, and Gait. On comparing the performance of all the aforementioned biometric, the fingerprint identity and the voice-based authentication system holds good in applying for securing the smart systems and the probability of duplication is also minimum.

The fingerprint is unique to each individual in the world and the probability of duplicating is very low. The various anomalies in the process of extracting the minutiae from the fingerprint are bifurcations of ridges and valleys. The various vital parameters of the human fingerprint are portrayed in Figure 14.4.

Figure 14.4 Vital parts of human fingerprint.

The minutiae extraction from the fingerprint [9] can be extracted through different types of algorithms, and all these algorithms concentrate on cross numbering systems at any point X, X' in the obtained fingerprint. The cross numbering systems and the mathematical expression to detect the cross numbering system is expressed as equation 14.1.

$$CrossNumberCN_i = 0.5\sum_{i=1}^{n} |X_i - X_i'| \tag{14.1}$$

The Peak signal-to-noise ratio in the extracted fingerprint minutiae is low as the noisy content in the fingerprint may be more. This increases the necessity of the introduction of a filter in minutiae extraction so that the authorized user of the smart system does not face the denial of service due to bad credentials. The filter employed in the process of removing the noise from the extracted minutiae may be a low pass or bandpass filter and the filtration process is expressed in equations 14.2 and 14.3.

$$BandPassFilter(F_{BP}) = exp\left\{-\frac{1}{2}\left(\frac{X^2}{\sigma_i^2}\right)\cos(\omega_0 x)\right\} \tag{14.2}$$

$$LowPassFilter(F_{LP}) = exp\left\{-\frac{1}{2}\left(\frac{Y^2}{\sigma_i^2}\right)\right\} \tag{14.3}$$

The extracted minutiae from the user fingerprint can be used as the second factor for the user authentication process.

The user voice-based authentication process is more popular in accessing the resources of smart systems. The voice-based authentication system [10] accepts the pre-registered voice content of the user and registers the users to provide the privilege of access. The user uses the same voice content as the authentication credential and has to undergo a credential verification process to enjoy the resources of smart systems. The user voice data is continuous in time signal and is represented as x(n) and the power spectrum of the input voice signal can be obtained by treating the input voice signal with the Discrete Fourier Transform as expressed in equation 14.4.

$$PowerSpectralDensity: P(s) = |DFT(x(n))|^2 \tag{14.4}$$

The feature of the input voice signal x(n) is mentioned as Cepstrum and is extracted from the power spectral density as expressed in equation 14.5.

$$C(s) = IDFT(\log P((s))) \tag{14.5}$$

To obtain the cepstrum, a logarithmic process is performed for the obtained power spectrum density of the input voice signal and then preceded with the inverse Discrete Fourier Transform (IDFT). The cepstrum is considered to be the authentication credential for the user and this system has optimized immunity towards the noise. The drawback of employing the voice-based authentication process in accessing the smart system is its poor level of immunity towards the noise, which affects the accessing process. The noise inclusive of background noise and problems related to vocal cords affects the authentication mechanism of smart systems.

14.4.3 Block Chain Technology

Blockchain technology is based on the bitcoin principle and is used for secured data transmission over the internet. The smart systems are connected with the backbone of internet connectivity and the inclusion of blockchain technology [11] in security the smart systems are due to its salient features of the improvised level of transparency and security, providing integrity over the authentication credentials and the data transferred over the internet, high level of security at reduced cost and energy consumption. The pre-requisite for employing blockchain technology in securing the smart systems are it requires multiple users to share the common data or resources, registration of users and resources, privacy and Security policy verification, removing the dependencies on users, and transaction interactions.

The role of blockchain in preserving the security of the smart system is to verify the authentication credentials of the users and to verify the legal documents, resources in the smart systems. The blockchain provides a secure access to the data resources of the smart systems at a reduced cost and minimal power consumption.

14.5 Machine Learning, Deep Learning, and Artificial Intelligence

14.5.1 Machine Learning in Smart Systems

The incorporation of Machine Learning technology in the smart systems assists the systems to be securely available to the users and to remove the dependencies of the smart systems to wait for the direct command from the users. The Machine Learning technology [12] enables the connective devices in the smart systems to perform the actions and to make certain decisions independently based on the program instructions fed to the systems. To enable this instructed independent working the smart systems, the sensors play a vital role in measuring the objective data and executing the instructions. The machine language works on the prediction concept and various predictor algorithms exist for performing predicting efficiently. The efficient machine learning prediction algorithms are genetic algorithm and decision tree algorithm.

14.5.2 Genetic Algorithm

The Genetic Algorithm employs chromosomes as the specific string to identify the cross-over points existing between the user and the smart device. The Genetic algorithm [13] incorporates three smart system administrators namely reproduction, cross-over, and mutation. Reproduction is the principle, in which the smart device identity multiplies itself as a reproduction process. The chromosome trades fewer bits of data communication in creating new chromosomes. The mutation process communicates the least bit of data in an individual chromosome. The genetic algorithm is classified into two type's namely single point genetic algorithm and two-point genetic algorithms and is portrayed in Figure 14.5.

The feature and the major pros of incorporating the genetic algorithm is, it is more efficient in detecting intrusion detection and thus protects the smart system environment.

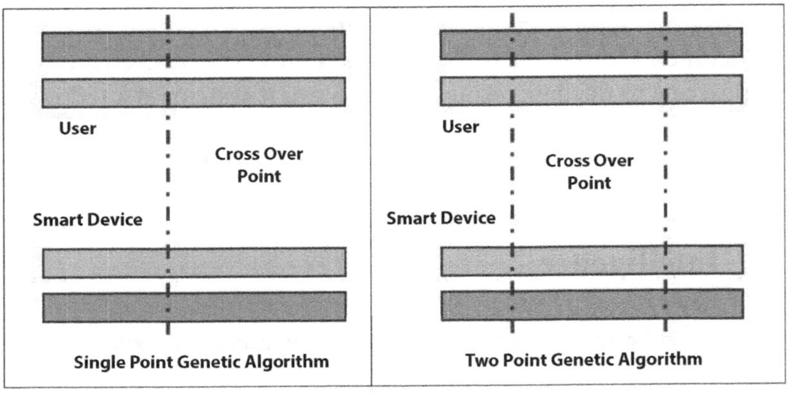

Figure 14.5 Genetic algorithm types.

14.5.3 Deep Learning in Smart Systems

The Deep Learning methods for handling the smart systems are exhibiting efficient performance in the feature extraction process and in handling a huge amount of data, which the machine learning fall shorts in providing expected performance. Smart systems when extended to a larger volume of devices and data, deep learning is considered to be the best technology for handling the data securely. Deep Learning employs Neural Networks in the authentication verification process and fingerprint verification or voice-based authentication is generally used in smart systems for which the neural networks work efficiently in terms of the authentication verification process. Some of the well-known performing algorithms of Deep Learning are Artificial Neural Networks, Convolution Neural Networks, and Support Vector Machines. The aforementioned algorithms outperform well in maintaining the security of the smart systems and overheads the performance when compared with the Machine learning algorithms.

14.5.4 Artificial Intelligence in Smart Systems

Artificial Intelligence (AI) plays an imperative role in maintaining the huge data stored in smart systems. Smart systems hold a variety of sensitive information like authentication credentials, sensor measured data, and instruction algorithms. The security breach in the smart systems leads to the attack of the entire smart systems and hence requires strong security and privacy policies. Artificial Intelligence optimizes the resource allocation in smart systems and in addition, restricts the unauthorized access.

14.6 Security Requirement for Smart Systems

The standard and norms for the security policy and privacy preservation policy of smart systems have been framed and published by European Union Agency for Network Information Society and the report states the results of threat analysis using STRIDE. The security requirements for maintaining high level of security in smart systems are as follows, thwarting of Data leakage and Falsification, identification and Prevention of Device Tampering, incorporation of Light Weight Encryption algorithm for Authentication, access Restrictions to Users and incident Response for entire systems.

14.6.1 Thwarting of Data Leakage and Falsifications

In smart systems, the different classes' data namely open data and closed data are handled together without any notable bifurcations. The best example for the open data is the place where we are planned to visit while the closed data are more sensitive as it covers the account details. There is no much harm if open data is disclosed to others while the exposure of closed data costs high. Care must be taken that the closed data must be not disclosed anywhere unnecessarily to others. Falsification is a hazard of duplicating the data in the central repository as it will create a serious concern of data duplication effects.

14.6.2 Identification and Prevention of Device Tampering

A smart system is a group of devices connected with a backbone of the internet. The smart device may be a surveillance camera, smart electric meter, air conditioner, washing machine, water heater, smart coffee machine, computer systems, printer, smart bulb, smart television, etc. The hacker targets any one of the weaker aforementioned smart devices, gaining access to any one of the devices, will start invading the entire smart systems. The hacker may perform any type of attack to access the directory of the smart devices, acquiring access will be followed by the data falsification process, which is also termed as device tampering. The administrator of the smart device has to frame a strong security policy and strengthening the gateway of all the access points of smart systems by installing gateways, Intrusion Prevention, and detection algorithms. In addition to that, the administrator has to review the security policies and privacy principles at constant time intervals to determine whether any attempt for security breach was performed.

14.6.3 Light Weight Encryption Algorithm for Authentication

The smart system administrator restricts the access of the smart systems by providing access privilege to the users based on the authentication credentials. The administrator forms a framework incorporating cryptographic algorithms such that the identity of the authorized person is verified before granting access to the smart system. The cryptographic algorithm may be symmetric key cryptography or public-key cryptography and in recent days the security mechanism is provided utilizing blockchain technology as described in the previous section. The usage of cryptographic algorithms encrypts the sensitive authentication credentials during the logging-in process and also transmits the control data in the form of ciphertext so that the adversary cannot interpret the meaning of the message even if acquired the ciphertext illegally. The dilemma of opting suitable and rigid cryptographic algorithm exists while concentrating much on the security aspect of the smart system. The incorporation of advanced or heavyweight (bigger key size and involving complex procedure) cryptographic algorithms reduces the performance of the smart system and it consumes more time for granting access to the authorized user. Hence, care must be taken while choosing a suitable cryptographic algorithm and it is feasible to opt for a light weight cryptographic algorithm with optimal key size and procedures for the encryption and decryption process.

14.6.4 Access Restrictions to Users

In smart systems, the user must be provided with the privilege of accessing the devices of smart systems. For example, the children in a home can be given the privilege to access smaller smart system devices like smart audio players, smart television, etc and privilege restriction may be provided for performing certain critical tasks like allowing a new user to access the resource, permitting new device into the directory, permitting new user into the cluster, etc. The access privilege provided to the users, monitors and records the activities of the authorized users and can identify the user who misuses the access privileges.

14.6.5 Incident Response for Entire Systems

Smart systems are the group of smart devices that are bound together to perform certain automatic applications [14]. There are numerous incident response algorithms. The incident response algorithms provide a federated authentication process, protecting the sensitive information stored in the smart system devices. The incident prevention and detection algorithm, which was readily available in the market, can be installed to determine the illegal access hoisted by the attacker.

14.7 Instruction to Build Strong Privacy Policy

14.7.1 Privacy Policy

A protection strategy is a lawful assertion or record that uncovers the manners in which a party will be gathering data from another party. Also, it should diagram how the strategies for utilizing this data and how the strategies for dealing with the data that they acquire. This assertion satisfies a lawful prerequisite to help ensure the security of a client or customer. Contingent upon what data is gathered, organizations may have to incorporate exceptional provisos that relate to that data explicitly. Security arrangements are utilized to ensure the two clients and the organization that is utilizing them.

14.7.2 Definition

A protection explanation might be an actual report or it could be housed on a site. In this day and age, notwithstanding, you are bound to see it on a site than elsewhere because the internet is so common and numerous individuals use it as their method of speaking with the organizations that they work with.

14.7.3 Key Reasons Why There Is a Need for Privacy Policy

There are numerous reasons why you ought to have a security strategy. These are the absolute broadest explanations behind the market, yet there are more explicit motivations to have one dependent on the idea of site are, gain trust from guests, assist your guests with unwinding, permits you to submit to outsider prerequisites, you might be lawfully needed to have one, you can all the more likely secure yourself, how to Create a Privacy Policy, utilize plain English, decide the sort of data you will gather, figure out who you will be offering data to, survey your nearby and government laws, update your approach consistently, try not to request data you don't require, ensure the information you store is secure, permit clients and guests to quit, ensure it is not difficult to track down.

14.8 Role of Internet in Smart Systems

The Internet of Things (IoT) now plays a significant role in daily life. Some IoT technologies are home automation, agriculture, industry, health, and lifestyle.

The word "Web of Things" refers to all technologies that allow the connection of a device to the Internet. These frameworks depend on the collection of data. The information is then used to search, monitor, and pass data to various devices via the internet. This allows for natural overt behaviors.

14.8.1 Home Automation

It is now feasible for a large part of the home to be associated with keen gadgets. There are brilliant attachments that consequently turn on/off gadgets. Keen cautions can play music when you awaken or even reveal to you the news. Voice aides can even run whole schedules where the lights, home apparatuses, indoor regulators, cautions, and different gadgets are controlled. Going ahead, this will be stretched out all through the home.

14.8.2 Agriculture

The applications, which are used in smart agriculture, are livestock and storage monitoring etc. in the future these smart farming technologies [14] will be used. There are plenty of benefits in agriculture while using IoT. Let's see some of the applications in real-time and their use cases. Farmers may be using IoT services to obtain information concerning their cattle location, fitness, and well-being. This information will assist them in evaluating the quality of their animals. For example, detecting creatures that are wiped out in this way and can be isolated from the rest of the group; helps to avoid the infection from spreading to the entire species of dairy cattle.

Climate stations fitted with smart sensors will collect data on the forecast and transfer it to a rancher. Furthermore, the data is dissected by unique programming, and the rancher prepares a personalized examination that facilitates him to develop an accurate description and avoid crop losses.

Drones can be used for anything from soil and harvest field investigation to planting and pesticide spraying in precision agriculture. Robots can be used in connection with different imaging modalities, such as hyperspectral, multispectral, warm, and others to support farmers in time and site-specific data on edit health, infectious diseases, construction bottlenecks, and so on. Robots can also identify drier areas in a region, allowing greater protocols to be introduced to flood those areas. Accuracy horticulture provides ranchers with such strong data that they can make informed decisions and make better use of their resources as depicted in Figure 14.6.

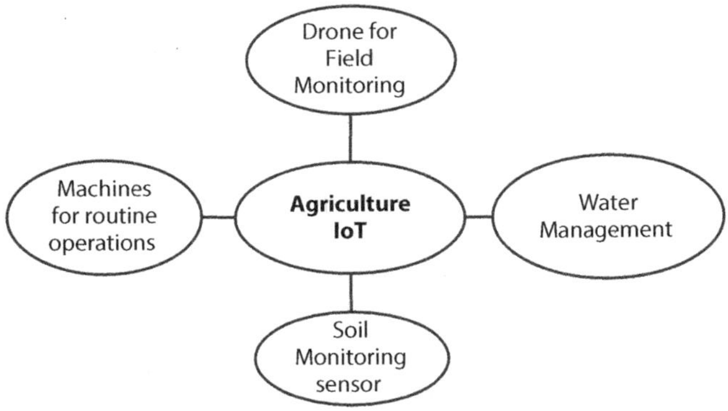

Figure 14.6 Role of IoT in agriculture.

14.8.3 Industry

IoT-enabled hardware can share operational data with partners, such as specific equipment manufacturers and management engineers. This will enable operation managers and manufacturing plant heads to deal with industrial facility units [15] from a distance and take advantage of measure mechanization and improvement. In addition, a closely linked unit will establish a superior line of orders and assist the supervisor in identifying main outcome regions (KRAs).

Condition-based maintenance warnings are allowed by the use of IoT sensors in assembly hardware. There are a variety of simple machine apparatuses designed to operate within specific temperature and vibration ranges. As hardware deviates from its recommended boundaries, IoT Sensors can effectively screen machines and send a ready signal. Makers can conserve resources, cut costs, eliminate machine downtime, and increase service quality by maintaining the desired working environment for machines.

Through the use of IoT in combination with large predictive analytics will surely enhance the health and welfare of the plant's workers. By keeping track of important measures of performance (KPIs) for health and safety, such as the number of wounds and disease rates, close calls, qualitative and mixed applying knowledge, automobile occurrences, and property damage or failure during day-to-day operations. Effective verification ensures greater protection in this way. Slack markers, if any, can be addressed in this manner, ensuring legitimately resolved health, protection, and climate (HSE) issues as mentioned in Figure 14.7.

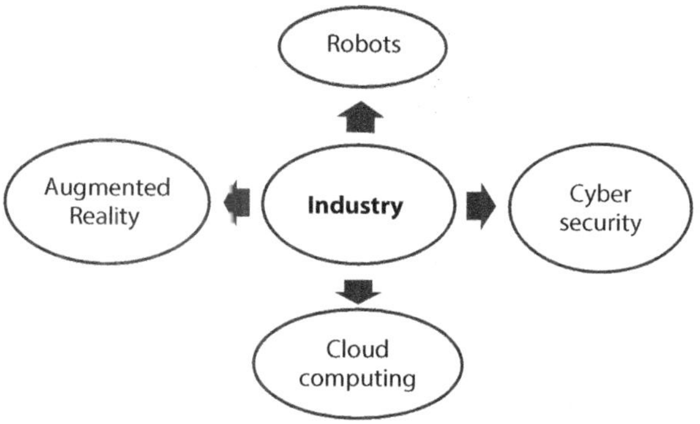

Figure 14.7 Role of IoT in the industry.

14.8.4 Health & Lifestyle

Patients suffering from the effects of diabetes will wear devices with sensors embedded under their skin. The sensors in the gadgets will send data to a patient's cell phone [16] when their glucose levels get excessively low and will record chronicled information for them as well. Thus, patients can likewise advise when they are doomed to suffer from lowered insulin levels throughout the coming years.

The best care for a cancer patient is typically determined by factors other than their weight and age. Their habits and levels of well-being also play a major role in deciding the right care plan for them. Action trackers also keep track of a patient's progress, fatigue levels, hunger, and other factors. Healthcare professionals would use the data collected from the tracker's initial treatment and after treatment has started to assess what improvements can be made to the recommended therapy plan.

Patients may wear devices that monitor their pulses and determine whether or not they have hypertension. When medical care providers need to pull respiratory and heart monitor statistics during testing and scans, they will disclose it. When patients experience arrhythmias, palpitations, strokes, or outright coronary episodes, the wearable devices may alert medical professionals.

People can wear anything that looks like a gem but is supposed to alert relatives or companions in the event of a crisis. For example, if a person is wearing a clinically equipped wrist band and falls asleep in the middle of the night, Individuals they appoint to assist in the event of a crisis will be informed via text message whether their assistance was needed as portrayed in Figure 14.8.

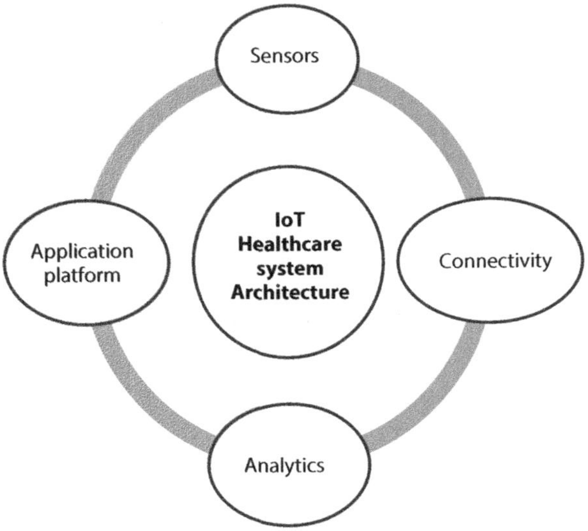

Figure 14.8 Architecture of IoT healthcare system.

14.9 Frameworks, Algorithms, and Protocols for Security Enhancements

The internet of things paves a delightful way in introducing the innovative novel challenges that discriminate ahead in reducing the consumption of energy and resources in many of the sectors. The prominent use of devices might cause certain security-related issues when dealing with some applications. Smart systems mainly use the internet of things for their connectivity. This contains some devices getting connected with vehicles, banks, homes, villas, etc. that can be accessed through the internet [17]. The network infrastructure enables the connection of such physical objects with IoT technology. Enhancing the security and privacy in intelligent systems it's necessary to focus on internet of things related environments. As the IoT enhances the most informative and descriptive form of discussions when dealing with the most sensitive data or information. Many types of research are developed for providing suitable well-equipped frameworks the smart systems for solving some of the interoperability problems. The framework design comprises certain required things, which accomplish the various standards and protocols that are used in the communication systems. Normally the hardware design and the web technologies use the framework for the various protocols and standards. By the security-related issues various newly identified challenges have been emerged and the most predominant of all these is the use of primitives called lightweight cryptography. This usage might limit the resources thus by reducing the consumption of energy for the smart systems.

14.9.1 Framework for the Internet of Things by Cryptography

Smart systems use the Internet of things to impose security-related aspects. When the same is done with the implementation of such devices then it can use the cryptographic framework, which is lightweight. The major properties that enable the use of lightweight cryptography are specified with some standards proposed by ISO 29192. This is marked in the form of projects specified in the target platform. Normally the energy consumption and the size of the chip is the major aspect to be considered in the properties of lightweight cryptography. The software-related implementation backs the hardware because of its RAM size nevertheless to the small code. In this chapter, a framework is discussed, which deals with the security aspects. The proposed architecture might contain a smart object in the Internet of things related network, which contains a material made of re-keying material that is managed by the implementation of transfer protocol in a safer manner. This could be a resource-friendly mechanism that is much fast and adorable for certain agreements related to the key. The verification of the authentication process is enhanced with the usage of certain parameters in the wireless sensor networks. This usage of the computation polynomials makes the system more secure and light weighted in case of dealing with the internet of things. As per the identification of certain compromises in the security-related aspects that get initiated from the internet applications the transport layer in the OSI model is just opposite to that of the layer that has low power. The end-to-end security is not identified to be safe in this part as it contains the constraint application protocol and the security associated with this is the datagram transport layer security. The secure end-to-end connection has a highly secured communication channel, which is much vulnerable to the attacks like flooding and amplification. The mapping of the various protocols for ensuring the end-to-end security to the first upper layer might not allow the data to transit.

For the effective implementation of the lightweight cryptographic protocols, the redistributions methods are used for the Internet of things environment. The most needed efforts used in the lightweight operations of the cryptographic-based system might get optimized with the security associated with the Internet of things. The novel identity-based establishment scheme might consist of an identifier in addition to the trusted third party that provides the device identifier and the common key usage efficiently. The architecture for the IP-based Internet of things might include certain parameters that enhance the security of the system. The designed architecture might contain a security network that contains medium-scale Adhoc security for deploying large-scale domains. The novel framework is not designed only for the resource-constrained aspect but also for the improvement of the computing modes that provide many resources and enormous spaces for storage purposes. Also, the security aspect is made higher in the network.

14.9.2 Protocols for Security Enhancements

The existing systems focus on the optimization strategies of various operations held in the internet of things network in the aspect of considering authentication and security. The comparison is made following the lightweight operations in the IoT-based network. The most challenging task is the comparison of the wireless sensor networks with the other IoT devices. Considering the wireless sensor networks the IoT might have other protocols like IPv4 and IPv6.The lightweight security protocols are implemented for unnatural environments like WSN. The new security protocols are designed, which can tolerate all the security needs that might exist on the Internet. The electronic security protocols are designed to satisfy the wireless sensor networks, which are not similar one just like the Internet protocol. For using such networks, the major requirement is the ratification that is done using the WSN protocols [18]. Based on the devices used on the internet the solution needed for the enhancement of the security needs some sort of modification, which is not made practical indeed. The adaptation of the lightweight protocols inside the architecture might be related to the security placements within the various layers like the application, data link, and the network layers in the Internet of things as the layers of unique nature processes different security needs and information sharing.

14.10 Design Principles of Privacy Enhancing Methodologies

In an undeniably associated world, a variety of information is gathered from the climate and its occupants. In light of the lavishness of the data, protection turns into a significant prerequisite. Even though there are standards and rules, there is as yet an absence of systems to direct the coordination of security rules into the advancement cycle. Approaches that are possible in the sense of protection utilizing design might get obscured, which may leave many open inquiries scheduled in the best way for applying them practical manner. Another form of approach is proposed which enhances the protection utilizing evidence that serves as a product advancement strategy to give protection mindfulness. Another mode is proposed for recording the most alleviations might gets equipped with a lot of issues thus by providing some complete protection by identifying the steadiness of authentication and privacy that is expected to expand in case of retaining the certainty. For approving the viability, the protection utilizing evidence is made utilized throughout the improvement of some contextual investigations like counting the people, meter related applications, and also the applications that involved in the observation of the energy effectiveness checking framework in addition to a couple of validation frameworks. Considering all the applications involved in the groups, it may have the option of giving some particular range of confirmations for the security, which is separately presumed with the protection by the form of evidence, which can be made powerful associated with the assistance of comprehensive approaches, which can address most of the protection authentication services and requests while generating programs.

Four contextual analyses executing the exercises depicted here in smart energy metering, individuals checking, energy proficiency, and two-factor validation applications to approve the viability of PbE. These are instances of uses that are filling in numbers, and some of them may bring security hazards, causing numerous individuals and even the media to show doubt about them. The fundamental target of these contextual analyses is to survey the capacity of PbE in delivering confirmations of security.

14.11 Conclusion

The advancement of technology thrives the human life to move towards the advanced standard by incorporation of smart systems, which is comparatively small when compared with the geographical area covered by the smart city and Internet of Things. The smart system assists the day-to-day processes of human life with the inclusion of sensors along with the devices. The smart system devices are highly vulnerable to attacks and the sensitive data in the hub of the smart system is highly prone to the launch of any type of attack. The adversary tries to invade the smart systems with a dual intention of accessing the data illegally and generating revenue by cracking the smart systems. This book chapter describes the various attacks of the smart systems, their security challenges, and the security principles of the smart systems. In future, the smart systems may be provided with a high level of security by the inclusion of lightweight cryptographic algorithms.

References

1. Yu, S., Gu, G., Barnawi, A., Guo, S., Stojmenovic, I., Malware propagationin large-scale networks. *IEEE Trans. Knowl. Data Eng.*, 27, 1, 170–179, 2015.
2. Bastug, E., Bennis, M., Médard, M., Debbah, M., Toward interconnected virtual reality: Opportunities, challenges, and enablers. *IEEE Commun. Mag.*, 55, 6, 110–117, 2017.
3. Angrishi, K., Turning internet of things (iot) into internet of vulnerabilities (iov): Iotbotnets, 2017, arXiv preprint, arXiv: 1702.03681.
4. Hasan, M., Biswas, P., Bilash, M.T.I., Dipto, M. A. Z., Smart Home Systems: Overview and Comparative Analysis. *Fourth International Conference on Research in Computational Intelligence and Communication Networks (ICRCICN)*, 2018.
5. Romero, M., Guédria, W., Panetto, H., Barafort, B., Towards a Characterisation of Smart Systems:A Systematic Literature Review. *Comput. Ind.*, 120, 1–23, Mar. 2020.
6. Abrishamchi, M., Abdullah, A., Cheok, A., Bielawski, K., *Side channel attacks on smart home systems: A short overview*, pp. 8144–8149, 2017, doi: 10.1109/IECON.2017.8217429.
7. Shyamala., R. and Prabakaran, D., A survey on security issues and solutions in virtual private network. *Int. J. Pure Appl. Math.*, 119, 1183–1191, 2018.
8. Prabakaran, D. and Shyamala, R., *A Review on Performance of Voice Feature Extraction Techniques*, pp. 221–231, 2019, doi: 10.1109/ICCCT2.2019.8824988.
9. Kaur, R., Sandhu, P.S., Kamra, A., A novel method for fingerprint feature extraction. *International Conference on Networking and Information Technology*, Manila, Philippines, 2010, pp. 1–5, 2010, doi: 10.1109/ICNIT.2010.5508569.
10. Prabakaran, D. and Sriuppili, S., Speech Processing: MFCC Based Feature Extraction Techniques- An Investigation. *J. Phys. Conf. Ser.*, 1717, 012009, 2021, doi: 10.1088/1742-6596/1717/1/012009.
11. Biswas, K. and Muthukkumarasamy, V., *Securing Smart Cities Using Blockchain Technology*, 2016, doi: 10.1109/HPCC-SmartCity-DSS.2016.0198.
12. Hussain, F., Hussain, R., Hassan, S., Hossain, E., Machine Learning in IoT Security: Current Solutions and Future Challenges. *IEEE Commun. Surv. Tut.*, 22, 3, 1686–1721, 2020, doi: 10.1109/COMST.2020.2986444.
13. Jain, V. and Agrawal, M., *Applying Genetic Algorithm in Intrusion Detection System of IoT Applications*, pp. 284–287, 2020, doi: 10.1109/ICOEI48184.2020.9143019.
14. Yang, C., Shen, W., Wang, X., Applications of Internet of Things in manufacturing. *IEEE 20th International Conference on Computer Supported Cooperative Work in Design (CSCWD)*, Nanchang, China, 2016, pp. 670–675, 2016, doi: 10.1109/CSCWD.2016.7566069.

15. Xu, L.D., He, W., Li, S., Internet of Things in Industries: A Survey. *IEEE Trans. Industr. Inform.*, 10, 4, 2233–2243, 2014Nov. 2014. doi: 10.1109/TII.2014.2300753.
16. Jerabandia, M. and Kodabagi, M.M., A review on home automation system. *International Conference on Smart Technologies for Smart Nation (SmartTechCon)*, Bengaluru, India, 2017, pp. 1411–1415, 2017, doi: 10.1109/SmartTechCon.2017.8358597.
17. Arul, L.J. and Senthil Kumaran, R., Development of Vulnerability Scanner. *Int. Res. J. Eng. Technol.*, 5, 7, 1278–1281, 2018.
18. Preethi, S. and Senthil Kumaran, R., Secure Key Establishment Scheme for Wireless Sensor Networks in Distributed IoT Applications. *IJMTES*, 3, 8, 90–93, 2016.

15

Artificial Intelligence and Blockchain Technologies for Smart City

Jagendra Singh[1*], Mohammad Sajid[2], Suneet Kumar Gupta[1] and Raza Abbas Haidri[3]

[1]School of Engineering and Applied Sciences Bennett University, Greater Noida, India
[2]Department of Computer Science, Aligarh Muslim University, Aligarh, India
[3]Department of Computer Science and Information Technology, Khwaja Moinuddin Chishti Language University, Lucknow

Abstract

In this digital age of rapid communication, the advances of emerging technologies can be used for making smart city as intelligent society. Specially, the speedy acceptance of artificial intelligence (AI) and blockchain technologies have guided a paradigm that is shifting to a new dimension called as digital ecosystem for smart city. A large number of AI and applications of blockchain guarantee resolutions for challenges in the fields varying from financial services, and threat management to cryptocurrency, and from social and public services to Internet of Things (IoT). Moreover, the conjunction of blockchain and AI technologies is transforming the network of smart city architecture for developing sustainable ecosystems. When we try to achieve the goal of developing smart cities, the innovations in technologies created both challenges and opportunities. This chapter presents a broad literature examination to the safety problems and the challenges, which influence the blockchain utilization in developing sustainable and smart societies. Our work represents a comprehensive dialogue of various vital issues for coming together for AI and blockchain knowledge, which help us to develop smart societies. Therefore, we talk about the solutions of security issues of blockchain and summarize the important concepts that need be utilized to develop many AI and blockchain-centered smart transportation techniques. Moreover, we review the problems that stay public and our forthcoming research directions, this contains new proposals for security and future regulations for developing a smart society with sustainable ecosystem.

**Corresponding author*: jagendra.singh@bennett.edu.in

Suman Lata Tripathi, Souvik Ganguli, Tengiz Magradze, and Abhishek Kumar (eds.) Intelligent Green Technologies for Sustainable Smart Cities, (317–330) © 2022 Scrivener Publishing LLC

Keywords: Blockchain technologies, artificial intelligence, smart city, Internet of thing, green computing, smart transport system

The major contribution of this chapter can be summarized as follows:

- This chapter introduces the important contextual understanding of blockchain and their conjunction with artificial intelligence and describes an artificial intelligence and blockchain cantered smart city design.
- This study scientifically depicts and examines the limitation of security, liabilities, challenges, and issues of Internet of Things–based smart society ecosystem.
- In adding, we as well consider a few business and technical problems in artificial intelligence and blockchain organization.
- The current solutions of security are discussed for AI and blockchain systems in the environment of smart city.

15.1 Introduction

In metropolitan areas, where density of people is high, smart city concepts brings better lifestyle by using intelligent services of high quality. Many social, environmental, and economic problems are created with speedy expansion of global inhabitants that can change life quality and the lifestyle of numerous people substantially. The concept of sustainable smart cities promises to assist boost understanding and make public green energy less utilization with safest habits.

As United Nations predicted, 66% of globe inhabitants (Bibri & Krogstie, 2017) will soon reside in large cities, it means, for social sustainability we need to face many issues or problems. Furthermore, this kind of the city form can be considered as environmental and social challenge. The municipalities are consuming nearby more than half of global resources. It creates problems in distributing these resources in the uniform way by using present technologies (Li *et al.*, 2016). Therefore, for the concept of smart cities, a vital role can be play by information and communication technology (ICT).

Now a days, artificial intelligence and blockchain technology are two well-known emerging technologies, that have various admiring features like security, fairness, decentralization and automation. From 2008, the attention in blockchain related technology is growing; the technology of blockchain is catching consideration and can be utilized as a dominant subject (Zhang *et al.*, 2016; Statista, 2015). Blockchain can be used as a decentralized technique, civic register of all communications in a distributed kind of network. Private and public keys are used in all the transactions that are mathematically related to each other, these transactions can be traced and recognized by the participants in place of using record preservation. There are many features of blockchain technology and one of that is distributed databases, the advantage of that is downloading of transaction copy from anywhere in the network, because each computer is assumed as node of the network. But this produces trust and privacy questions; we will confer these issues and solutions of these issues in this chapter.

Blockchain is an interesting technology that provides us various features like data integrity, anonymity, and security without controlling by third party for transaction controlling. These features makes this technology appropriate for handling limitations and challenges of smart cities. Public security and safety have become a main concern in the development of smart cities specially for handing terrorism, traditional crime, cybercrime, and natural disasters. IoT world hearsays that in 2016, nearby 750 million dollars were disbursed for privacy and security issues of IoT and assume to increase 4.5 billion up-to 2021 (Statista, 2021) that is depicted in Figure 15.1. While foreseeing privacy and security issues of smart city, it must stay pointed that many same types of problems exists today but often they do not occur, while fully interconnection of these technologies.

For solving the privacy and security issues of smart cities, the investors must solve the whole scenario to safeguard from these problems not to remain in this intelligent type of network. For attaining this giant goal, the

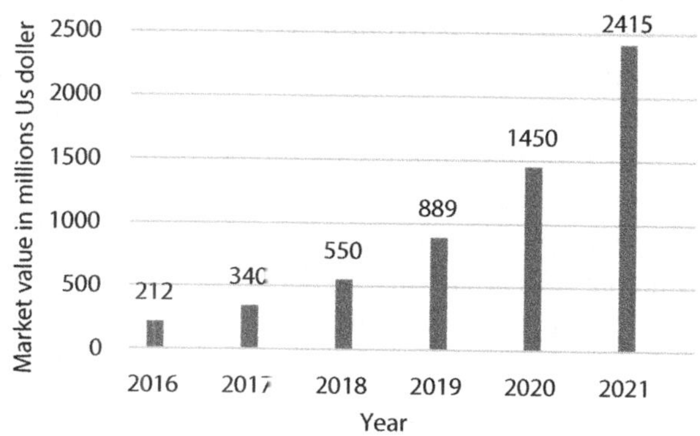

Figure 15.1 Blockchain market from 2016 to 2021.

planner of smart city and professionals of security want to pull related projects and should encourage making of required smart system for smart cities. To address these challenges and issues, recent attempts and considerations will guide that how securely we can build system for sustainable smart towns in the future. Therefore, blockchain is one of the reliable and safe technological architecture for developing transport system with intelligence.

For developing smart towns, the transport system application can be develop using blockchain technology that is now facing revolution in its usage and has potential to develop such kind of applications. For building a reliable, secure, autonomous, and decentralized transport system using blockchain technology, we need to use current available resources and infrastructure (B-IoT, 2021). In smart cities, for controlling the autonomous vehicles, transport system will be a vast connected network of vehicles. The demand of smart transport system is increasing in our society because road resources are consuming by current system at high rate (Tu *et al.*, 2018).

Smart transport system (STS) will change a lot in human factor that is an important role in all kinds of technologies, it considers management of traveller information and design of vehicle (Zhou *et al.*, 2018). And it is a legal sector for shifting items from one place to another. Over time, the sector of transport sees numerous problems and issues, such as congestion of traffic, high rate of accidents, emissions of carbon, and air pollution. Further, the transport sector has public nature, so the use of wireless technology creates various privacy and security issues (Behrens, 2021).

These issues include authentication of privacy of location, privacy of identity, nonrepudiation, integrity, anonymity, authentication, confidentiality, revocation of certificate, and verification of certificate (Tang *et al.*, 2018). Developing environment for future smart network, the blockchain technology can play a significant effect, and while using this, several technical issues will generate, most of them are related to privacy, security, business exposures and scalability. Many researchers focused on liabilities of smart contract, such as wallet contracts, bitcoin, and Ethereum (Singh, 2020), and with some other challenges of security (Wang *et al.*, 2017). Smart cities and smart homes are the beginning of smart networks, Internet of Things, also called IoT, is an emerging area with huge impact with potential use, as per the survey up to 2020, around 55 billion linked devices are available (Cai & Huang, 2017).

In the last few years, connectivity has been improved by IoT using blockchain technology. The network representation of futured smart city presented in Figure 15.2. Intelligent vehicle system is an example of distributed system monitor each connected device, and it can find what kind of action is performed by which device. Smart contract in IoT devices get

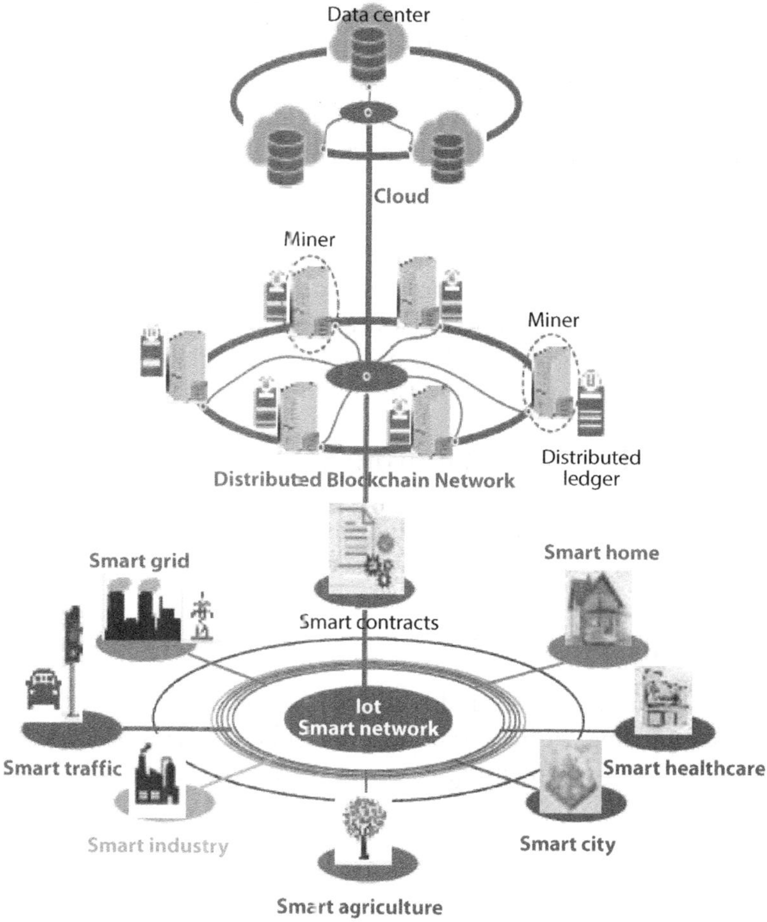

Figure 15.2 Smart city network design based on distributed blockchain.

advantage to communicate with each other's device and connect with the Internet because of this interaction between two devices need not any kind of central authority to handle it. And big enterprises can get help of blockchain and IoT technology for enhancing security infrastructure and can also be used for data analysis and extraction (Singh, 2017).

By the help of IoT and blockchain technology, the integration of fault proof connected devices can be maintained by open ledgers and can help in eliminating following issues:

- network issues of IoT devices,
- in real-time scenario exchange of data throughout the network,

- data transparency issues,
- process tracking of end-to-end process,
- transaction verification and automation issues,
- IoT platforms with cloud shortcoming issues,
- data analytics issues of IoT devices.

Many opportunities has opened by using the extraordinary evolution of IoT technologies for developing techniques to share and access information and open data is an important part of this development. Various issues and challenges are faced by business platforms for adopting IoT because many gaps are there in integration and security of data (Singh & Srivastava, 2021).

15.2 Standard for Designing Smart City and Society

The prerequisites required to design an efficient, scalable, secure, and enabled with artificial intelligence smart city are as follows.

15.2.1 Scalability

A huge amount of software and hardware devices are to be used in smart cities for providing services to citizens for collecting and processing data at real time. For implementing it, we need a blockchain and IoT enabled infrastructure.

15.2.2 Intelligent Health Care

The healthcare sector can get advantage of IoT technology for bringing together intelligent system, big data, and critical analysis.

15.2.3 Flexible and Interoperable

Physical items can be interconnected using communication approaches and technologies. IoT platform can be a communication approach for home related applications that can be handled and controlled from any other place and data can be managed by cloud services.

15.2.4 Safeguard Infrastructure

For protecting network and information from numerous types of attacks, entity authorization, authentication, malicious activities monitoring, and policies for controlling network development for protecting and securing smart towns and cities.

15.2.5 Robust Environment

The urbanization and globalization convergence creates unexpected issues and problems in smart cities, so for making smart cities sustainable the used infrastructure should be robust type.

15.2.6 Distribution and Sources of Energy

Supplied energy quality and resources of renewable energy is a vital requirement of smart cities, and there should be guarantee of this type of energy. Therefore, power grid kind of distributive technology can be a good solution in this area for efficient use.

15.2.7 Intelligent Infrastructure

Intelligent system can bring various individual items together, therefore, all the individuals should work together for developing a smart city that can sustain for a long time.

15.2.8 Choice-Based Backing System

Smart city has a combination of diverse and complex system that need a balance in environmental, industrialization, and economic development essentials. It is a big challenge for policy-making firms, for finding optimal solutions a supportive decision system should be implemented.

15.2.9 Monitoring of Behavior

Human adaptive behavior can be simulated by computer system simulation in the smart cities, owing to the going web, and relationship cycle of reciprocal behavior produces pattern of behavior data, which is exposed in the collaboration between physical, social, and economic systems that signify webs of the networks.

15.3 Blockchain and Artificial Intelligence

Blockchain (Prasad *et al*, 2015) and artificial intelligence (Makhija, 2021; Singh & Sharan, 2017) technologies disagree with each other in different

Table 15.1 Blockchain and AI with various aspects.

Aspects	Blockchain	Artificial Intelligence
Approach	Deterministic	Probabilistic
Access	Open	Closed
Nature	Decentralized	Centralized
Transparency	Transparent	Black box

and separate type paradigms, that is depicted in Table 15.1, but in various field mixture of both, the technologies are used. Blockchain and artificial technology can seem like cutting-edge technology, and these has possibility of various new opportunities.

For designing self-driving vehicle, there is a good opportunity of using smart transportation system. Blockchain and artificial intelligence combinedly are used in self-driving cars, learning vehicles, and in intelligent system for transport management (Sgantzos & Grigg, 2019; Park & Park, 2017). Independent vehicles are now in conventional trend, but only few can use cars using artificial intelligence techniques. Renting a car is also a good option. So, the rental process can be solved to simpler form by blockchain technology by minimizing the numbers of people engaged, audit trails providing, and by synchronizing available data. There are various advantages of this kind of technology, I am listing few of that as follows:

- share and interact various available vehicles to with AI-blockchain technology,
- vehicle training by AI-blockchain technology will improving efficiency,
- tumbling human effort, cost reduction and time saving,
- learning process get modularity.

15.4 Contributions and Literature Study

Many challenges and issues have been focussed on in recent years, and several type issues of blockchain technology and their combination with cloud system and IoT system are also addressed (Singh & Bohat, 2021). The authors addressed blockchain system for IoT called B-IoT, and blockchain system for smart transport approaches (Joshi *et al.*, 2018). In Panarello *et al.* (2018), the authors reviewed artificial technology for next generation and congregated the blockchain to distribute services and speed up healthcare study. In Singh & Sharan (2017), the authors talked about smart grid with its security tasks and its applicability for healthcare development that can sustain. The current research work represented in Table 15.2.

For blockchain healthcare–related applications, latest state of work was reported and for medical record automation, HDG used that is a model for mobile app (Banerjee *et al.*, 2018). Simon suggested a sensor framework for applications in data domain to design smart cities with IoT in future using environment sustainability (Halim *et al.*, 2017). The authors studied prior works for focusing on applications of big data approach and additionally discover the chances to increase the knowledge environment for smart cities that can sustain. The authors likewise concentrating on applications, specialized technologies, and research that are more abstract and their innovative treatments.

The authors in Singh (2020) analyzed the existing investigation in blockchain knowledge; the purpose of this study was to identify the existing topics on challenges, research, and future developments of the current technology. The authors in Ande *et al.* (2019) considered adoption of blockchain technology and various issues like security applied in environment of cloud computing. The authors deliberated the structure of block and worked on case studies in security case studies comprising attacks also. In Sharma & Park (2018), the authors studied security in blockchain with measuring real attack and threats in it. The research article précises the analysis of the allied liabilities exploited and the real outbreaks on blockchain approach.

The authors moreover conversed solution for security and its achievements for academic side and research future scope prediction. Though we studied a lot of advantage of blockchain-related technology, the absence of zeal in research topic investigation displays that investigator working on addressing to mitigate risks of security in the blockchain technology in place of solving the complexities of potential claims. Thus, this is authoritative for addressing security glitches in blockchain approaches and environments and make them safer, and additional protected applications of blockchain technology crossways industries.

This chapter addresses an inclusive taxonomical study of security and safety in blockchain-related technology to a smart city that can sustain. Contrasting in existing current research work, this analysis addresses a comprehensive argument with numerous important features for the AI and blockchain approaches for a smart and intelligent cities and society. The impact and effect of AI and blockchain approaches discussed in this research article for developing a smart societies and environments.

Table 15.2 Summary of existing surveys conducted.

Reference/ year	Security Challenges	Focus	Blockchain-AI Opportunities	Existing Solutions	Future Suggestion	Approach	Threats and Attacks	Classific-ation	Open Issues
Xiaoqi Li *et al.*, [6]	Common challenges and attacks are defined but with lack technical details	Blockchain-IoT	Scope of the technology in various domains	Solutions according to few attacks	Directed to limited scopes	Descriptive	Yes	Yes	No
Jesse Yli-Huumo *et al.* [36]	Challenges are highlighted for technical perspective	Blockchian	No	Limited	Yes	Analytic	No	Yes	Well-directed
Jin Ho Park [37]	Security and privacy issues regarding elements of a smart city connecting to IoT using different network applications	IoT	No	Regarding anonymity, security, privacy, and transparency concerns	No	Descriptive	Yes	No	No

(*Continued*)

Table 15.2 Summary of existing surveys conducted. (*Continued*)

Reference/ year	Security Challenges	Focus	Blockchain-AI Opportunities	Existing Solutions	Future Suggestion	Approach	Threats and Attacks	Classific-ation	Open Issues
Alfonso Panarello *et al.* [40]	Challenges of blockchain and IoT integration in different application areas	Blockchain-IoT	No	Solutions regarding scalability	No	Analytic	No	Yes	No
Madrita Banarjee *et al.* [41]	No	Blockchain-IoT	No	Solutions are regarding the challenges of the applications	Yes	Descriptive	No	Limited	No
Md Arafatur Rahman [42]	Security issues are defined by approaches, network structure, applications in terms of IoT environment	IoT	Limited	Limited	No	Analytic	No	No	No
This survey	Yes (in detail)	Blockchain IoT and AI	Yes	Yes (in detail)	Yes	Analytic	Yes	Yes	Well-directed

The key methodical contribution of this research work is concluded as follows:

- This chapter presents the required contextual information of artificial technology and its conjunction with blockchain technology and produces AI- and blockchain-based architectures of smart city.
- This chapter moreover enables the researchers to get knowledge and familiarity in blockchain architecture and technologies.
- This study methodically discussed and presented the limitation of security concern, challenges, liabilities, and problems of IoT for developing smart cities. Moreover, we consider certain business and technical problems in AI and blockchain-based industries.
- This study describes broad elements of smart structure for smart cities that can sustain. Additionally, it involves numerous current projects that combine the AI and blockchain technologies.
- Present solutions for security are discussed for AI and blockchain-related technology in developing smart cities. After all, in our study, this chapter discusses relations among urban problems in the direction of sustain abilities and existence in studies of life cycle, and feasible forthcoming directions.

15.5 Conclusion

The artificial intelligence (AI) and blockchain merging concept is transforming concept of smart cities. It brings governments, companies, and countries together. AI and blockchain technologies are commonly recognized, accepted, and highly valued because of its nature that is based on peer to peer and decentralized attributes. In this study, we demonstrated many challenges and security issues that hamper the implementation and acceptance of AI and blockchain technology, discovering the numerous trials in diverse characteristics of smart cities that can sustain. This study additionally take on a glance at numerous ongoing AI- and blockchain-based projects and their advantages at a company or business level. In the end, we conversed security solutions of blockchain technology for transportation systems of smart cities. After all, this study confronted problems connected to ability of sustain and existence in the life sequence investigations based on AI and blockchain comprising certain forthcoming possible directions and tips of the technology.

References

Ande, R., Adebisi, B., Hammoudeh, M., Saleem, J., Internet of Things: Evolution and technologies from a security perspective. *Sustainable Cities Soc.*, 54, 101–128, 2019.

Banerjee, M., Lee, J., Choo, K.K.R., A Blockchain future for internet of things security: a position paper. *Digital Commun. Networks*, 4, 3, 149–160, 2018.

Behrens, A., 2021, March 23, https://spendmatters.com/2019/08/22/blockchain-ai-offer-big-savings-in-transportation-and-logistics-pwc-surveyfinds/.

Bibri, S.E. and Krogstie, J., Smart sustainable cities of the future: An extensive interdisciplinary literature review. *Sustainable Cities Soc.*, *31*, 183–212, 2017.

B-IoT, Blockchain Technology for IoT in Intelligent Transportation Systems, 2021, January 10, http://iot.ed.ac.uk/projects/b-iot/.

Cai, Z. and Huang, L., Finite-time stabilization of delayed memristive neural networks: Discontinuous state-feedback and adaptive control approach. *IEEE Trans. Neural Networks Learn. Syst.*, 29, 4, 856–868, 2017.

Halim, N.S.A., Rahman, M.A., Azad, S., Kabir, M.N., Blockchain security hole: issues and solutions. *Proc. of the International Conference of Reliable Information and Communication Technology*, pp. 739–746, 2017.

Joshi, A.P., Han, M., Wang, Y., A survey on security and privacy issues of Blockchain technology. *Math. Found. Comput.*, 1, 2, 121–147, 2018.

Li, X., Jiang, P., Chen, T., Luo, X., Wen, Q., A survey on the security of Blockchain systems. *Future Gener. Comput. Syst.*, 9604, 106–125, 2016.

Makhija, P., Top 5 Reasons: How Blockchain and Artificial Intelligence is Revolutionizing the Transport Industry, 2021, January 18, https://it.toolbox.com/guest-article/top-5-reasons-how-blockchain-and-artificial-intelligence-is-revolutionizing-the-transport-industry.

Panarello, A., Tapas, N., Merlino, G., Longo, F., Puliafito, A., Blockchain and IoT integration: a systematic survey. *Sensors*, 18, 8, 1–37, 2018.

Park, J. and Park, J.H., Blockchain security in cloud computing: use cases, challenges, and solutions. *Symmetry*, 9, 8, 164–177, 2017.

Prasad, M., Lin, C.T., Prasad, O.K., Mohanty, M., Singh, J., Novel Data Knowledge Representation with TSK-Type Pre-processed Collaborative Fuzzy Rule Based System. *IEEE Symposium Series on Computational Intelligence*, pp. 14–21, 2015.

Sgantzos, K. and Grigg, I., Artificial Intelligence Implementations on the Blockchain: Use Cases and Future Applications. *Future Internet*, 11, 8, 170–182, 2019.

Sharma, P.K. and Park, J.H., Blockchain based hybrid network architecture for the smart city. *Future Gener. Comput. Syst.*, 86, 650–655, 2018.

Singh, J. and Bohat, V.K., Neural network model for recommending music based on music genres. *International Conference on Computer Communication and Informatics (ICCCI -2021)*, pp. 1073–1078, 2021.

Singh, J., Ranks aggregation and semantic genetic approach based hybrid model for query expansion. *Int. J. Comput. Intell. Syst.*, 10, 1, 34–55, 2017.

Singh, J., Collaborative Filtering based Hybrid Music Recommendation System, in: *proceedings of the 3rd International Conference on Intelligent Sustainable Systems (ICISS 2020)*, pp. 1183–1086, 2020.

Singh, J., Learning based Driver Drowsiness Detection Model, in: *proceedings of the 3rd International Conference on Intelligent Sustainable Systems (ICISS 2020)*, pp. 1084–1087, 2020.

Singh, J. and Sharan, A., Rank fusion and semantic genetic notion based automatic query expansion model. *Swarm Evol. Comput.*, 38, 295–308, 2017.

Singh, J. and Sharan, A., Context window based co-occurrence approach for improving feedback based query expansion in information retrieval. *Int. J. Inf. Retr. Res.*, 5, 4, 31–45, 2018.

Statista, Size of the Blockchain technology market worldwide from 2016 to 2021, 2021, Feb 15, https://www.statista.com/statistics/647231/worldwide-Blockchain-technology-market-size/.

Tang, Z., Ding, X., Zhong, Y., Yang, L., Li, K., A Self-Adaptive Bell–LaPadula Model Based on Model Training With Historical Access Logs. *IEEE Trans. Inf. Forensics Secur.*, 13, 8, 2047–2061, 2018.

Tu, Y., Lin, Y., Wang, J., Kim, J.U., Semi-supervised learning with generative adversarial networks on digital signal modulation classification. *Comput. Mater. Continua*, 55, 2, 243–254, 2018.

Wang, D., Huang, L., Tang, L., Synchronization criteria for discontinuous neural networks with mixed delays via functional differential inclusions. *IEEE Trans. Neural Networks Learn. Syst.*, 29, 5, 1809–1821, 2017.

Zhang, L., Zhipeng, C., Xiaoming, W., FakeMask: a novel privacy preserving approach for smartphones. *IEEE Trans. Netw. Serv. Manage.*, 13, 2, 335–348, 2016.

Zhou, S., Liang, W., Li, J., Kim, J.U., Improved VGG model for road traffic sign recognition. *CMC Comput. Mater. Continua*, 57, 1, 11–24, 2018.

16

Android Application for School Bus Tracking System

S. Sriram

Dept. of Computer Engineering, Sri Ramakrishna Polytechnic College, Coimbatore, Tamil Nadu, India

Abstract

Android is becoming very popular and reason for Android platform it requires an open source development, which is probably the most feasible and a user-friendly approach. This represents an approach using android SDK to develop application for intelligent mobile devices mainly GPS. This application uses the GPS function, available in most of smartphones today, to pinpoint current location accurately. The bus location will update the location of a bus on server frequently. The coordinates of location will be sent to the server, and the server will send current location of bus to student, which will provide additional services through notification if the bus reaches near to the student location. This system presents an adaptive location information sharing system that allows for a user to know the exact location of a vehicle. So the passenger does not need to wait long to catch the vehicle. In this system, the vehicle GPS location is updated to the web server. The web service receives the location information of a vehicle continuously. That information is updated in the database. While the vehicle location is near to the passenger location, the system sends a notification alert to the passenger mobile.

Keywords: School bus application, school bus monitor, location tracker, ip camera school bus monitor

16.1 Introduction

Among all public transportation services, the bus service is a major transportation used by the public. Especially in a busy town or city, a bus is the

Email: srsriramenggl@gmail.com

Suman Lata Tripathi, Souvik Ganguli, Tengiz Magradze, and Abhishek Kumar (eds.) Intelligent Green Technologies for Sustainable Smart Cities, (331–340) © 2022 Scrivener Publishing LLC

most easy, convenient, and cheaper transportation. There are various reasons that people take bus instead of driving their own vehicle, such as traffic jam, heavy parking fee, and lack of parking slot in destination. However, bus transportation service has very poor transportation information system nowadays. Bus users do not know the exact arrival time for a bus, but only know the scheduled arrival time [1–5]. Compared with train or flight transportation system, bus transportation service does not have a proper system to track all buses position and the actual arrival time in every bus stops.

These problems occur because current bus service system do not apply real-time tracking technology to track each bus on the road and also lack of a platform to update latest bus traffic information to bus users [5]. In order to solve these problems and enhance current bus service system, real-time school bus tracking system has to develop and implement With real-time school bus tracking system [6, 7]. Bus position data is connected real time and transmitted to client mobile and extracting transit information. The main technology used to develop this system is Global Positioning System (GPS). GPS technology is able to receive the position of an object from space-based satellite navigation system [9, 10]. The developed school bus tracking system will be able to provide bus users a real-time platform to check on updated bus traffic information, for examples, bus arrival or departure time [11]. Besides, this system is also able to reduce workload for school bus management team and provide an immediate platform to update latest and accurate location and estimate time information to the client's mobile.

An IP camera is set up inside the school bus to project the monitoring of the students. The images captured by an IP camera may be viewed from anywhere in the world via the Internet, whether via pc, laptop or mobile phone. In many cases, as well as being able to view video footage and listen to audio streaming, the camera can also be controlled remotely. IP cameras capture images in much the same way as a digital camera and compress the files to transmit over the network. IP cameras may be used with a wired network connected via Ethernet cable to a broadband modem or router, or wirelessly via a Wi-Fi router [18]. The school bus tracking system, which is being fitted in every school vehicles, allows school authorities to view the location of the vehicles from their smart phone, [12]. Additionally, Parents can likewise track the School Bus through Mobile Application, they can see their bus live location.

16.2 Application Methods for Access

This application is specially designed for school children using the school bus. The student does not know exactly what time a bus will arrive, but only know the scheduled arrival time. The student needs to wait for a bus without knowing what time the bus will actually arrive. Sometime, student might feel anxious and impatient when they waiting for a bus if they do not know what time the bus will arrive especially when student rushing of time for certain reason. Additionally, this situation wastes a lot of time when waiting in bus stop because the time wasted can actually be spent on another matter [13, 14].

Figure 16.1 School bus tracking applications.

16.2.1 Driver Portal Screen

While the school bus starts, the driver needs to log in the application, and he is able to share the current location using Google maps to all registered mobile numbers as shown in Figure 16.2. A notification message is also sent immediately to the corresponding registered mobile numbers of the parents. When the school student enter into the school bus, daily smart attendance can be taken using QR scanner methods daily report will be submitted through e-mail in an Excel file format and the driver is able to send an alert message to the school management in any emergency situation.

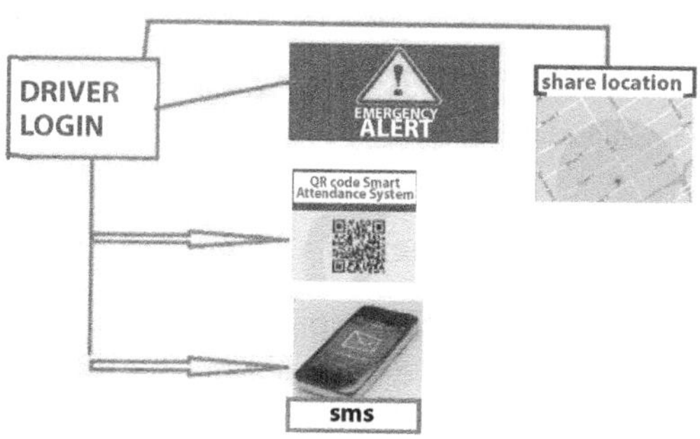

Figure 16.2 Driver login portals.

16.2.2 Parent Portal Screen

Through these School bus Tracking application Parents are able to get a notification message from the school bus driver at the start that the bus inside the school campus [6]. School student waiting for a school bus if they do not know what time the bus will arrive especially when student rushing of time for certain reason. Additionally, this situation wasted a lot of time when waiting in bus stop. The estimate time for their school bus stop could be display in the application for time management purpose that helps the student to be present in their corresponding allotted bus stop. The School bus driver shares the bus location through the application while stating the school bus inside the school bus campus [8]. The location link also send by notification message to each parent mobile number. When the school bus reaches the school campus, the school bus driver terminates the current location sharing point at the same time a notification text message could be sent to parent mobile that school bus reaches destination. Parents are able to watch their children inside the bus at any time through an IP camera by online process as shown in Figure 16.3.

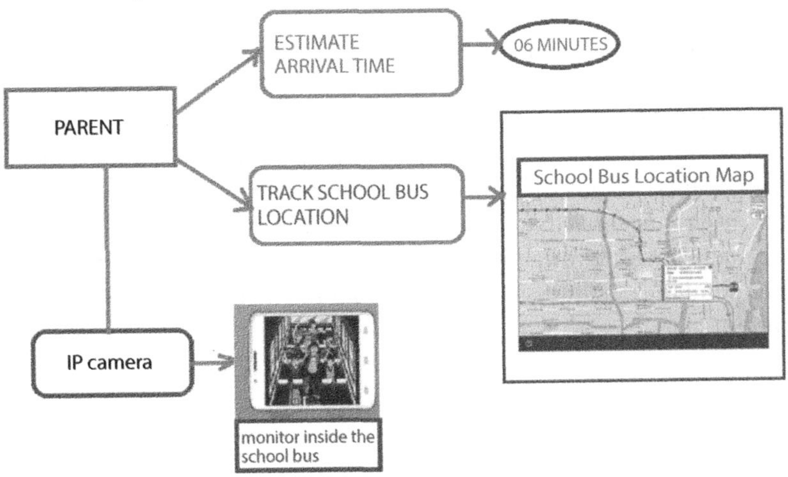

Figure 16.3 Parent login portals.

16.2.3 Teachers Portal Screen

School Teachers also able to use this application through teachers login portal. The main three categories every school teachers are able to get access. Estimate school bus arrival time, location tracking and Message to Parent, whether if a school student get absent in their corresponding bus stop as shown in Figure 16.4.

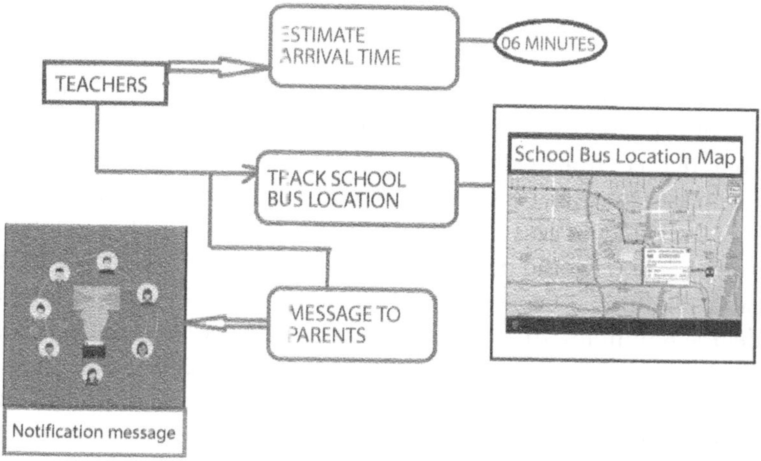

Figure 16.4 Teachers login portal.

16.3 GPS Data Processing Methodology

GPS stands for "Global Positioning System" by which anyone is able to identify the position information anywhere in the world. The GPS location sharing school bus makes your fleet, environment-friendly given in Figure 16.5. Through optimization of routes and monitoring hard breaking and acceleration, the menace of pollution can be kept under control [15].

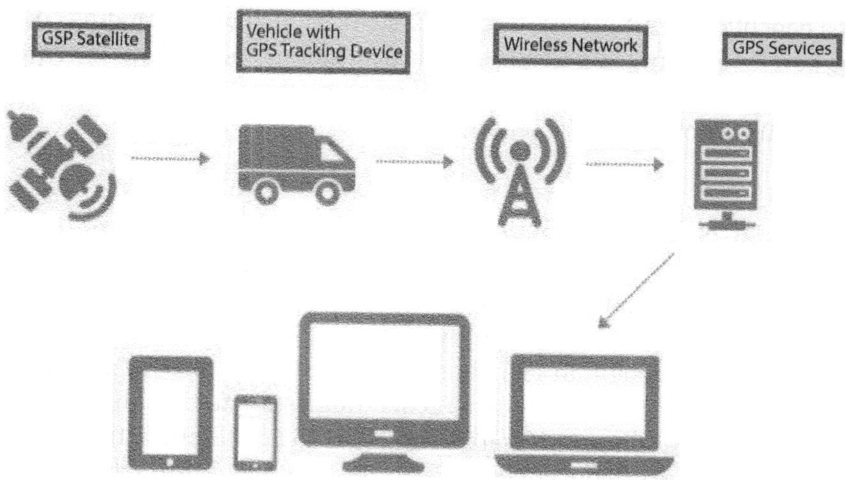

Figure 16.5 GPS data processing.

16.4 GPS Working Process

Latitude and longitude are primary mention in degrees. However, the data increment is expressed in timing, such as minutes and seconds when less than one degree. It also can convert to decimal for calculation. Latitude lines is measure north and south between the pole, while longitude lines is measure the west and east position. For example, the latitude and longitude position of School would be 5 ° 18' 34.1972" N, 104 ° 8 ' 40.7544" W. It can be read as latitude 5 degree 18 minutes with 34.1972 seconds, longitude 104 degree 8 minutes 40.7544 seconds. It can also convert to decimal, latitude 4.326277407216858 and longitude 101.14465355873108. In school bus tracking system, School bus current positioning module will install in a smart phone with GPS feature. It will be an active GPS receiver in the client's smart phone and continuously receive the current location point data from satellite system by the GPS receiver and send to user system with interval 2 seconds.

16.5 System Implementation

System implementation is performed by installing the school bus tracking system (SBTS). Before using this application, every user should enroll their register mobile number for authentication process. This application helps the school students to be present and on time in their respective places in school bus arrival to their stop. Parent and teachers are also satisfied with the benefit of the school bus tracking system. Mutual understanding could be developed between parents and teachers among the students' activities in the corresponding bus stops. At any moment, the parent can watch their children's action inside the school bus.

16.6 Result and Discussion

The important achievement of the proposed system is work is to develop the school bus tracking system by including extra facility, like reporting bus time through an alert bus timings, location sharing to clients, student smart attendance system, generating estimate time display projection and by adding an IP camera to focus the children's activity given in Figure 16.6. This concept accepts entirely designed through a mobile application develop by Android-based platform.This application makes friendly access for parent to monitoring their children's inside the school bus. School buses are allotted as the safest means of school commute for children. Android application for school bus tracking system helps schools to increase efficiency and profitability.

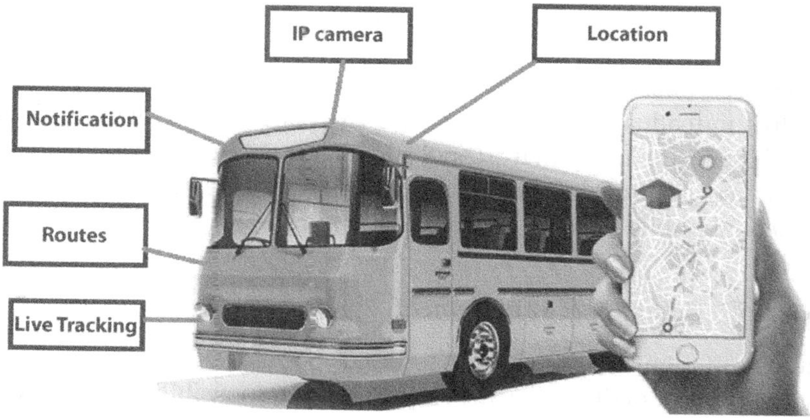

Figure 16.6 School bus tracking system.

16.6.1 Reasons to Utilize Android Application for School Bus Tracking System

16.6.1.1 Perfect Child Security

Android application for school bus tracking system enables real-time tracking of the school bus for school authorities, parents as well as transport managers. Emergency situations can be quickly responded, and parents are constantly updated about the expected time of arrival of the school buses so little children do not need to wait at bus stops and get exposed to the various dangers. Internet Protocol camera systems can cover the entire interior, as well as the exterior perimeter of the bus, ensuring that unruly behavior inside the bus is monitored.

16.6.1.2 Elaborate Operational Efficiency

The entire system enables optimizes the routes in real time, which translates into fuel as well as time savings. It accurately monitors driver work hours eliminating the possibility of drivers claiming excess hours.

16.6.1.3 *Valid Timely Maintenance*

Android application for school bus tracking system alerts the clients to be ready at their corresponding stop for pickup the students in perfect time that manage to reach the correct destination punctual in all times.

16.6.1.4 *Automating Attendance Management*

QR code scanner is embedded in the school ID cards, which helps to track children individually at every stage of their commute, as well as inside the school premises. In the smart school attendance system, precise attendance reports are generated. Parents and school authorities get instant alerts based on the child misses a school bus.

16.6.1.5 *Better Staff Management*

The system monitors driver behavior sends out violation alerts for speeding, aggressive braking, etc, which instills more responsibility and accountability among the drivers.

16.6.1.6 *Addressing Environmental Concerns*

Along with transport authorities and schools also have a responsibility to ensure a cleaner environment for children as well as people at large. By optimizing bus routes and monitoring mileage, excessive idling and unauthorized vehicle usage, the system keeps harmful emissions in check.

16.7 Conclusion

School bus tracking system provided a real-time platform for school bus user to check on bus status in anytime and anywhere. It also provided a platform for school bus service provider to monitor bus status and update latest information to user. By implementing the system, parents are able to monitor their children activities inside the school bus through IP camera functioning inside the school bus. Parents can be notifying by a text message that school bus reaches the destination. Smart attendance system helps the client to know whether the children presence or absence inside the school bus. This application may help parents in a user-friendly manner who have busy at their entire work. The application equipped with it helps to improve safety and enhance the operational efficiency for better route planning. The idea is to make small kids realize that sometimes they need to look after themselves as elders will not be with them all the time.

References

1. Chen, H.W., Chiang, Y.T., Chang, F.R., Wang, H.S., Toward real-time precise point positioning: Differential GPS based on IGS ultra rapid product, in: *Proceedings of SICE Annual Conference 2010*, 2010, August, IEEE, pp. 355–358.
2. Al-Hindawi, A.M. and Talib, I., Experimentally Evaluation of GPS/GSM Based System Design. *J. Electron. Syst.*, 2, 2, 67, 2012.
3. Peijiang, C. and Xuehua, J., Design and Implementation of Remote monitoring system based on GSM, in: *2008 IEEE Pacific-Asia workshop on computational intelligence and industrial application*, 2008, December, vol. 1, IEEE, pp. 678–681.
4. Ramya, V., Palaniappan, B. Karthick, K., Embedded controller for vehicle In-Front obstacle detection and cabin safety alert system. *Int. J. Comput. Sci. Inf. Technol.*, 4, 2, 117, 2012.
5. Chheda, G., Gajra, N., Chhaya, M., Deshpande, J., Gharge, S., Real time bus monitoring and passenger information system. *Int. J. Soft Comput. Eng.*, 1, 6, 34–38, 2012.
6. Maruthi, R. and Jayakumari, C., SMS based bus tracking system using open source technologies. *Int. J. Comput. Appl.*, 86, 9, 44–46, 2014.
7. Salim, K.A. and Idrees, I.M., Design and implementation of web-based GPS-GPRS vehicle tracking system. *Int. J. Sci. Eng. Comput. Technol.*, 3, 12, 443, 2013.
8. Sardey, Y., Deshmukh, P., Mandlik, P., Shelar, S., Nerkar, M., A mobile application for bus information system and location tracking using client-server technology. *Int. J. Emerg. Technol. Adv. Eng.*, 4, 4, 86–91, 2014.
9. Yang, J.S., Travel time prediction using the GPS test vehicle and Kalman filtering techniques, in: *Proceedings of the 2005, American Control Conference, 2005*, 2005, June, IEEE, pp. 2128–2133.
10. Singla, L. and Bhatia, P., GPS based bus tracking system, in: *2015 International Conference on Computer, Communication and Control (IC4)*, 2015, September, IEEE, pp. 1–6.
11. Williams, B.M. and Hoel, L.A., Modeling and forecasting vehicular traffic flow as a seasonal ARIMA process: Theoretical basis and empirical results. *J. Transp. Eng.*, 129, 6, 664–672, 2003.
12. Shinde, P.A., Mane, Y.B., Tarange, P.H., Real time vehicle monitoring and tracking system based on embedded Linux board and android application, in: *2015 International Conference on Circuits, Power and Computing Technologies [ICCPCT-2015]*, 2015, March, IEEE, pp. 1–7.

13. Gleason, J.M., Bus stop location on school bus routes: Gleason JM (1974) Paper available from the author, College of Business Administration, Texas Tech University, Lubbock, Texas 79409, USA. *Omega*, 3, 2, 242–243, 1975.
14. Goodboy, A.K., Martin, M.M., Brown, E., Bullying on the school bus: Deleterious effects on public school bus drivers. *J. Appl. Commun. Res.*, 44, 4, 434–452, 2016.
15. Kinable, J., Spieksma, F.C., Vanden Berghe, G., School bus routing—a column generation approach. *Int. Trans. Oper. Res.*, 21, 3, 453–478, 2014.
16. Baltes, M.R., Measuring motorist comprehension of Florida's school bus stop law and school bus signalization devices. *Transp. Res. Rec.*, 1640, 1, 10–16, 1998.
17. Park, J. and Kim, S., Security requirements analysis on IP camera via threat modeling and common criteria. *KIPS Trans. Comput. Commun. Syst.*, 6, 3, 121–134, 2017.
18. Mugisha, E., Ogwang, B., Ongoma, V., Zhang, G., A secured automatic notification system based on short message service. *Int. J. Inf. Secur. Res.*, 5, 1, 546–550, 2015.
19. Yeop, M.K., Aziz, M.A., Shah, M.M., Abd Kadir, M.F., Smart Attendance System by suing RFID, in: *2007 Asia-Pacific Conference on Applied Electromagnetics*, 2007, December, IEEE, pp. 1–4.

About the Editors

Suman Lata Tripathi, PhD, is a professor at Lovely Professional with more than seventeen years of experience in academics. She has published more than 45 research papers in refereed journals and conferences. She has organized several workshops, summer internships, and expert lectures for students, and she has worked as a session chair, conference steering committee member, editorial board member, and reviewer for IEEE journals and conferences. She has published one edited book and currently has multiple volumes scheduled for publication, including volumes available from Wiley-Scrivener.

Souvik Ganguli, PhD, is an assistant professor and received his PhD from Thapar Institute of Engineering and Technology, Patiala. With fourteen years of experience in academics and several years in industry, he has been a session chair, keynote speaker, and conference organizer for scholarly conferences, and he has published over 50 papers in academic journals. He also has coveted grants to his credit and has published a number of book chapters in edited volumes.

Abhishek Kumar, PhD, is an associate professor at and obtained his PhD in the area of VLSI Design for Low Power and Secured Architecture from Lovely Professional University, India. With over 11 years of academic experience, he has published more than 30 research papers and proceedings in scholarly journals. He has also published five book chapters and one authored book. He has worked as a reviewer and program committee member and editorial board member for academic and scholarly conferences and journals.

Tengiz Magradze, PhD, is an electrical design advisor for WINDTHINK, head of power transmission lines projects with "Georgian State Electrosystem," and an adjunct professor of Electrical/Power Engineering/Management at Ballsbridge University, Dominica. He has published 14 journal articles and one book and is an editorial board member for a number of academic journals.

About the Editors

Index

Acid rain, 54
Agriculture, 307
AI, 303
Anaerobic, 4
Android, 331
Architecture, 54
Audit, 89
Authentication, 304
Axis, 92

Battery, 103
Bias, 117
Biomass, 55
Biometric, 299
Blockchain, 301

Charger controller, 179
Chlorofluorocarbons, 2
Clean energy, 55
Connectivity, 270
Conversion, 20
Cryptography, 297

Database, 45
Deep learning, 303
Detection, 46
Digestion, 65
Driver portal, 333
DSM, 94
Dust, 104
Dust remover, 183
Dynamic, 196

Energy, 198
Energy audit, 98
Energy conservation, 104
E-waste management, 75

Falsifications, 304
Flood, 227
Fossil fuel, 112
Frequency, 229
Fuel, 247

Gasification, 63
Genetic algorithm, 275
Geothermal energy, 92
GPS, 332
Greenhouse gas, 58

Hacking, 295
Hetrojunction, 205
HEMT, 255
Home automation, 307
Hotspot, 104
HVAC, 20
Hybrid, 41
Hydrogen vehicles, 24
Hypothesis,

Intelligent Health Care, 322
Inspection, 102
Intelligent infrastructure, 323
Internet of Things, 35
ITRS, 196

Kane's Model, 208

Machine learning, 301
MEMS, 270

Mixer, 241
Modulation, 180
MOSFET, 195
Multithreshold, 235
Municipal, 34

Nonrenewable energy, 52

Parent portal, 334
Photovoltaic, 166
Power optimization, 230
Privacy, 305
Privacy challenges, 292
Pulse width modulation, 180
PV cell, 113
Pyrolysis, 61

RAM, 236
Recycle, 57
Recycling, 79
Renewable energy, 23
Remover, 183
Reserve, 23
Routing, 271

Sensors, 45
Smart city, 322
Smart garbage management, 41
Smart governance, 8
Smart waste bin, 40
Solar energy, 91
Solar tracking, 131
Solid waste, 57
Subthreshold, 195
Subthreshold slope, 199
Support vector machine, 27
SDG, 3

Tampering, 304
Teachers portal, 334
TFET, 195
Tracking topology, 121
Transformer, 249
Transistor, 255
Transmission, 271
Tunneling, 203

VCO, 229
Vulnerabilities, 293

Waste, 333
Wind, 19
Waste management, 6
Web of Things, 306
Wildlife, 55

Also of Interest

By the same editors

DESIGN AND DEVELOPMENT OF EFFICIENT ENERGY SYSTEMS, edited by Suman Lata Tripathi, Dushyant Kumar Singh, Sanjeevikumar Padmanaban, and P. Raja, ISBN 9781119761631. Covering the concepts and fundamentals of efficient energy systems, this volume, written and edited by a global team of experts, also goes into the practical applications that can be utilized across multiple industries, for both the engineer and the student. *NOW AVAILABLE!*

Electrical and Electronic Devices, Circuits, and Materials: Technical Challenges and Solutions, edited by Suman Lata Tripathi, Parvej Ahmad Alvi, and Umashankar Subramaniam, ISBN 9781119750369. Covering every aspect of the design and improvement needed for solid-state electronic devices and circuit and their reliability issues, this new volume also includes overall system design for all kinds of analog and digital applications and developments in power systems. *NOW AVAILABLE!*

Green Energy: Solar Energy, Photovoltaics, and Smart Cities, edited by Suman Lata Tripathi and Sanjeevikumar Padmanaban, ISBN 9781119760764. Covering the concepts and fundamentals of green energy, this volume, written and edited by a global team of experts, also goes into the practical applications that can be utilized across multiple industries, for both the engineer and the student. *NOW AVAILABLE!*

Check out these other related titles from Scrivener Publishing

INTELLIGENT RENEWABLE ENERGY SYSTEMS: Integrating Artificial Intelligence Techniques and Optimization Algorithms, edited by Neeraj Priyadarshi, Akash Kumar Bhoi, Sanjeevikumar Padmanaban, S. Balamurugan, and Jens Bo Holm-Nielsen, ISBN 9781119786276. This collection of papers on artificial intelligence and other methods for improving renewable energy systems, written by industry experts, is a reflection of the state of the art, a must-have for engineers, maintenance personnel, students, and anyone else wanting to stay abreast with current energy systems concepts and technology. *NOW AVAILABLE!*

SMART CHARGING SOLUTIONS FOR HYBRID AND ELECTRIC VEHICLES, edited by Sulabh Sachan, Sanjeevikumar Padmanaban, and Sanchari Deb, ISBN 9781119768951. Written and edited by a team of experts in the field, this is the most comprehensive and up to date study of smart charging solutions for hybrid and electric vehicles for engineers, scientists, students, and other professionals. *NOW AVAILABLE!*

Microgrid Technologies, edited by C. Sharmeela, P. Sivaraman, P. Sanjeevikumar, and Jens Bo Holm-Nielsen, ISBN 9781119710790. Covering the concepts and fundamentals of microgrid technologies, this volume, written and edited by a global team of experts, also goes into the practical applications that can be utilized across multiple industries, for both the engineer and the student. *NOW AVAILABLE!*

Progress in Solar Energy Technology and Applications, edited by Umakanta Sahoo, ISBN 9781119555605. This first volume in the new groundbreaking series, Advances in Renewable Energy, covers the latest concepts, trends, techniques, processes, and materials in solar energy, focusing on the state-of-the-art for the field and written by a group of world-renowned experts. *NOW AVAILABLE!*

Energy Storage 2nd Edition, by Ralph Zito and Haleh Ardibili, ISBN 9781119083597. A revision of the groundbreaking study of methods for storing energy on a massive scale to be used in wind, solar, and other renewable energy systems. *NOW AVAILABLE!*

Nuclear Power: Policies, Practices, and the Future, by Darryl Siemer, ISBN 9781119657781. Written from an engineer's perspective, this is a treatise on the state of nuclear power today, its benefits, and its future, focusing on both policy and technological issues. *NOW AVAILABLE!*

Zero-Waste Engineering 2nd Edition: A New Era of Sustainable Technology Development, by M. M. Kahn and M. R. Islam, ISBN 9781119184898. This book outlines how to develop zero-waste engineering following natural pathways that are truly sustainable using methods that have been developed for sustainability, such as solar air conditioning, natural desalination, green building, chemical-free biofuel, fuel cells, scientifically renewable energy, and new mathematical and economic models. *NOW AVAILABLE!*

Printed and bound by CPI Group (UK) Ltd, Croydon, CR0 4YY
17/09/2026
14964247-0006